Environmental Endocrine Disrupters

Environmental Endocrine Disrupters:
An Evolutionary Perspective

Edited by
Louis J. Guillette, Jr.
and
D. Andrew Crain

TAYLOR & FRANCIS
ALERE FLAMMAM
· Founded 1798 ·

USA	Publishing Office	Taylor & Francis 29 West 35th Street New York, NY 10001-2299 Tel: (212) 216-7800
	Distribution Center	Taylor & Francis 47 Runway Road, Suite G Levittown, PA 19057-4700 Tel: (215) 269-0400 Fax: (215) 269-0363
UK		Taylor & Francis 11 New Fetter Lane London EC4P 4EE Tel: 011 44 207 583 9855 Fax: 011 44 207 842 2298

ENVIRONMENTAL ENDOCRINE DISRUPTERS

1 2 3 4 5 6 7 8 9 0

A CIP catalog record for this book is available from the British Library.

Library of Congress Cataloging-in-Publication Data

Environmental endocrine disrupters : an evolutionary perpective /
    edited by Louis J. Guillette, Jr., and D. Andrew Crain.
        p.   cm
    Includes bibliographical references and index.
    ISBN 1-56032-571-2 (alk. paper)
    1. Reproductive toxicology.   2. Endocrine toxicology.
3. Environmental toxicology.   I. Guillette, Louis J., 1954– .
II. Crain, D. Andrew, 1970– .
RA1224.2.E68   1999
616.4'07—dc21

                                                        99-35847
                                                        CIP

Printed on acid-free, 250-year–life paper.
Manufactured in the United States of America.

Contents

Preface

Do environmental contaminants, released by anthropogenic activities, pose "new" threats to ecosystem health? Do wildlife studies from various environments worldwide implicate environmental contaminants as probable causal agents in various widespread human diseases? These questions and others are being addressed by governments, policy advisors, and research scientists in response to new research indicating that contaminants can alter chemical signaling among cells. Environmental contaminants continue to adversely influence wildlife by altering normal reproduction, growth, and immunity. Another system in the body, the endocrine system, orchestrates these systems and appears to be the primary target of many environmental toxicants. Thus, a great deal of the current research on environmental pollution focuses on the endocrine-altering actions of various chemical contaminants. Alteration of hormonal action is not an evolutionarily new phenomenon; for example, many plants make compounds (phytoestrogens) that mimic estrogenic sex steroids in vertebrates. What is novel is the increasing number of xenobiotic chemicals that interact with and modify vertebrate and invertebrate endocrine systems. These chemicals have been observed to mimic hormones, act as antihormones, or alter the synthesis and/or degradation of hormones. Thus, they have been termed *endocrine-disrupting contaminants*.

In her eloquent presentation *Silent Spring* (1962) more than 30 years ago, Rachel Carson demonstrated that wildlife are excellent sentinels of ecosystems and human health. This concept was further elaborated by Theo Colborn and colleagues recently in *Our Stolen Future* (1996). Humans are but one of many species living on earth; we are, however, unique in that we modify our environment rapidly through technology, either negatively or positively. It is now apparent that the intentional release of compounds through agricultural, industrial, and municipal activities has influenced the health of wildlife and human populations. In many cases, we have altered the survival, biodiversity, and genetic diversity of populations. For example, many scientists working with wildlife must now address whether their research subjects represent "normal" or "altered" populations. Indeed, field biologists can no longer use the term *control* when referring to a population or site but rather use the term *reference*. Comparative endocrinologists can no longer ignore the possible influences of pesticides and industrial contaminants on various endocrine endpoints of interest, such as circulating steroid or thyroid hormone concentrations. Further, laboratory-based endocrinologists are concerned with the endocrine-altering potential of plastisizers and dyes. These concerns of both field- and laboratory-based researchers are not limited to any particular group of animals, because all species in the kingdom Animalia (and also the kingdom Plantae) utilize endocrine signaling to communicate vital information.

This book does not cover all of the topics of current concern in the field of endocrine-disrupting contaminants. Quite simply, that is beyond the scope of any single volume. It does address a number of questions, such as the mode of disruption and effects of endocrine-altering contaminants on behavior, embryonic development, reproduction, and immune function in various invertebrates and vertebrates (including humans). These discussions raise many new questions and indicate that future research must continue to focus on interdisciplinary and cross-disciplinary studies. Examples of such studies are the endocrine/immune interactions in a variety of species (see Chap. 7) and anthropological studies that provide a practical approach to examining the impact of endocrine-altering compounds in human development and performance (see Chap. 12).

Resolution of any problem is a multistep process. First, the problem must be recognized, and it was just several decades ago that studies began to emphasize the potential detrimental effects of endocrine-altering compounds. Second, the scope of the problem must be characterized. In this book, we illustrate that the effects of endocrine-altering compounds are not isolated to one mechanism or several species but are extremely ecologically and mechanistically complex with evolutionary implications. Third, solutions to the problem are found. We do not present solutions here, but we have raised many of the problems that must be addressed in the coming years if we are to maximize ecosystem health and minimize the effects of endocrine-disrupting contaminants on future generations. There is no question that there is a problem, and there is no question that the scope of this problem is vast. The only question is whether or not we will truly be stewards of our environment and seek solutions.

Dedication/Acknowledgments

This book is dedicated to our academic grandfathers, Howard Bern and Richard E. Jones, for their friendship, mentoring, and infectious enthusiasm for science and life; their contributions continue to influence our professional and personal lives. We thank the many colleagues and students who have contributed to our professional and personal development—in particular, Theo Colborn, Cathy Cox, Mark Gunderson, Greg Masson, John Matter, Matt Milnes, Ed Orlando, Dan Pickford, Andy Rooney, and Allan Woodward. Most of all we thank our families for their love, sacrifice, and constant support. They make all the work and dedication worthwhile.

Louis J. Guillette, Jr.
D. Andrew Crain

Chapter 1

ENDOCRINE-DISRUPTING CONTAMINANTS AND HORMONE DYNAMICS: LESSONS FROM WILDLIFE

D. Andrew Crain,[1] Andrew A. Rooney,[2] Edward F. Orlando,[2] and Louis J. Guillette, Jr.[2]

[1] Department of Biology
Maryville College
Maryville, TN 37804

[2] Department of Zoology
University of Florida
Gainesville, FL 32611

Introduction

Over the past few decades, accumulating evidence has suggested that many environmental chemicals can alter reproduction, growth, and survival by changing the normal function of the endocrine system. The endocrine system is complex, with many organs contributing to a multifaceted regulatory system. This complexity obstructs the identification of the specific mechanisms through which endocrine-disrupting contaminants (EDCs) elicit their responses. The lack of an organized, mechanistic understanding of EDC effects provides perhaps the greatest impediment to this area of research (Rudel, 1997). Before considering how EDCs can alter the endocrine system, we must first have an understanding of normal hormone dynamics. Figure 1-1 presents a simplistic model of hormone dynamics, following hormones from production to excretion. Accurate assessment of an environmental chemical's potential to alter the endocrine system depends on consideration of the entire hormone dynamic pathway.

Wildlife studies have contributed greatly to our understanding of contaminant-induced endocrine disruption. In a sense, studies of wildlife are natural experiments, testing the effects of anthropogenic perturbations on populations with natural genetic variation. Studies of wildlife are especially useful for considering one of the great dilemmas in evolutionary and developmental biology:

how animals maintain a level of uniformity while experimenting with diversity. For instance, vertebrates are limited in the number of hormones they have to perform biological functions, but the study of vertebrate wildlife has revealed that the same hormone can have different functions in various species. Prolactin, for instance, is used for osmoregulation in fish and for milk production in mammals (Norris, 1997).

The purpose of this chapter is to review the various mechanisms through which EDCs elicit their effects, specifically focusing on the insight gained from wildlife studies. This review will be organized based on the model presented in Fig. 1-1; we will examine the potential for EDCs to alter each point in the steroid hormone pathway. We believe that this method will both provide an organizational framework to the apparently disparate mechanisms of endocrine disruption and introduce an evolutionary perspective to the effects of endocrine-altering contamimants.

EDCs and Hormone Production

Steroidogenesis is the biochemical synthesis of steroid hormones. It is a system of interlinked pathways consisting of precursors, enzymes, and their products (Fig. 1-2). In general, the first step in steroid production is the conversion of cholesterol into pregnenolone by the side-chain cleavage enzyme. In the

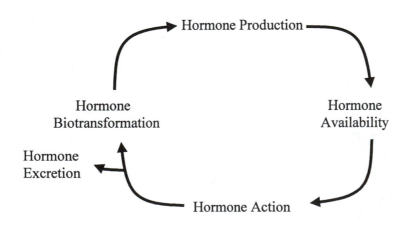

Figure 1-1.
A simplistic model of hormone dynamics. After a hormone is produced, circulating and intracellular binding proteins regulate the hormone's bioavailibility. Then the hormone elicits action by binding to a specific receptor. The hormone is then either excreted in the urine after hepatic conjugation reactions or biotransformed into another hormone, which will begin the cycle of bioavailability, action, and excretion/biotransformation.

gonads, pregnenolone is converted by the enzyme 3β-hydroxysteroid dehydrogenase to progesterone. Either pregnenolone or progesterone can be converted via a series of enzymes and intermediary substrates into testosterone (T). Finally, testosterone can be aromatized into estradiol-17β (E_2) (Miller, 1988; Felig et al., 1995). Steroid hormones move from the site of their production into the bloodstream where they usually bind to plasma-binding proteins, such as α-fetoprotein, albumin, and sex-hormone–binding globulin, which protect them from degradation and excretion (Martin, 1975; Norris, 1997).

An intricate system of receptor-mediated feedback mechanisms controls the timing and duration of steroid production. At each level of the hypothalamic-pituitary-gonadal (H-P-G) axis, the hormones released downstream in the axis usually provide negative feedback to the producing tissue. This results in a decrease not only of the particular hormones synthesis but also of any hormone downstream in the axis. For example, gonadotropic hormone (GtH) released from the pituitary stimulates the testis to produce T, which is transported via the circulatory system back to the pituitary, as well as the hypothalamus, causing a decrease in the synthesis of gonadotropin-releasing hormone (GnRH), GtHs, and thereby a reduction in T. Although less common, positive feedback control loops do exist. For example in mammals during ovulation, E_2 causes an increase in the release of luteinizing hormone (LH) from the pituitary gland; this feedback is a positive relationship, resulting in an increased release of E_2 (Gilbert, 1997; Norris, 1997; Peter and Yu, 1997).

Exposure of animals to many environmental contaminants results in altered circulating concentrations of hormones (Table 1-1). These alterations could either be due to changes in hormone production or to changes in hormone

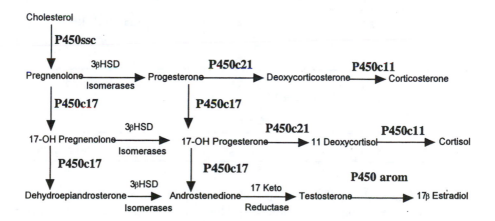

Figure 1-2.
The steroidogenic pathway illustrating both the steroids and steroidogenic enzymes involved. Enzymes in the P450 family are in boldface.

Table 1-1. Representative studies showing altered circulating hormone concentrations in wildlife species after exposure to environmental contaminants

ORGANISM	DISRUPTION	CONTAMINANT	REFERENCE
		Fishes	
Roach White sucker Lake whitefish	↓plasma T ↓plasma E_2, GtH-II, 17,20β-P ↓plasma E_2, T	Paper mill effluent	McMaster, 1991 Van Der Kraak et al., 1992 Munkittrick et al., 1992
Salmon	↓E_2, ↓DHP ↓in vitro E_2	PCBs, DDT, PCDDs, PCDFs, and heavy metals β-naphthoflavone 20-methylcholanthrene	Leatherland,1992 Alfonso et al., 1997
Common carp	↓serum T	Sewage plant effluent	Folmar et al., 1996
Atlantic croaker	↑plasma E_2, T ↓plasma E_2, T	Cadmium, lead, benzo[a]pyrene, Aroclor 1254	Thomas, 1990
Monopterus albus	↓plasma E_2, T	Malathion, cadmium, 3-methylcholanthrene	Singh,1989
Heteropneustes fossilis	↓plasma E_2, T, 17αOHP (reproductive stage dependent)	Malathion, γ-hexachlorocyclohexane	Singh and Singh, 1992
Common carp Rainbow trout	↓plasma androgens, estrogens, and corticoids	Arochlor, 1254	Sivarajah et al., 1978
		Reptiles	
American alligator	↑T in chorioallantoic fluid ↑T in plasma ↓plasma T	Tamoxifen Organochlorines or modern-use herbicides	Crain and Guillette, 1997 Crain et al., 1998a
		Birds	
Japanese quail	↓plasma E_2 (before sexual maturity)	PCBs	Biessmann, 1982
		Mammals	
Dall's porpoise	↓circulating T	*p,p'*-DDE, PCBs	Subramanian et al., 1987

excretion/metabolism. This section will consider those endocrine alterations due to altered hormone production rates. There are a variety of ways that environmental contaminants could alter the production of hormones, including: (1) altered availability of cholesterol to begin steroidogenesis, (2) altered steroidogenic enzyme activity, and (3) alterations in feedback loops.

EDCs can effect steroidogenesis by changing the availability of cholesterol to the side-chain cleavage enzyme ($P450_{scc}$) in the mitochondria or by decreasing the activity of the $P450_{scc}$. Dioxin (TCDD) was shown to inhibit steroidogenesis in the rat testis by inhibiting the mobilization of cholesterol to the $P450_{scc}$ (Moore et al., 1991). This reduction in activity of the $P450_{scc}$ leads to a decrease in the conversion of cholesterol to pregnenolone, which is the precursor of progestogens, androgens, and estrogens. Thus, dioxin elicits a wholesale decrease in all steroid production. In addition, many EDCs with known antisteroidogenic effects inhibit the binding of the mitochondrial peripheral-type benzodiazepine receptor (PBR) to its ligand (Papadopoulos et al., 1997). Since the formation of the PBR-ligand complex promotes the loading of cholesterol to the $P450_{scc}$, interference with this binding could explain the mechanism underlying the observed reduction in steroidogenesis.

EDCs can also increase or decrease steroidogenic enzyme expression or activity, thereby altering circulating hormone concentrations. In alligators, *in ovo* treatment with the herbicide atrazine stimulated aromatase ($P450_{arom}$) in testes such that there was no significant difference between treated testes and control ovaries in *in vitro* gonadal aromatase activity (Crain et al., 1997). $P450_{arom}$ is the steroidogenic enzyme that converts T to E_2 in gonads as well as other tissues in vertebrates (Simpson et al., 1994). Therefore, stimulating or depressing the normal activity of $P450_{arom}$ could alter the relative balance of androgens and estrogens within an organism.

A third way that EDCs can cause altered hormone production is through disruption of normal feedback loops in the H-P-G axis. As explained in detail later, many EDCs directly bind to native hormone receptors. These EDCs can act in an agonistic or antagonistic fashion at any place along the H-P-G axis, thereby altering normal feedback mechanisms. As a result, xenobiotics that act as steroid agonists or antagonists can effectively create an internal hormonal environment very different from an organism's normal endocrine profile. This hormonal profile is dependent on the timing of exposure to xenobiotics.

One of the best-studied examples of disruption of normal feedback mechanisms in the pituitary-gonadal axis has been documented in the white suckers, *Catostomus commersoni*, of Lake Superior. Populations of these fish exposed to bleached kraft mill effluent (BKME) have demonstrated a broad array of reproductive abnormalities, including delayed sexual maturity and reduced primary (gonad size) and secondary sexual characteristics (McMaster, 1991; McMaster et al., 1991; Munkittrick et al., 1991; McMaster et al., 1992). These fish have reduced circulating levels of reproductive steroids and pituitary peptides, which suggest a disruption to the normal function of the pituitary-gonadal axis (Van Der Kraak et al., 1992). Exposed fish of both sexes showed significant reductions in plasma levels of GtH-II, T, $17\alpha,20\beta$–dihydroxy-4-pregnen-3-one (17,20βP), and 11-ketotestosterone (11-KT) in males. Following stimulation by injection of salmon gonadotropin-releasing hormone (sGnRH), exposed white suckers of both sexes exhibited a smaller increase in GtH release than fish from the reference site. These same BKME-exposed fish did not show an increase in 17,20βP, or in 11-KT in males on stimulation. Interestingly, the exposed white suckers of both sexes did have a transitory increase in plasma T following stimulation. *In vitro* incubations of ovarian follicles removed from fish exposed to BKME revealed depressed basal and stimulated secretions of T and 17,20βP (Van Der Kraak et al., 1992). Results from these studies suggest that alterations of circulating steroid levels as well as steroidogenic capability of fish ovaries can be caused by disruption of the pituitary-gonadal axis by exposure to BKME.

EDCs and Hormone Availability

The ability of contaminants to disrupt endocrine processes depends not only on their ability to alter hormone production but also on their effects on the bioavailability of the hormones. Bioavailability is controlled by several factors, including plasma or tissue concentration, sequestration by binding proteins, clearance, and hepatic metabolism. For native steroid hormones such as E_2, T, and cortocosterone/cortisol, binding proteins in the blood and cellular cytoplasm regulate metabolism, clearance, and bioavailability. Serum or extracellular binding proteins interact with steroids and protect them from renal and hepatic clearance, whereas intracellular proteins reduce steroid metabolism and serve as molecular chaperones.

Two well-studied serum binding proteins are sex hormone–binding globulin (SHBG) and corticosteroid-binding globulin (CBG). When bound to these proteins, steroid hormones are unable to interact with receptors and thus exhibit reduced biological effects (Miller, 1988; Felig et al., 1995; Norris, 1997). The synthesis of both CBG and SHBG are stimulated by estrogens in the mammals studied to date, but SHBG preferentially binds T and E_2, whereas CBG preferentially binds cortisol and progesterone (Baxter et al., 1995). Circulating binding proteins related to CBG and/or SHBG have been reported in fish (Martin, 1975), amphibians (Ozon et al., 1971; Martin and Ozon, 1975), reptiles (Salhanick and Callard, 1980; Ho et al., 1987; Paolucci and Di Fiore, 1992), birds (Wingfield et al., 1984), and mammals (Siiteri et al., 1982), indicating a conserved role for these proteins.

Intracellular steroid-binding proteins are also thought to regulate the amount of free hormone available for receptor binding. Fox (1975) characterized a protein that binds estradiol in the brains of neonatal rats. This neonatal binding protein (NBP) is believed to protect the brain cells from high levels of maternal estrogen. Without the binding protein, the rats likely would be sterilized due to brain exposure to estrogens. NBP-like proteins have been reported in the cytosol of turtle oviduct (Salhanick et al., 1979), the alligator oviduct (Crain et al., 1998b), and avian liver (Dower and Ryan, 1976). Studies have also characterized the function of α-fetoprotein, a protein found in the developing reproductive and nervous tissues of animals; α-fetoprotein regulates the amount of unbound estradiol available for cellular activity and is influenced by local production of growth factors (Nunez, 1994).

The effects of EDCs on binding-protein concentrations remain largely unstudied. However, it is likely that EDCs that act as "estrogen mimics" alter binding-protein concentrations, as SHBG concentrations can be regulated by endogenous estrogens (Loukovaara et al., 1995) and ingested phytoestrogens (Mousavi and Aldercreutz, 1993; Loukovaara et al., 1995). Future studies should examine the effects of EDCs on intracellular and extracellular binding-protein concentrations.

Besides regulating the bioavailability of endogenous steroid hormones, steroid-binding proteins could regulate the bioavailability of the EDCs. Numerous studies have shown that many EDCs bind avidly to steroid receptors (see the following section), and therefore it is possible that the EDCs also bind to the less-specific binding proteins. To test this hypothesis, a recent study examined the binding affinity of cytosolic steroid-binding proteins from the oviduct of the American alligator (*Alligator mississippiensis*; Crain et al., 1998b). The oviductal cytosolic binding proteins had a significantly lower affinity for most environmental chemicals compared to native steroids (Fig. 1-3). These data support a previous study showing that the activity (and thus the bioavailability) of estradiol was decreased more than 25% on incubation with a physiological level of SHBG, whereas the estrogenic activity of the ecoestrogens octylphenol and *o,p'*-DDT was unchanged on incubation with SHBG (Arnold et al., 1996). These data suggest that a lack of affinity between serum and/or cytosolic binding proteins and various steroid-mimicking contaminants can increase the availability of these chemicals to intracellular steroid receptors and, therefore, increase the cellular potency of such contaminants relative to native hormones. Future studies should focus on intracellular availability as an important variable in evaluating the potency of EDCs.

EDCs and Hormone Action

After steroids arrive at target cells, the hormones cross the cell membrane and elicit actions by binding to specific protein receptors. The receptor-steroid complex then binds to chromatin (DNA and associated proteins) in the cell nucleus, stimulating the synthesis of specific RNAs and proteins. Steroid hormone receptors have high affinity for ligands, but the receptors do not display absolute specificity, as the same receptor can bind to many ligands. Some of these ligands act in an agonistic manner to mimic the actions of the native hormone, whereas other ligands act as antagonists to block the effects of native hormones. As an example, consider the human estrogen receptor (hER).

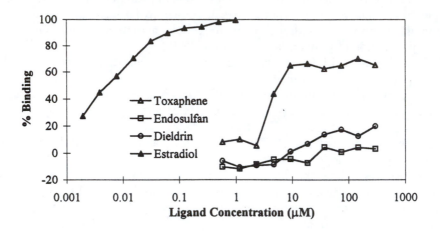

Figure 1-3.
Binding of estradiol and several environmental contaminants to cytosolic binding proteins from the oviduct of the American alligator. Of the four contaminants only one, the pesticide toxaphene, exhibited an affinity for the cytosolic steroid binding proteins. Therefore, it appears that relative to endogenous steroid hormones, many environmental contaminants may be more readily available for intracellular actions (from Crain et al., 1998b, with permission).

In one study using *in vitro* assays (varying studies often show differing binding affinities, see Chap. 4), purified hER exhibits a binding affinity for estradiol 1.57 times more potent than diethylstilbestrol (DES, a synthetic, pharmacudical estrogen), 77 times more potent than coumestrol (a phytoestrogen), 273 times more potent than estriol (a native estrogen), and 2,000 times more potent than dihydrotestosterone (DHT, a native androgen; Gaido et al., 1997). Many environmental chemicals released by human activities also bind to purified hERs, although with much less potency; for instance, estradiol is 5,000 times more potent than p-nonylphenol (a sewage treatment biodegradation product of nonionic surfactants), and 8 million times more potent than *o,p'*-DDT (a persistant organochlorine pesticide; Gaido et al., 1997). While the affinity of these contaminants for the hER seems extremely low, the actual potency of these contaminants may be increased due to alterations induced at other endocrine axes (receptor-binding affinity is only one factor in the determination of endocrine-disrupting potency; see Fig. 1-1). As discussed previously, increased bioavailability resulting from weak interaction of EDCs with binding proteins is a major factor augmenting EDC action. Additionally, the potency of many EDCs (such as kepone, a persistent organochlorine insecticide) are increased by a relatively long biological half-life. For example, in the short time frame of *in vitro* assays, kepone is 100,000 to 1 million times less potent than estradiol-17β (Soto et al., 1995). However, in a rat uterine weight assay, kepone is only 1,000 to 5,000 times less potent than estradiol-17β (Hammond et al., 1979). The duration of a ligand/receptor interaction can also affect EDC effects. Receptor dissociation rates differ between natural ligands such as testosterone and dihydrotestosterone bound to AR (Zhou et al., 1995) as well as between EDCs such as kepone and *o,p'*-DDT bound to ER (Hammond et al., 1979). The combination of increased bioavailability and prolonged receptor association can lead to increased potency of many endocrine-altering compounds.

ER binding dynamics are extremely complex and not completely understood. For instance, while *o,p'*-DDT binds to rainbow trout and alligator ERs, *p,p'*-DDD exhibits no binding affinity for the ERs (Donohoe and Curtis, 1996; Vonier et al., 1996). However, a mixture of *p,p'*-DDD and *o,p'*-DDT exhibits significantly more ER binding than *o,p'*-DDT alone, indicating complex interactions between ERs and compound mixture (Vonier et al., 1996).

Few studies have pursued EDC steroid receptor interactions within wildlife species. Most evidence from nontraditional research species or wildlife is evidence of a correlation between EDC exposure and altered endocrine parameters. For example, Bergeron et al. (1994) demonstrated PCB-induced sex reversal in the red-eared slider turtle, *Trachemys scripta elegans*, a species in which exogenous estrogens can induce ovarian development in animals that would otherwise develop as males. Two of 11 PCBs (2',4',6'-trichloro-4-biphenylol   and   2',3',4',5'-tetrachloro-4-biphenylol) applied to turtle eggs induced sex reversal in *T. scripta elegans* (Bergeron et al., 1994). Although they have not been tested against isolated *T. scripta* estrogen receptor (ER), the same two PCBs bind to mouse estrogen receptor *in vitro*. PCB exposure in wild mink (Larsson et al., 1990), and ranch mink (Sundqvist et al., 1989) was used to select significant PCB congeners and doses for lab studies of PCB effects on ER and progesterone receptors (PR). Using a major PCB found in wild mink (2,2',4,4',5,5'-hexachlorobiphenyl), Patnode and Curtis (1994) were able to show that a PCB impaired normal estrogen-induced increases in nuclear ER in anestrous, but not pregnant, mink. The same PCB was also found to increase PR dissociation constants in pregnant mink, indicating that receptor number and function can both be affected by EDCs (Patnode and Curtis, 1994). The American alligator is another of the few wildlife species that have been studied under field conditions of EDC exposure and laboratory conditions, including receptor-binding studies. A combination of field and laboratory studies have demonstrated abnormal plasma concentrations of estrogen, testosterone, thyroxine, and triiodothyronine, as well as reduced phallus size associated with EDCs in alligators (Guillette et al., 1994, 1995, 1996; Crain et al., 1997, 1998a). Recently, the alligator ER and PR were also studied directly, and binding to several EDCs was demonstrated (Vonier et al., 1996). In fact, Vonier et al. (1996) demonstrated that several EDCs such as *o,p'*-DDD, *o,p'*-DDT, *o,p'*-DDE, dicofol, atrazine, and *trans*-Nonachlor that were implicated by previous studies did bind alligator steroid receptors. Both the alligator and mink studies illustrate the benefits of an experimental approach to studying EDCs that includes coupled wildlife and laboratory studies.

There is no guarantee that a compound that binds to a receptor in one species will bind to that receptor in another species. Three factors determine whether or not a ligand will cause endocrine disruption by binding to receptors in a particular species: the amount of receptor expressed in the species, the developmental or reproductive stage at which the animal is exposed to the compound, and the structure of the receptor in the species. First, the expression of more receptor in a species is expected to cause that species to be more susceptable to receptor agonists and antagonists, but do species express different quantities of steroid receptors? Apparently so, because two closely related lizard species (*Cnemidophorus uniparens* and *Cnemidophorus inornatus*) express different amounts of estrogen and progesterone receptors (Young et al., 1995), suggesting that native steroids and steroidal agonists/antagonists can affect similar species to different extents.

Second, consider the effects of the developmental or reproductive stage. Throughout embryonic development, steroid receptors are temporally up- and down-regulated to control normal development of the reproductive system. Similarly, tissue-dependent steroid receptor concentrations are dependent on reproductive status. For instance, androgen receptors (ARs) in Leydig cells, which are responsible for testosterone production, are intermediate during immaturity, highest during puberty, and lowest during adulthood (Shan et al., 1995). However, in Sertoli cells, which aid in sperm maturation, ARs are lowest during immaturity, intermediate during puberty, and highest during adulthood (Shan et al., 1995). Thus compounds such as vinclozolin and *p,p'*-DDE that would cause endocrine alteration via interaction with the AR (Kelce et al., 1997) are expected to be more disruptive to Leydig cells during puberty, but more disruptive to Sertoli cells during adulthood. The timing of steroid receptor up-and down-regulation during development and reproduction is species specific and poorly characterized in almost all species. Therefore, it is difficult to predict with certainty the potential of a compound to alter the endocrine system at the level of the receptor unless the normal temporal pattern of receptor abundance has been elucidated. For instance, consider a pesticide that interacts with the estrogen receptor. If you needed to apply the pesticide to a lake or riverine system, it would seem reasonable to apply the pesticide during a time of low reproductive activity such as the fall. However, a recent study indicates that reproduction in the female American alligator is initiated in the fall, when estrogen receptor mediated responsivity is high (Guillette et al., 1997). Thus, application of the pesticide during the fall months would cause endocrine alterations in the female alligators.

The third factor that determines the binding affinity of a ligand for a steroid receptor is the structure of the receptor in the species of interest. Steroid receptors are proteins, and proteins evolve just as morphological characteristics do. Thus, evolutionary divergence between species indicates structural divergence between receptors. For instance, the estrogen receptor (ER) from the whiptail lizard *Cnemidophorous uniparens* shares 89% of the amino acid sequence of rainbow trout ER and 88% of the amino acid sequence of chicken ER (Krust et al., 1986; Pakdel et al., 1989; Young et al., 1995). As expected, the DNA-binding domains and steroid-binding domains of estrogen, progesterone, and androgen receptors share a high degree of sequence homology among vertebrates, with most of the amino acid variations occurring in the hinge region of the receptors (Young et al., 1995). The function of the non-ligand–binding/non-DNA–binding regions is unknown; and the consequences of the variation in these regions on ER binding or transcription is also unclear (Nimrod and Benson, 1996). However, even slight differences in amino acid sequences have the potential to change the tertiary folding of the receptor, affecting ligand binding. Evidence for this is demonstrated in the alligator estrogen receptor. ER from the oviduct of female alligators binds to dihydrotestosterne (Vonier et al., 1997), whereas no DHT binding is noted for ERs from turtle testes (Mak et al., 1983), turtle oviducts (Salhanick et al., 1979), or humans (Obourn et al., 1993). Therefore, it is possible that species specificity may exist in receptor-binding affinity for steroids and steroidal agonists and antagonists. Future studies should explore this species specificity.

EDCs and Hormone Excretion and Biotransformation

A mechanism that likely is involved in contaminant-induced endocrine disruption is an alteration in the rate of steroid excretion and steroid biotransformation. Enzymes in the cytochrome P450 family catalyze both steroid excretion and biotransformation. Cytochrome P450 is a generic term for a superfamily of more than 230 enzymes, all of which contain a heme group and approximately 500 amino acids (Miller, 1988). P450 enzymes have been studied intensively due to their action in phase I detoxification of xenobiotics, whereby the P450 adds a hydroxyl moiety to the xenobiotic, making the xenobiotic hydrophilic and an acceptable substrate for phase II conjugation reactions. Two lesser-studied actions of the P450 enzymes are their roles in the synthesis and excretion of steroids. Steroid hormone biosynthesis is catalyzed by specific P450 steroidogenic enzymes (Fig.1-2) that transform one steroid into another by cleaving side chains and adding hydroxyl groups. For instance, the androgen testosterone is converted into the estrogen 17β-estradiol by P450 aromatase, which aromatizes the phenolic ring by hydroxylating the C18 carbon (Fig. 1-4). Other P450 enzymes have a metabolic (rather than a biosynthetic) effect on steroid hormones, catalyzing site-specific hydroxylation reactions that may inactivate circulating steroids or target steroids for conjugation and elimination (Waxman, 1996). Based on the variety of functions performed by members of the P450 enzyme family, it is clear that the evolution of P450 enzymes has provided two critical adaptations for vertebrates: the ability to detoxify contaminants and the ability to exhibit sexual dimorphism. The detoxifying and steroidogenic actions of P450s are normally treated as separate and distinct functions, but it is proposed that these activities are closely related and, perhaps, regulated by similar mechanisms.

Many of the same P450 enzymes that are responsible for the detoxification of xenobiotics are involved in the metabolic and biosynthetic conversion of steroids (Waxman, 1996), suggesting that exposure to environmental contaminants could alter endogenous steroid concentrations by inducing P450 enzymes. Early studies in fish indicated that exposure to PCBs caused an increase in hepatic enzyme activity and a simultaneous decrease in circulating steroid concentrations (Sivarajah et al., 1978), suggesting that P450 induction increased the metabolic conversion of endogenous steroids. Recent studies on the effects of bleached kraft pulp mill effluent (BKME) indicate that BKME causes elevated hepatic P450IA-dependent activity and decreased plasma sex steroid concentrations in lake whitefish (*Coregonus clupeaformis*; Munkittrick et al., 1992) and white sucker (*Catostomus commersoni*; Munkittrick et al., 1992, 1994). The effects on steroids may be independent of the P450IA induction, but the fact that P450 enzymes are involved in xenobiotic response and steroid production suggests that the two effects are linked. Interestingly, BKME exposure reduces follicular P450 aromatase activity in the white sucker (McMaster et al., 1995), and this reduction in P450 aromatase is also seen in alligators *(Alligator mississippiensis)* exposed to persistent organochlorines and

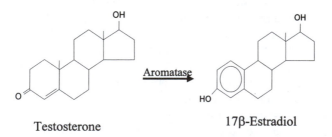

Figure 1-4.
Structures for the androgen testosterone and the estrogen 17β-estradiol; conversion of testosterone into 17β-estradiol is catalyzed by P450 aromatase.

agricultural compounds at Lake Apopka, Fl. (Crain et al., 1997). Scientists are just beginning to appreciate the potential for EDCs to cause changes in endogenous hormones by altering P450 activity. For instance, genes encoding the mitochondrial P450 11β-hydroxylase (P450c11) were thought to only be involved in steroidogenesis of the cortisol hormones (Fig. 1-2), but a recent study has shown that P450c11 metabolizes the environmental contaminant MeSO2-DDE (Lund and Lund, 1995). MeSO2-DDE is a metabolite of the global, persistent environmental pollutant DDT, and if MeSO2-DDE induces P450c11 activity, this could be a major mechanism of endocrine disruption in vertebrate wildlife.

Vertebrates may be especially sensitive to exposure to EDCs during embryonic development, as EDCs can change P450 enzyme activity that, in turn, alters reproductive development and function. For instance, maternal exposure to the environmental estrogen 4-octylphenol causes a reduction in the expression of P450 17α-hydroxylase/C17-20 lyase (P450c17) in the fetal rat testis (Majdic et al., 1996). P450c17 is a highly regulated steroidogenic P450 responsible for testosterone production (Fig. 1-2) and, therefore, reductions in this enzyme could alter normal development of the male reproductive tract. Regulation of the P450 enzyme responsible for 17β-estradiol production can also be altered following embryonic exposure to environmental contaminants. P450 aromatase (Figs. 1-2, 1-4) is elevated in hatchling alligators exposed to the herbicide atrazine during embryonic development (Crain et al., 1997). Studies in rats indicate that males and females exhibit different gender-dependent expression of various P450 enzymes (Waxman, 1996); these differences allow normal metabolism of testosterone or estradiol-17β. Because these sexually dimorphic enzyme patterns are organized during embryogenesis (Gustafsson, 1994), exposure of embryos to EDCs has the potential to permanently alter the normal sex-specific steroid metabolism (Guillette et al., 1995).

The observation that P450 induction can lead to enhanced detoxification and altered steroidogenesis leads to an evolutionary dilemma. If an animal is to survive exposure to xenobiotics, that animal requires a well-developed P450 detoxification system. Evolution is not driven by survival, however, but by reproduction. In terms of evolution, it matters not how long an individual survives; what matters is how many viable offspring that individual contributes to the population. For example, consider an individual that is born, by chance, with a superior P450 detoxification system. This individual may live a long life but never successfully reproduce because of altered steroidogenesis. Another individual with a less-effective detoxification system but normal reproduction has higher fitness, because this individual produces viable offspring. Therefore, evolution would favor individuals having efficient detoxification systems (to promote survival so that future reproduction can occur) that do not alter reproduction. Millions of years of evolution have produced such a hepatic P450 detoxification system in vertebrates. However, during the last 100 years, animals have been exposed to an increased number of novel, persistent anthropogenic compounds. This brief time period, a matter of several generations in most vertebrate species, would not allow evolution to generate a balance between survival and reproduction. Therefore, many of the reproductive abnormalities currently seen in wildlife and humans could be due to the balance shifting toward survival, with reproduction being compromised. An increase in hepatic P450 induction is but one potential mechanism that could lead to increased survival but decreased reproduction in vertebrates.

Summary and Conclusions

Figure 1-5 revisits the model presented in Fig.1-1, citing representative wildlife studies that describe the effects of EDCs on particular sites in the cycle of steroid dynamics. By far, the most thoroughly studied aspects of this cycle is the effect of EDCs on steroid production and action. Numerous studies have noted that many EDCs bind with steroid receptors, eliciting either an agonistic or antagonistic effect. Because of the prevalence of these studies and the availability of numerous techniques to measure steroid receptor-binding, the phenomenon of receptor interaction has become synonymous with endocrine disruption. As is apparent in Fig. 1-1, however, there are many sites independent of the receptor that should be considered when assessing endocrine disruption.

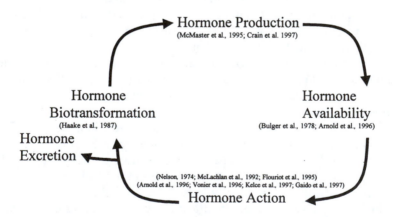

Figure 1-5.
Figure 1-1 revisited. Selected references are given that illustrate how environmental contaminants can affect hormone dynamics. Numerous studies have documented changes in hormone activity (e.g., through receptor interactions), but few studies have explored the interaction of environmental contaminants with hormone production, binding proteins, excretion, and biotransformation.

The study of the adverse effects of xenobiotics dates back to the earliest humans. We are still far from understanding all the ways that xenobiotics alter our physiology, but our increased understanding of physiological mechanisms has elucidated many of the mechanisms through which xenobiotics can elicit unfavorable effects. For instance, 30 years ago the interaction of contaminants with estrogen receptors was unknown, mainly because basic steroid-receptor dynamics were not understood. Therefore, our future understanding of xenobiotic actions is dependent on both basic and applied research. In a day when humans and wildlife are exposed to increasing amounts of novel compounds, these basic and applied research efforts are both timely and essential.

References

Alfonso, L.O.B., Campbell, P.M., Iwama, G.G., Devlin, R.H., and Donaldson, E. M. (1997). The effect of the aromatase inhibitor on fadrozole and two polynuclear aromatic hydrocarbons on sex steroid secretion by ovarian follicles of coho salmon. *Gen. Comp. Endocrinol.* 106: 169–174.

Arnold, S.F., Robinson, M.K., Notides, A.C., Guillette, L.J., Jr., and McLachlan, J.A. (1996). A yeast estrogen screen for examining the relative exposure of cells to natural and xenoestrogens. *Environ. Health Perspec.* 104: 544–548.

Baxter, J.D., Frohman, L.A., and Felig, P. (1995). Introduction to the endocrine system. *Endocrinology and Metabolism, 3d ed.* (P. Felig, J. Baxter, and L. Frohman), pp. 3–22. McGraw-Hill, New York.

Bergeron, J. M., Crews, D., and McLachlan, J. A. (1994). PCBs as environmental estrogens: Turtle sex determination as a biomarker of environmental contaminantion. *Environ. Health Perspec.* 102: 780–781.

Biessmann, A. (1982). Effects of PCBs on gonads, sex hormone balance and reproduction processes of Japanese quail *Coturnix coturnix japonica* after ingestion during sexual maturation. *Environ. Pollution, Series A* 27: 15–30.

Crain, D.A., and Guillette, L.J., Jr. (1997). Endocrine-disrupting contaminants and reproduction in vertebrate wildlife. *Reviews Toxicol.* 1: 207–231.

———, ———, Rooney, A.A., and Pickford, D.B. (1997). Alterations in steroidogenesis in alligators (*Alligator mississippiensis*) exposed naturally and experimentally to environmental contaminants. *Environ. Health Perspec.* 105(5): 528–533.

———, ———, Pickford, D.B., Percival, H.F., and Woodward, A.R. (1998a). Sex-steroid and thyroid hormone concentrations in juvenile alligators (*Alligator mississippiensis*) from contaminated and reference lakes in Florida, USA. *Environ. Toxicol. Chem.* 17: 446–452.

———, Noriega, N., Vonier, P.M., Arnold, S.F., McLachlan, J.A., and Guillette,

L.J., Jr. (1998b). Cellular bioavailability of natural hormones and environ-mental contaminants as a function of serum and cytosolic binding factors. *Toxicol. Industrial Health* 14: 261–273.

Donohoe, R.M., and Curtis, L.R. (1996). Estrogenic activity of chlordecone, o,p'-DDT and o,p'-DDE in juvenile rainbow trout: Induction of vitellogen-sis and interaction with hepatic estrogen binding sites. *Aquatic Toxicol.* 36: 31–52.

Dower, W.J., and Ryan, K.J. (1976). A cytoplasmic estrone-specific binding protein (E1BP) in hen liver. *Fed. Proc.* 35: 1366.

Felig, P., Baxter, J.D., and Frohman, L.A. (1995). *Endocrinology and Metabolism.* McGraw-Hill, New York.

Folmar, L.C., Denslow, N.D., Rao, V., Chow, M., Crain, D.A., Enblom, J., Marcino, J., and Guillette, L.J., Jr. (1996). Vitellogenin induction and reduced serum testosterone concentrations in feral male carp (*Cyprinus carpio*) captured near a major metropolitan sewage treatment plant. *Environ. Health Perspec.* 104: 1096–1101.

Fox, T.O. (1975). Oestradiol receptor of the neonatal mouse brain. *Nature* 258: 441–444.

Gaido, K.W., Leonard, L.S., Lovell, S., Gould, J.C., Babai, D., Portier, C.J., and McDonnell, D.P. (1997). Evaluation of chemicals with endocrine modulat-ing activity in a yeast-based steroid hormone receptor gene transcription assay. *Toxicol. Applied Pharmacol,* 143: 205–212.

Gilbert, S.F. (1997). *Developmental Biology.* Sinauer, Sunderland.

Guillette, L.J., Jr., Crain, D.A., Rooney, A.A., and Pickford, D.B. (1995). Orga-nization versus activation: The role of endocrine-disrupting contaminants (EDCs) during embryonic development in wildlife. *Environ. Health Perspec.* 103 (Suppl. 7): 157–164.

———, Gross, T.S., Gross, D.A., Rooney, A.A., and Percival, H.F. (1995). Gonadal steroidogenesis in vitro from juvenile alligators obtained from con-taminated or control lakes. *Environ. Health Perspec.* 103 (Suppl. 4): 31–36.

———, ———, Gross, T.S., Masson, G.R., Matter, J.M., Percival, H.F., and Woodward, A.R. (1994). Developmental abnormalities of the gonad and abnormal sex hormone concentrations in juvenile alligators from contami-nated and control lakes in Florida. *Environ. Health Perspec.* 102(8): 680–688.

———, ———, Pickford, D.B., Crain, D.A., Rooney, A.A., and Percival, H.F. (1996). Reduction in penis size and plasma testosterone concentrations in juvenile alligators living in a contaminated environment. *Gen. Comp. Endocrinol.* 101: 32–42.

———, ———, Woodward, A.R., Crain, D.A., Masson, G.R., Palmer, B.D., Cox, M.C., You-Xiang Q., and Orlando, E.F. (1997). The reproductive cycle of the female American alligator (*Alligator mississippiensis*). *Gen. Comp.*

Endocrinol. 108: 87–101.

Gustafsson, J.-A. (1994). Regulation of sexual dimorphism in the rat liver. *The Differences Between the Sexes.* (R.V. Short and E. Balaban), pp. 231–242. Cambridge University Press, Cambridge.

Hammond, B., Katzenellengogen, B.S., Krauthammer, N., and McConnell, J. (1979). Estrogenic activity of the insecticide chlordecone (Kepone) and interaction with uterine estrogen receptors. *Proc. Nat. Acad. Sci. U.S.A.* 76: 6641–6645.

Ho, S.-M., Lance, V., and Megaloudis, M. (1987). Plasma sex-binding protein in a seasonally breeding reptile, *Alligator mississippiensis. Gen. Comp. Endocrinol.* 65: 121–132.

Jansson, J.-O., Ekberg, S., Isaksson, O., Gustafsson M.A, and Gustafsson, J.-A. (1985). Imprinting of growth hormone secretion, body growth and hepatic steroid metabolism by neonatal testosterone. *Endocrinol.* 106: 306–316.

Kelce, W.R., Lambright, C.R., Gray, L.E., Jr., and Roberts, K.P. (1997). Vinclozolin and p,p'-DDE alter androgen-dependent gene expression: In vivo confirmation of an androgen receptor-mediated mechanism. *Toxicol. Applied Pharmacol.* 142: 192–200.

Krust, A., Green,S., Argos, P., Kumar, V., Walter, P., Bornert, J., and Chambon, P. (1986). The chicken oestrogen receptor sequence: Homology with v-erbA and the human oestrogen and glucocorticoid receptors. *EMBO* 5: 891–897.

Larsson, P., Woin, P., and Knulst, J. (1990). Differences in uptake of persistant pollutants for preditors feeding in aquatic and terrestrial habitats. *Holartic Ecol.* 13: 149–155.

Leatherland, J.F. (1992). Endocrine and reproductive function in Great Lakes salmon. In *Chemically-induced Alterations in Sexual and Functional Development: The Wildlife/Human Connection, Vol. XXI.* (T. Colborn and C. Clement), pp. 129–146. Science Publishing, Princeton.

Loukovaara, M., Carson, M., and Adlercreutz, H. (1995). Regulation of sex-hormone-binding globulin production by endogenous estrogens *in vitro. Biochem. Biophys. Res. Comm.* 206: 895–901.

———, ———, Palotie, A., and Adlercreutz, H. (1995). Regulation of sex hormone-binding globulin production by isoflavonoids and patterns of isoflavonoid conjugation in HepG2 cell cultures. *Steroids* 60: 656–661.

Lund, B., and Lund, J. (1995). Novel involvement of a mitochondrial steroid hydroxylase (P450c11) in xenobiotic metabolism. *J. Biol. Chem.* 270: 20895–20897.

MacLatchy, D.L., and Van Der Kraak, G.J. (1995). The phytoestrogen β-sitosterol alters the reproductive endocrine status of goldfish. *Toxicol. Appl. Pharmacol.* 134: 305–312.

Majdic, G., Sharpe, R.M., O'shaughnessy, P.J., and Saunders, P.T.K. (1996). Expression of cytochrome P450 17α-hydroxylase/C17-20 lyase in the fetal rat testis is reduced by maternal exposure to exogenous estrogens. *Endocrinology* 137: 1063–1070.

Mak, P., Ho, S.M., and Callard, I.P. (1983). Characterization of an estrogen receptor in the turtle testis. *Gen. Comp. Endocrinol.* 52: 182–189.

Martin, B. (1975). Steroid-protein interactions in nonmammalian vertebrates. *Gen. Comp. Endocrinol.* 25: 42–51.

———, and Ozon, R. (1975). Steroid-protein interactions in nonmammalian vertebrates. II. Steroids binding proteins in the serum of amphibians: A physiological approach. *Biol. Reprod.* 13: 371–380.

McMaster, M.E. (1991). *Impact of Bleached Kraft Pulp Mill Effluent on Fish Populations in Jackfish Bay, Lake Superior.* University of Waterloo, Waterloo, Ontario.

———, Portt, C.B., Munkittrick, K.R., and Dixon, D.G. (1992). Milt characteristics, reproductive performance, and larval survival and development of white sucker exposed to bleached kraft mill effluent. *Ecotox. Environ. Safety* 23: 103–117.

———, Van Der Kraak, G.J., and Munkittrick, K.R. (1995). Exposure to bleached kraft pulp mill effluent reduces the steroid biosynthetic capacity of white sucker ovarian follicles. *Comp. Biochem. Phys.* 112C: 169–178.

———, ———, Portt, C.B., Munkittrick, K.R., Sibley, P.K., Smith, I.R., and Dixon, D.G. (1991). Changes in hepatic mixed function oxygenase (MFO) activity, plasma steroid levels and age at maturity of a white sicker (*Catastomus commersoni*) population exposed to bleached kraft pulp mill effluent. *Aquatic Toxicol.* 21: 199–218.

Miller, W.L. (1988). Molecular biology of steroid hormone synthesis. *Endocrine Reviews* 9: 295–318.

Moore, R.W., Jefcoate, C.R., and Peterson, R.E. (1991). 2,3,7,3-Tetrachlorodibenzo-p-dioxin inhibits steroidogenesis in the rat testis by inhibiting the mobilization of cholesterol to cytochrome P450$_{scc}$. *Toxicol. Applied Pharmacol.* 109: 85–97.

Mousavi, Y., and Aldercreutz, H. (1993). Genistein is an effective stimulator of sex hormone-binding globulin production in hepatocarcinoma human liver cancer cells and suppresses proliferation of those cells in culture. *Steroids* 58: 301–304.

Munkittrick, K.R., McMaster, M.E., Portt, C.B., Van Der Kraak, G.J., Smith, I.R., and Dixon, D.G. (1992). Changes in maturity, plasma sex steroid levels, hepatic mixed-function oxygenase activity, and the presence of external lesions in lake whitefish (*Coregonus clupeaformis*) exposed to bleached kraft mill effluent. *Can. J. Fish Aquat. Sci.* 49: 1560–1569.

————, Portt, C.B., Van Der Kraak, G.J., Smith, I.R., and Rokosh, D.A. (1991). Impact of bleached kraft mill effluent on population characteristics, liver MFO activity, and serum steroids of the Lake Superior white sucker (*Catostomus commersoni*) population. *Can. J. Fish Aquat. Sci.* 48: 1–10.

————, Van Der Kraak, G.J., McMaster, M.E., and Portt, C.B. (1992). Response of hepatic MFO activity and plasma sex steroids to secondary treatment of bleached kraft pulp mill effluent and mill shutdown. *Environ. Toxicol. Chem.* 11: 1427–1439.

————, ————, ————, ————, van den Heuvel, M.R., and Servos, M.R. (1994). Survey of receiving-water environmental impacts associated with discharges from pulp mills. 2. Gonad size, live size, hepatic EROD activity and plasma sex steroid levels in white sucker. *Environ. Toxicol. Chem.* 13: 1089–1101.

Nimrod, A. C. and Benson, W.H. (1996). Environmental estrogenic effects of alkylphenol ethoxylates. *Crit. Rev. Toxicol.* 26(3): 335–364.

Norris, D.O. (1997). *Vertebrate Endocrinology.* Academic, San Diego.

Nunez, E.A. (1994). Biological role of alpha-fetoprotein in the endocrinological field: Data and hypotheses. *Tumor Biol.* 15: 63–72.

Obourn, J.D., Koszewski, N.J., and Notides, A.C. (1993). Hormone- and DNA-binding mechanisms of the recombinant human estrogen receptor. *Biochemistry* 32: 6229–6236.

Ozon, R., Martin, B., and Boffa, G.A. (1971). Protein-binding of estradiol and testosterone in newt serum (*Pleurodeles waltlii* Michah). *Gen. Comp. Endocrinol.* 17: 566–570.

Pakdel, F., Guellec, C., Vaillant, C., Roux, M., and Valotaire, Y. (1989). Identification and estrogen induction of two estrogen receptors (ER) messenger ribonucleic acids in the rainbow trout liver: Sequence homology with other ERs. *Mol. Endocrinol.* 3: 44–51.

Paolucci, M., and Di Fiore, M.M. (1992). Putative steroid-binding receptors and nonreceptor components and testicular activity in the lizard *Podarcis sicula sicula. J. Repro. Fert.* 96: 471–481.

Papadopoulos, V., Amri, H., Boujrad, N., Cascio, C., Culty, M., Garnier, M., Hardwick, M., Li, H., Vidic, B., Brown, A.S., Reversa, J.L., and Drieu, K. (1997). Peripheral benzodiazepine receptor in cholesterol transport and steroidogenesis. *Steroids* 62: 21-28.

Patnode, K.A., and Curtis, L.R. (1994). 2,2′,4,4′, 5,5′- and 3,3′,4,4′,5,5′- hexachlorobiphenyl alteration of uterine progesterone and estrogen receptors coincides with embryotoxicity in mink (*Mustela vision*). *Toxicol. Appl. Pharmacol.* 127: 9–18.

Peter, R.E., and Yu, K.L. (1997). Neuroendocrine regulation of ovulation in fishes: Basic and applied aspects. *Rev. Fish Biol. Fish* 7: 173–197.

Rudel, R. (1997). Predicting health effects of exposures to compounds with estrogenic activity: Methodolical issues. *Environ. Health. Perspec.* (105, Suppl.) 3: 655–663.

Salhanick, A.C.R., and Callard, I.P. (1980). A sex steroid-binding protein in the plasma of the fresh water turtle, *Chrysemys picta. Gen. Comp. Endocrinol.* 42: 163–166.

————, Vito, C.C., Fox, T.O., and Callard, I.P. (1979). Estrogen-binding proteins in the oviduct of the turtle, *Chrysemys picta:* Evidence for a receptor species. *Endocrinology* 105: 1388–1395.

Shan, L., Zhu, L., Bardin, C.W., and Hardy, M.P. (1995). Quantitative analysis of androgen receptor messenger ribonucleic acid in developing Leydig cells and Sertoli cells by in situ hybridization. *Endocrinology* 136: 3856–3862.

Siiteri, P.K., Murai, J., Hammond, G.L., Nisker, J.A., Raymoure, W.J., and Kuhn, R.W. (1982). The serum transport of steroid hormones. *Rec. Prog. Hormone Res.* 38: 457–510.

Simpson, E.R., Mahendroo, M.S., Means, G.D., Kilgore, M.W., Hinshelwood, M.M., Graham-Lorence, S., Amarneh, B., Ito, Y., Fisher, C.R., Michael, M.D., Mendelson, C.R., and Bulun, S.E. (1994). Aromatase cytochrome P450, the enzyme responsible for estrogen biosynthesis. *Endocrine Reviews* 15: 342–355.

Singh, H. (1989). Interaction of xenobiotics with reproductive endocrine functions in a protogynous teleost, *Monopterus albus. Marine Environ. Res.* 28: 285–289.

Singh, P.B., and Singh, T.P. (1992). Impact of malathion and g-BHC on steroidogenesis in the freshwater catfish, *Heteropneustes fossilis. Aquatic Toxicology* 22: 69–80.

Sivarajah, K., Franklin, C.S., and Williams, W.P. (1978). The effects of polychlorinated biphenyls on plasma steroid levels and hepatic microsomal enzymes in fish. *J. Fish. Biol.* 13: 401–409.

Soto, A.M., Sonnenschein, C., Chung, K.L., Fernandez, M.F., Olea, N., and Olea Serrano, F. (1995). The E-SCREEN assay as a tool to identify estrogens: An update on estrogenic environmental pollutants. *Environ. Health. Perspec.* 103 (Suppl. 7): 113–122.

Subramanian, A.N., Tanabe, S., Tatsukawa, R., Saito, S., and Miyazaki, N. (1987). Reduction in the testosterone levels by PCBs and DDE in Dall's porpoises of Northwestern North Pacific. *Mar. Pollution Bull.* 18: 643–646.

Sundqvist, C., Amador, A.G., and Bartke, A. (1989). Reproduction and fertility in the mink *(Mustela vison). J. Reprod. Fert.* 85: 413–441.

Thomas, P. (1990). Teleost model for studying the effects of chemicals on female reproductive endocrine function. *J. Exp. Zool. Suppl.* 4: 126–128.

Van Der Kraak, G.J., Munkittrick, K.R., McMaster, M.E., Portt, C.B., and

Chang, J.P. (1992). Exposure to bleached kraft pulp mill effluent disrupts the pituitary-gonadal axis of white sucker at multiple sites. *Toxicol. Appl. Pharmacol.* 115: 224–233.

Vonier, P.M., Crain, D.A., McLachlan, J.A., Guillette, L.J., Jr., and Arnold, S.F. (1996). Interaction of environmental chemicals with the estrogen and progesterone receptors from the oviduct of the American alligator. *Environ. Health Perspec.* 104(12): 1318–1322.

———, Guillette, L.J., Jr., McLachlan, J.A., and Arnold, S.F. (1997). Identification and characterization of an estrogen receptor from the oviduct of the American alligator *(Alligator mississippiensis). Biochem. Biophys. Res. Commun.* 232: 308–312.

Waxman, D.J. (1996). Steroid hormones and other physiologic regulators of liver cytochromes P450: Metabolic reactions and regulatory pathways. *Advan. Mol. Cell. Biol.* 14: 341–374.

Wingfield, J.C., Matt, K.S., and Farner, D.S. (1984). Physiologic properties of steroid hormone-binding proteins in avian blood. *Gen. Comp. Endocrinol.* 53: 281–292.

Young, L.J., Godwin, J., Grammer, M., Gahr, M., and Crews, D. (1995). Reptilian sex steroid receptors: amplification, sequence and expression analysis. *J. Steroid Biochem. Mol. Biol.* 55(2): 261–269.

———, Nag, P.K., and Crews, D. (1995). Species differences in estrogen receptor and progesterone receptor-mRNA expression in the brain of sexual and unisexual whiptail lizards. *J. Neuroendocrinol.* 7: 567–576.

Zhou, Z.-X., Lane, M.V., Kemppainen, J.A., French, F.S., and Wilson, E.M. (1995). Specificity of ligand-dependent androgen receptor stabilization: Receptor domain interactions influence ligand dissociation and receptor stability. *Mol. Endocrinol.* 9: 208–218.

Chapter 2

THE IMPORTANCE OF COMPARATIVE ENDOCRINOLOGY IN EXAMINING THE ENDOCRINE DISRUPTER PROBLEM

Tyrone B. Hayes

Laboratory for Integrative Studies in Amphibian Biology
Department of Integrative Biology, Group in Endocrinology,
    and Museum of Vertebrate Zoology
University of California, Berkeley, CA 94720-3140

Introduction

Increasing evidence shows that many environmental contaminants can interact with the endocrine system of vertebrates. New evidence showing that many chemicals can disrupt developmental processes at contaminant levels that are not lethal or carcinogenic creates a sinister threat to wildlife and humans alike. Many environmental contaminants can cause developmental abnormalities that alter developmental patterns or impair reproductive capabilities. Although not necessarily fatal to the individual, these developmental effects can result in dramatic declines, and even complete losses of populations as recruitment decreases. Several environmental contaminants may interact with endogenous hormonal systems, and through these interactions induce developmental abnormalities. To date, the main hormonal systems noted to be affected are those of the sex steroids, glucocorticoids, and thyroid hormones.

Evidence for Endocrine Disruption

Sex Steroids

Pesticides and other environmental contaminants have estrogen-like effects in a variety of vertebrates. Exposure to pesticides (such as DDT) and polychlorinated biphenyls (PCBs) can induce vitellogenin in fish (Anderson et al., 1996a,b; Flouriot et al., 1995), at least one amphibian (*Xenopus laevis*; Palmer and Palmer, 1996), and one reptile (*Trachemys scripta*; Palmer and Palmer, 1996). Treatment with PCBs also produced skewed sex ratios in hatchling turtles (*T. scripta*; Crews et al., 1995), similar to the effects of natural estrogens (Janzen and Paukstis, 1991; Crews et al., 1994). Exposure to environmental contaminants also resulted in abnormal gonadal development, abnormal plasma steroid levels, and decreased phallus size in male alligators, at a contaminated site in Florida (Guillette et al., 1994). In mammals, DDT treatment results in hypertrophy of the myometrium in rats (Bustos et al., 1996) and initiates implantation and maintenance of pregnancy in mice (Johnson et al., 1992), and may be associated with estrogen-related cancers in humans (Adami et al., 1995; Soto et al., 1994). The described effects are likely due to the binding of contaminants to the estrogen receptor: Several contaminants bind and activate the estrogen receptor as shown in a yeast estrogen screen (Arnold et al., 1996b) and in estrogen-sensitive cancer cell lines (Soto et al., 1995).

A variety of contaminants also affect male secondary sex differentiation and reproductive physiology: DDT inhibits testicular growth and development of male secondary sex characters (Hayes, 1982), but other contaminants such as PCBs increase testis size and sperm production (Cooke et al., 1996). In humans, dioxin exposure results in increased gonadotropin and decreased testosterone in the plasma of males (Egeland et al., 1992, 1994). The effect of contaminants on the function of the testes varies with the compounds, however. Although contaminants such as dioxins (Moore et al., 1991b; Kleeman et al., 1990) and PCBs (Kovacevic et al., 1995) inhibit steroidogenesis in the testes, others induce testosterone synthesis (Johnson et al., 1994).

Glucocorticoids

Environmental contaminants may also interfere with adrenal corticoid action. In particular, two metabolites of DDT (*o,p'*-DDD and *o,p'*-DDE) are potent glucocorticoid synthesis inhibitors and have been used to treat adrenal disorders (Benecke, 1991). Thus, these compounds may affect any physiological or developmental processes regulated by glucocorticoids. PCBs can also induce (turtles and birds: Goldman and Yawetz, 1991; rats: Miller et al., 1993) or inhibit (guinea pigs: Goldman and Yawetz, 1990, 1991, 1992) glucocorticoid synthesis.

Hayes (1997a) and Hayes et al. (1997) also showed that corticosterone treatment in larvae of the African walking frog *(Kassina senegalensis)* can result in erosion of the upper mandible, similar to the effects of DDT on anuran larvae (Cooke, 1970). Although DDT produced the effect in several species of anuran larvae, corticosterone only produced the effect in *K. senegalensis*. Recent data suggest that the effect is produced because corticosterone causes the release of DDT stored in the fat (Hayes, 1996; Hayes and Guzman, in preparation).

Thyroid Hormones

Several environmental contaminants interact with the thyroid axis. In rats, PCBs caused decreases in plasma thyroid hormones in pregnant mothers, fetuses, and weanlings (Morse et al., 1992, 1993, 1996). PCBs can also bind thyroid hormone binding proteins in pregnant mice and be transferred to fetal mice, resulting in a reduction in plasma thyroid hormone in the fetuses (Darnerud et al., 1996a,b). Also, Seo et al. (1995), Barter and Klaasen (1994), and Van Birgelen et al. (1995) showed that the decrease in plasma thyroid hormone in response to PCBs is in part due to increases in glucoronidation and elimination of thyroid hormone by the liver. PCBs and dioxins also bind thyroid hormone binding proteins in humans, as shown in the studies in rodents (Lans et al., 1993, 1994). Studies in humans also showed that decreased thyroid hormone levels in human mothers and infants is correlated with exposure to dioxins and PCBs (Koopman-Esseboom et al., 1994).

Mechanisms of Action

The aforementioned studies show that DDT and its metabolites, PCBs, and dioxins all have the potential to interact with sex steroids, corticoids, and thyroid hormones in several vertebrate classes. The mechanisms of action for contaminant interference may differ for the different hormone classes, however. Estrogen-like effects of environmental contaminants appear to be due to compounds binding directly to the estrogen receptor. Compounds that interfere with androgen, corticoid, or thyroid hormone–mediated processes, however, may act by disrupting hormone synthesis, metabolism, or transport in the blood. There is, of course, potential for interactions with corticoid and thyroid receptors.

The reported and potential effects of environmental contaminants on all steps in hormone action (synthesis, transport, receptor binding, postreceptor activity, and degradation) and the potential effects of endocrine disrupters on individuals and populations (wildlife and humans) create an urgent need to understand the mechanisms of action of endocrine-disrupting environmental contaminants. A clearer understanding of the mechanisms of action will better prepare researchers to predict their effects on individuals and their impacts on populations. Using our knowledge of endogenous hormonal effects and mechanisms as models to predict the behavior of endocrine disrupters is the most efficient approach in examining the problem.

In examining the effects and mechanisms of endocrine disrupters, it is also important not to examine detailed mechanisms of action in model species at the expense of conducting comparative studies. Although mechanisms of hormone/endocrine disrupter action may be similar across species, comparative studies are necessary, because the roles that these hormones play in development and physiology can be quite distinct in any given vertebrate class, family, or species. In addition, the effects of endocrine disrupters likely vary between species, and even within a species, depending on environmental factors or developmental state, as the natural hormones show such variability (Hayes and Licht, 1995). Thus, examinations of model systems may not allow the prediction of effects in a diversity of species. There are some general similarities in the effects and roles of sex steroids, corticoids, and thyroid hormones in development across vertebrates, however, creating a useful starting point for understanding the effects and mechanisms of endocrine disrupters.

Role of Sex Steroids, Glucocortioids, and Thyroid Hormones

Sex Steroids

Sex steroids (estrogens and androgens) are vital in regulating reproductive physiology and behavior across all vertebrates; however, the specific roles may vary between species. In oviparous and ovoviparous vertebrates (fish, amphibians, most reptiles, and monotremes), estrogens regulate vitellogenesis (production of yolk protein) and egg development (Ho, 1991). The role of sex steroids in these processes includes directly inducing the vitellogenin gene (Gupta and Kanungo, 1996) as well as increasing oviducal growth and inducing oviducal proteins important for egg production (Laugier et al., 1988; Simmen, 1987; Pawar and Pancharatna, 1995). Sex steroids are also involved in preparing the oviduct for ovulation, and estrogens increase the sensitivity of the oviduct to oxytocin, which stimulates ovulation (Ko et al., 1989; Ayad et al., 1996; Witt et al., 1991).

In viviparous animals, progesterone prepares the uterus for implantation, whereas estrogens increase myometrial excitability at parturition (Nathanielsz et al., 1995; Lye, 1996). Estrogens increase oxytocin receptors in the uterus (Liu et al., 1996) and increase oxytocin secretion from the pars nervosa (Wang et al., 1995; Thomas and Amico, 1996), prior to the stimulation of myometrial contractions by oxytocin. After birth, estrogens are involved in stimulating fat accretion and mammary gland growth (Snedeker, 1996) necessary for milk production.

Androgens are involved in several reproductive behaviors including aggressive (Hayes and Licht, 1992; Van Goozen et al., 1994, 1995; Ogawa et al., 1996), territorial (Hunt et al., 1995; Pankhuurst and Carragher, 1995), and reproductive behaviors in males (Justice and Logan, 1995; Mendonça et al., 1996). Estrogen and progesterone stimulate reproductive behaviors in females (Vailes et al., 1992; Young et al., 1995; Mech et al., 1996), and progesterone induces parental care in males and females (Brown, 1993; Matt, 1990; Kelley, 1988).

Sex steroids also play important roles in vertebrate development. Most secondary sex characteristics are dependent on sex steroids. For example, sexually dimorphic structures such as antlers in deer (Kolle et al., 1993) and thumb pads in frogs (Kanamadi and Saidapur, 1993) are androgen-dependent. Sexually dimorphic growth is dependent on sex steroids (Hayes and Licht, 1992; Chowen et al., 1996; Ho et al., 1996) as are sexual differences in color (Hews and Moore, 1995). Sex behaviors, although activated by steroids at reproductive maturity, are also dependent on brain exposure to sex steroids during early development (Carter et al., 1989; Balthazart et al., 1996). Although sex differentiation does not appear to be affected by sex steroids in mammals, a number of fish (Van den Hurk et al., 1989), amphibians (Gallien, 1974; Hayes and Licht, 1995), reptiles (Bull et al., 1988), and some birds (Maraud et al., 1990) are "sex reversed" when exposed to sex steroids during embryonic development.

Glucocorticoids

Although typically viewed as stress hormones, the glucocorticoids serve many roles in physiology and development. Glucocorticoids regulate intermediary metabolism, increasing proteolysis, lipolysis, and glycogenolysis, and stimulate gluconeogenesis from amino acids and free fatty acids. Most of the catabolic actions of glucocorticoids likely result from the inhibition of glucose uptake by cells, but glucocortioids also increase glycogenolysis and gluconeogenesis in the liver (Goldstein et al., 1993). As a result of effects on intermediary metabolism, glucocorticoids typically inhibit somatic growth when plasma glucocorticoid levels are high.

Although mineralocorticoids, such as aldosterone, are the main osmoregulatory hormones, glucocorticoids can also cause sodium retention and increase the osmolarity of the blood (Whitworth et al., 1995). Glucocorticoids also inhibit inflammatory and allergic reactions and are generally immunosuppressive and increase vulnerability to illness (Ottaviani and Franceschi, 1996).

During fetal development, glucocorticoids are responsible for lung maturation (Pierce et al., 1995). Glucocorticoids also signal the start of parturition, as glucocorticoids increase near term in the fetus of mammals. The increase in glucocorticoids results in an increase in placental production of estrogens (although the mechanisms is not completely understood), which induces uterine sensitivity to oxytocin. However, the role of glucocorticoids in oviposition of nonmammalian vertebrates is unknown. Glucocorticoids may also serve a role in oviparous species, as it enhances estrogen-induced vitellogenesis in *Xenopus laevis* (Rabelo et al., 1994).

Thyroid Hormones

Like glucocorticoids, thyroid hormones (thyroxine T_4 and triiodothyronine T_3) are also involved in intermediary metabolism and favor lipolysis and glycogenolysis (McNabb, 1995). Thyroid hormone levels are altered with nutritional status and are generally increased with short-term overfeeding, but decreased during fasting (Morley, 1995). Hypothyroidism results in a decreased basal metabolic rate and increased fat storage (Drent and Van Der Veen, 1995).

In mammals, thyroid hormones are also involved in regulating temperature and stimulate nonshivering thermogenesis (Silva, 1993; Harper and Brand, 1995). Thyroid hormones serve an analogous role in some reptiles where they induce basking behavior that increases the animal's body temperature (Hulbert and Williams, 1988). Thyroid hormones are also necessary for growth, stimulate growth hormone release from the pituitary (Giustina and Wehrenberger, 1995), and are necessary for bone elongation and maturation (Klaushofer et al., 1995).

Thyroid hormones are also well known for their role in amphibian (Etkin and Gilbert, 1968; Dodd and Dodd, 1976; Kikuyama et al., 1993) and fish metamorphosis (Tagawa et al., 1994; Inui et al., 1995). These hormones directly regulate most of the metamorphic genes at metamorphic climax in amphibians (Tata, 1996) and stimulate most of the tissue changes that occur during metamorphosis. Thyroid hormone is also involved in fetal growth (Wallace et al., 95) and is necessary during mammalian fetal development, controlling important events such as brain development (Calikogu et al., 1996) and auditory function (Forrest et al., 1996).

Alternative Mechanisms of Endocrine Disrupter Action

Although many endocrine-disrupting compounds bind to hormone receptors, there are several points along endocrine axes where contaminants may interfere with the functioning of sex steroids, adrenal corticoids, and thyroid hormones. Secretion of all three classes of hormones is controlled by the pituitary, which is in turn regulated by the hypothalamus (Fig. 2-1). The steroids

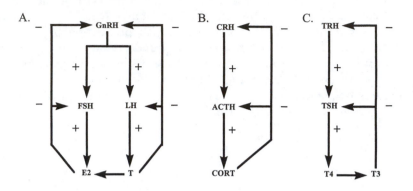

Figure 2-1.
(A) Hypothalamo-pituitary-gonadal, (B) hypothalamo-pituitary-adrenal, and (C) hypo-thalamo-pituitary-thyroid axes. Positive (+) and negative (-) influences are shown. ACTH = adrenal corticotrop(h)ic hormone; CRH = corticotrop(h)in releasing hormone; CORT = corticosterone; E_2 = estradiol, FSH = follicle-stimulating hormone; GnRH = gonadotrop(h)in-releasing hormone; LH = luteinizing hormone; T_3 = triiodothyronine; T_4 = tetraiodothyronine (thyroxine); T = testosterone; TSH = thyroid-stimulating hormone; TRH = thyrotrop(h)in-releasing hormone.

and thyroid hormones are all synthesized in specific tissues, circulate bound to globular binding proteins, and each has a specific receptor that is a zinc finger DNA binding protein that acts as a transcription factor (Beato et al., 1996; Freedman, 1992).

Regulation of Sex Steroid, Glucocorticoid, and Thyroid Hormone Synthesis, Secretion, Action, and Degradation

Hormonal Axes

The steroids and thyroid hormones are all part of hormonal axes that involve the hypothalamus and anterior pituitary. The hypothalamus has a blood portal system that allows secretion of its hormones (or releasing factors) directly onto the anterior pituitary, with only negligible release into the general circulation. The anterior pituitary releases its trop(h)ic hormones into the blood, which stimulate both growth (trophic effects) of the targeted peripheral glands as well as the secretion (tropic effects) of the peripheral gland's hormonal products. There is typically a feedback loop where the hormonal products of the peripheral endocrine glands (gonads, adrenal tissue, and thyroid gland) have a suppressive effect on both the releasing hormones of the hypothalamus as well as the trop(h)ic hormones of the anterior pituitary (Fig. 2-1).

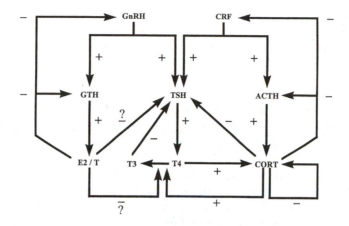

Figure 2-2.
Potential interactions between the classic endocrine axes (Fig. 2-1). The interactions shown were gathered from literature on diverse taxa. All interactions may not occur in a single species. References are cited in the text, and relationships where the data are not completely established are indicated by (?). ACTH = adrenal corticotrop(h)ic hormone; CRH = corticotrop(h)in-releasing hormone; CORT = corticosterone; E_2 = estradiol, FSH = follicle-stimulating hormone; GnRH = gonadotrop(h)in-releasing hormone; LH = luteinizing hormone; T_3 = triiodothyronine; T_4 = tetraiodothyronine (thyroxine); T = testosterone; TSH = thyroid-stimulating hormone; TRH = thyrotrop(h)in-releasing hormone.

Sex Steroids

Gonadotrop(h)in-releasing hormone (GnRH) is a decapeptide secreted by the hypothalamus. This hormone acts on anterior pituitary gonadotrop(h)es to stimulate synthesis and secretion of the gonadotrop(h)ins, luteinizing hormone (LH), and follicle-stimulating hormone (FSH), from the pituitary and into the blood. Luteinizing hormone and FSH are both glycoproteins. Each contains two subunits (α and β) that are synthesized independently. Within a species, the α subunits are the same and the variable β subunits confer specificity.

In males, LH binds to interstitial cells of Leydig and stimulates androgen synthesis and secretion. Follicle-stimulating hormone is primarily responsible for sperm maturation but may enhance the ability of LH to stimulate steroidogenic enzyme activity. In females, both gonadotrop(h)ins are involved in steroidogenesis: Luteinizing hormone stimulates androgen production in the thecal tissue of the follicles. Follicle-stimulating hormone then acts on the granulosa cells to stimulate aromatization of the androgens to estrogens.

All steroids are derived from cholesterol. Tissues that produce steroids are typically mesoderm-derived and have a high content of smooth endoplasmic reticulum. To produce androgens, cholesterol must first be converted to progestins, and androgen synthesis is required to produce estrogens. Estrogens are produced by the aromatase enzyme in an irreversible reaction that aromatizes the A ring of androgens (androstenedione to estrone, or testosterone to estradiol), resulting in the loss of carbon-19.

Steroids are lipophilic molecules, and unlike peptide hormones (which are synthesized and stored before their release is stimulated), sex steroids are generally released immediately following synthesis. However, androgens may be concentrated in the testes by bonding to a large protein (androgen-binding protein) that does not allow the hormone to traverse the blood testis barrier.

Also due to their hydrophobicity, steroid hormones do not dissolve well for travel in the blood. In addition to a loose association with albumen and prealbumen, there are several specific steroid-binding proteins that transport steroids to their target tissues. These binding proteins are large amphipathic globulins and include sex hormone–binding globulin that binds estrogens and androgens. There is a pool of free steroid, and the relative amounts of bound versus free steroid in the blood, however, depends both on the amount of binding protein in circulation and the amount of steroid secreted. In fact, there is some debate regarding whether steroid bound to the binding protein is available to target tissues. Some data argue for delivery of steroid to the target tissue by the binding protein, whereas other data suggest that only free steroid can enter target tissue and that uptake of free hormone causes further release of bound steroid from the binding proteins to maintain an equilibrium (Rosner, 1990).

In fact, the role of binding proteins varies with the hormone, and binding proteins can decrease or increase bioavailability. For example, during fetal development, α-fetoprotein binds to the E_2 produced by the mother and placenta. Because the E_2 is bound to this large globular hormone, it is unable to cross the blood brain barrier and expose the brain of the fetus. Thus, in this case, the binding protein limits bioavailability. In other cases, binding proteins increase bioavailability. Androgen-binding protein binds and prevents androgens from crossing the blood-testis barrier, thus increasing bioavailability in the testes where androgens are necessary for sperm maturation.

Because of their hydrophobic nature, steroids readily traverse the lipid membranes of cells in their target tissues. Once inside the cell, the steroid can also traverse the lipid nuclear membrane, where the sex steroid receptor resides. The steroid receptor belongs to a family of "zinc finger" proteins. These proteins are transcription factors and contain a distinct steroid-binding region and a DNA-binding region that contains the zinc finger. The steroid-binding region actually inhibits the protein from binding DNA and inducing transcription, as experimentally removing the steroid binding region results in constitutive binding to DNA and transcription by the receptor (Carson-Jurica et al., 1990). Thus, the binding of steroid hormones to the hormone-binding region activates the DNA-binding region by removing inhibition conveyed by the steroid-binding domain.

Individual receptors are considered specific for individual steroids, although a receptor is essentially defined by the steroid that it binds with the highest affinity, and no one receptor displays 100% fidelity. Also, the zinc finger domains recognize specific sequences in the promoters of genes (recognition elements). Thus, steroid-regulated genes are defined by their possession of such a recognition element in their promotor. After binding the receptor, the steroid-receptor complex regulates (there is both up and down regulation) specific genes. In many cases, the regulated genes are themselves transcription factors, and the activation by the steroid-receptor complex often sets off a cascade of changes in gene transcription. Testosterone directly regulates genes in the testes and epididymis (Maiti et al., 1996) and the foreskin of the penis (Nitsche et al., 1996). Androgens also up-regulate epidermal growth factor in the submaxillary gland (Sheflin et al., 1996) and may regulate their own receptors (Varialle and Serino, 1994). Estrogens regulate many egg yolk proteins (Lazier et al., 1994) and stimulate gene transcription in the uterus (Hyder et al., 1996).

There are also other possible mechanisms of steroid action, including interactions with cell surface receptors involving second messengers. In fact, some actions of steroids may be exclusively due to these alternative mechanisms. Possible mechanisms include interactions through cell surface receptors, changes in membrane fluidity, interactions with GABA receptors, and possible interactions between steroids and DNA directly without a receptor (Brann et al., 1994).

Once a steroid reaches its target tissue, it may function directly or it may be further metabolized to its active form. For example, testosterone is metabolized (irreversibly) to 5α-dihydrotestosterone (DHT) in many peripheral tissues by the enzyme 5α-reductase. Also, many tissues metabolize androgens (such as testosterone) to estrogens using the aromatase enzyme (also irreversible) After performing its function, the steroid reenters the blood. Most likely, because of the expense of breaking carbon bonds and because many of the active hormones are produced by irreversible reactions, steroids are not typically degraded and recycled but are rather targeted for excretion. This targeting involves the attachment of SO_4 and sugar moieties to oxygens on carbon-3 and carbon-17. The sulfations and glucoronidations, respectively, are regulated by sulfases and glucoronidases in the liver and kidney. Once sulfated and glucoronidated, steroids become hydrophilic, cannot traverse lipid membranes in target tissues, and can be filtered by the blood and excreted in the urine.

Glucocorticoids

Corticotrop(h)in-releasing hormone, secreted by the hypothalamus, is the releasing hormone that ultimately regulates the hypothalmo-pituitary-adrenal axis. Corticotrop(h)in-releasing hormone is under neural control, and most likely cholinergic neurons regulate its secretion. Corticotrop(h)in-releasing hormone stimulates the release of adrenal corticotrop(h)in (ACTH), which stimulates growth of the adrenal tissue and regulates secretion of the gluco-corticoids. Although there is typically a diurnal rhythm of CRH release, the axis is also activated by stress.

The structure of the adrenal tissue varies throughout vertebrates. In mammals, there is a distinct adrenal gland that has a steroid-producing adrenal cortex, surrounding an adrenal medulla that produces cate-cholamines. The adrenal cortex of mammals is divided into three zones: the zona reticularis, zona fasciculata, and zona glomerulosa. These zones possess different steroid-producing enzymes, and only the zona fasciculata produces glucocorticoids. The zona reticularis produces androgens, and the zona glomerulosa produces mineralocorticoids.

The adrenal tissue is not present in the cortex-medulla arrangement across all vertebrates. Similarly, the steroidogenic tissues are typically not arranged in zones in nonmammalian vertebrates. For example, in chondricthian fishes, steroidogenic (interrenal) is completely separate from the catecholaminergic (chromaffin) tissue. In teleosts, the steroidogenic tissue may be associated with the chromaffin tissue or not (depending on the species) and is embedded in the head kidney. In amphibians, the interrenal tissue is dispersed throughout the kidney. Chromaffin and interrenal tissue are intermingled in a distinct adrenal gland in many reptiles. In lizards, the catecholaminergic tissue surrounds the steroidogenic tissue (the reverse of mammals). Finally, in birds, the interrenal tissue is intermingled with the chromaffin tissue in a distinct adrenal gland.

The chemical nature of the active glucocorticoid also varies between species. All glucocorticoids contain 21 carbons (C21) and are synthesized from progestins. The main glucocorticoid in humans is cortisol, whereas corti-costerone is important in most other vertebrates. The key enzyme in corticoid production is 21β-hydroxylase that directs steroid metabolism away from sex steroid production. Other enzymes are important, and, in fact, deficiencies in the enzymes 3β-OH-dehydrogenase, 21β-hydroxylase, or 11β-hydroxylase result in adrenal production of androgens that in early development can androgenize female fetuses.

Once synthesized and secreted, glucocorticoids typically circulate in the blood bound to specific binding proteins such as corticoid-binding globulin (CBG). Glucocorticoids plasma levels can change very rapidly (in minutes, or even seconds) in response to acute stress such as handling (Wingfield et al., 1995). This rapid change suggests that there is some storage of glucocorticoids in the adrenal tissue, as it is unlikely that such a rapid increase in plasma levels could be achieved by newly synthesizing glucocorticoids at the time of the acute stress. Storage of steroids is unusual, however, because they are lipophilic.

Like the sex steroids, glucocorticoids bind to zinc finger proteins that act as transcription factors. Unlike the sex steroids, however, the glucocorticoid receptor resides in the cytoplasm. The glucocorticoid receptor is also unique in that it binds heat shock protein 90 (hsp 90). On binding, glucocorticoids induce the release of hsp 90 and initiate nuclear translocation. The glucocorticoid-receptor complex then binds to specific glucocorticoid recognition elements in the promoters of specific genes. After binding to its receptor, glucocortioids can both up- and down-regulate genes. Important genes regulated by glucocorticoids include those involved in the AP1 complex, which is itself a transcription factor (Karin et al., 1993; Vig et al., 1994) and which is involved in apoptosis (Zhou and Thompson, 1996). Once released from the receptor, glucocortioids return to the blood where they are glucoronidated and sulfanated in the liver and kidney, after which they are secreted as described for the sex steroids.

Thyroid Hormones

In most vertebrates, thyrotropin-releasing hormone (TRH) from the hypothalamus stimulates the release of thyrotrop(h)in (thyroid-stimulating hormone, TSH) from the pituitary, which in turn stimulates the release of thyroid hormone (T_4) from the thyroid gland. In anuran larvae, TRH does not regulate the thyroid axis, but rather CRH stimulates TSH (Denver and Licht, 1989). Thyrotropin is a glycoprotein in the same family as LH and FSH. The thyroid gland is comprised of multiple follicles. The cells of the follicles synthesize and secrete a glycoprotein, thyroglobulin, into the lumen of the follicle. The cells of the follicle also concentrate iodine that is used in the lumen to iodinate tyrosine residues in thyroglobulin. Molecules of thyroid hormone are cleaved from the glycoprotein and secreted by exocytosis.

Thyroxine is the major thyroid hormone secreted from the thyroid gland. Most thyroid hormone circulating in the plasma is bound to thyroid-hormone–binding globulin, along with some binding to albumin and prealbumin. Thyroxine is converted to triiodothyronine (T_3) in peripheral tissues by the enzyme monodeiodinase, and T_3 is the active form of thyroid hormone. Thus, tissue sensitivity to thyroid hormone may be determined by both the amount of available receptor as well as the amount of monodeiodinase.

Thyroid hormone has a specific receptor that is in the same family of zinc finger–binding proteins, as is the steroid receptors. The genes regulated by thyroid hormone during amphibian development and metamorphosis have been studied extensively in *Xenopus laevis*. Thyroid hormone regulates several genes involved in metamorphosis, including genes involved in tail resorption (Brown et al., 1996), limb development (Buckbinder and Brown, 1992), and gut regression and restructuring (Shi, 1996). Thyroid hormone also regulates genes important in brain development in mammals (Poddar et al., 1996).

Following binding and activation of the receptor, thyroid hormones are degraded by a system of monodeiodinase enzymes that progressively remove iodines from the thyroid hormones. The iodines are recycled to the thyroid gland where they are used to synthesize new thyroid hormone. The tyrosine may also be metabolized further and recycled. In addition, thyroid hormones are also glucoronidated and sulfated in the liver similar to steroids (Finnson and Eales, 1996).

Interactions Between Hormonal Axes

In addition to the classical hormonal axes described, there is a great deal of interaction between axes (Fig. 2-2). This interaction is extensive, and, in fact, it is an underestimate to present the regulation of steroids and thyroid hormones as individual axes. For example, regulation of TSH by CRF has already been discussed. In addition, GnRH also regulates TSH in anurans (Denver, 1988). The sex steroids have a negative feedback on the thyroid axis (Sekuli et al., 1995), inhibiting the thyroid gland's response to TSH (Leatherland, 1985) and decreasing monodeiodinase activity (Maclatchy et al., 1986). There are also many interactions between the thyroid hormones and glucocorticoids. Thyroxine can stimulate the production of corticoids in anuran larvae (Krug et al., 1983; Hayes and Wu, 1995a). In turn, corticosterone stimulates the conversion of T_4 to T_3 (Galton, 1990; Hayes and Wu, 1995a) and can further enhance thyroid hormone activity by increasing the binding of T_3 to its receptor (Kikuyama et al., 1983). Corticosterone may also decrease thyroid hormone activity by negative feedback effects on the thyroid axis (Hayes et al., 1993; Hayes and Wu, 1995a).

In addition to cross-regulation of hormone synthesis and release, there is also regulation at the receptor level. Thyroid hormone up-regulates the E_2 receptor, in addition to regulating its own receptor (Rabelo et al., 1993, 1994) in *Xenopus laevis*. Glucocortioids enhance T_3's binding to its receptor (Kikuyama et al., 1983) and may also enhance E_2 binding to its receptor (Ulisse and Tata, 1994). Similarly, thyroid hormone regulates the androgen receptor and determines androgen sensitivity in the larynx of *X. laevis* (Robertson and Kelley, 1996).

Potentially, other complex interactions are also possible: Because T_4 stimulates corticosterone production, many developmental effects in response to T_4 may actually be due to the subsequent increase in corticosterone. In addition, the increase in T_4 to T_3 conversions as a result of exposure to corticosterone may also increase T_3-like effects in response to increases in glucocortioids. Finally, because T_3 induces the estradiol receptor, E_2 effects may be enhanced. In fact, these complex interactions may explain why corticosterone appears to enhance estrogen activity in *Xenopus laevis*. Corticosterone may enhance the ability of T_3 to induce the estrogen receptor.

Interactions of Environmental Contaminants: Potential Mechanisms

The reported mechanisms of endocrine disrupter actions described earlier may represent a small number of ways that endocrine disrupters may function. Environmental contaminants may disrupt endocrine functioning by disrupting hormonal axes or hormone synthesis at any point in the sequences described above. Interference with hypothalamic releasing factors or pituitary trop(h)ic hormones can alter secretion of steroids and thyroid hormone. In addition, interference with binding-protein production or function may alter available pools of hormones, as can altering the half-lives of hormones by altering degradation and excretion rates.

Given the many interactions between thyroid hormones, glucocorticoids, and sex steroids, the true mechanisms of endocrine disrupter action, for any given effect in any given species, may be difficult to discern. For example, a compound that produces corticoid-like effects may do so because it is a corticoid mimic, or because the compound is like thyroxine hormone, and exposure subsequently results in the increased secretion of endogenous corticoids (as seen with thyroxine treatment: Hayes, 1995; Hayes and Gill, 1995; Hayes and Wu, 1995b). In another example, an estrogen mimic could appear to have antithyroid-like effects due to possible inhibition of the thyroid axis or inhibition of monodeiodinase (T_4 to T_3 conversions), as observed with natural estrogens. Thus, understanding the interactions and many levels at which endocrine disrupters may act is vital, and this understanding will come only through careful comparative endocrinological examinations.

The Need for Comparative Studies

Complex Problems

Several aspects of endocrine disruption make the problems more difficult compared to past concerns of environmental contaminants. Problems with environmental pollution that were faced in the past—mortality, deformities, or egg shell thinning—were more obvious effects, for which only the mechanisms of action were unclear. In addition, these obvious effects are direct, and the doses producing such effects are easily measured in the environment and in animal tissues.

In our current problems, not only are the mechanisms unclear but the effects are often hidden as well. Seemingly healthy populations of animals (no obvious deformities or unusual mortality) may disappear as a result of low fertility or decreased recruitment (because of skewed sex ratios, for example). Furthermore, the effects that have been described for vertebrates can vary drastically among species, as described above. Also, it is unclear if these diverse effects are caused by different mechanisms of action between species or the same mechanisms, but manifested in different ways, in different species. Likewise, the effects of compounds on different species may not differ, but apparent differences in effects may simply reflect the fact that different researchers study different phenomena in different animals. For example, the role of thyroid hormone in mammalian physiology has been examined extensively, whereas the role of this hormone in adult amphibian physiology has been ignored (although the role of thyroid hormones has been well studied in larval amphibian development). But how do we predict the effects of thyroid disruption in adult amphibians? The role that thyroid hormones play in mammals is most likely not transferrable to amphibians, nor is the role that these hormones play in larval amphibian development (most studies of thyroid hormone in anuran larvae focus on effects on the tail, a structure that is absent in the adult).

Exposure Routes

Studies of comparative biology are needed to better understand routes of contaminant exposure. Obvious major routes of contaminant exposure are direct exposure to contaminants or consumption of contaminated food and water. Increasing evidence suggests, however, that maternal exposure may be very important, and the mechanisms of maternal exposure will obviously be different for viviparous and oviparous organisms. Recent studies have shown that contaminants can travel across the placenta from the maternal circulation (Bjerke et al., 1994; Huisman et al., 1995; Alaluusua et al., 1996; Darnerud et al., 1996a, b; Li et al., 1995; Schantz, 1996; Schantz et al., 1996) and alter plasma hormone levels in developing fetuses. The resulting alteration in hormone levels can also affect fetal development (Guilette et al., 1996). Further exposure from maternal sources can occur in mammals, where breast milk is contaminated as a result of maternal exposure/consumption by the female (Brunetto et al., 1993; Bjerke et al., 1994; Becher et al., 1995; Granjean et al., 1995; Li et al., 1995; Quinsey et al., 1995; Schlaud et al., 1995; Scheele et al., 1995; Albers et al., 1996; Schantz, 1996; Sinjari et al., 1996; Waliszewski et al., 1996). Similar studies have not been conducted in nonplacental viviparous vertebrates where exposure rates and levels likely differ.

In oviparous animals, exposure via contaminated yolk may expose the developing young as they reabsorb the the yolk. The presence of hormones of maternal origin, such as thyroid hormone (Brown et al., 1987) and sex steroids (Schwabl, 1993) have been shown in the yolk of a number of animals, and these maternal hormones may serve an important role in embryonic development (Lam, 1980; Nacario, 1983; Schwabl, 1993, 1996; Janz et al., 1996). Similarly, contaminants can affect embryonic development and may also be transferred in the yolk (Spear et al., 1990; Nosek et al., 1992; Bishop et al., 1995; Pastor et al., 1996; Walker et al., 1996). Once again, comparative studies are lacking, and maternal transfer of hormones has not been addressed in amphibians or most other oviparous species. Recent studies have shown, however, that contaminants can be transferred from the female to the eggs (Hayes and Noriega, in preparation). In fact, estrogen-mimicking contaminants may ensure their transference in oviparous species by inducing/enhancing vitellogenesis, binding to the resulting vitellogenin protein, and eventually being stored in the developing embryo.

Conclusion

These examples make clear the need for comparative studies across a variety of species and the need to examine any species under study for previously described effects. We must also examine the effects of combined endocrine-disrupting contaminants, given that it is unlikely that organisms are exposed to individual compounds and are more likely bombarded with a mixture of contaminants in their lifetime. Clearly, the concept of the "model" animal will prove inadequate in solving today's problem, and must be replaced by comparative analyses.

References

Adami, H.-O., Lipworth, L., Titus-Ernstoff, L., Hsieh, C.-C., Hanberg, A., Ahlborg, U., Baron, J., and Trichopoulos, D. (1995), Organochlorine compounds and estrogen-related cancers in women. *Cancer Causes & Control* 6(6): 551–566.

Alaluusua, S., Lukinmaa, P.-L., Vartiainen, T., Partanen, M., Torppa, J., and Tuomisto, J. (1996). Polychlorinated dibenzo-p-dioxins and dibenzofurans via mother's milk may cause developmental defects in the child's teeth. *Environ. Toxicol. Pharmacol.* 1(3): 193–197.

Albers, J.M.C., Kreis, I.A., Liem, A.K.D., and Van Zoonen, P. (1996). Factors that influence the level of contamination human milk with poly-chlorinated organic compounds. *Arch. Environ. Contam. Toxicol.* 30(2): 285–291.

Anderson, M.J., Miller, M.R., and Hinton, D.E. (1996a). *In vitro* modulation of 17β-estradiol-induced vitellogenin synthesis: Effects of cytochrome P4501A1 inducing compounds on rainbow trout *(Oncorhynchus mykiss)* liver cells. *Aquatic Toxicol.* (Amsterdam) 34(4): 327–350.

————, Olsen, H., Matsumura, F., and Hinton, D.E. (1996b). *In vivo* modulation of 17-estradiol-induced vitellogenin synthesis and estrogen receptor in rainbow trout *(Oncorhynchus mykiss). Toxicol. Appl. Pharmacol.* 137(2): 210–218.

Arnold, S.F., Robinson, M.K., Notides, A.C., Guillette, L.J., Jr., and McLachlan, J.A. (1996b). A yeast estrogen screen for examining the relative exposure of cells to natural and xenoestrogens. *Environ. Health Perspec.* 104(5): 544-548.

Ayad, V.J., Gilbert, C.L., McGoff, S.A., Matthews, E.L., and Wathes, D.C. (1996). Actions of oxytocin and vasopressin on oestrogen-induced electromyographic activity recorded from the uterus and oviduct of anoestrous ewes. *Reprod. Fert. Develop.* 6(2): 203–209.

Balthazart, J., Foidart, A., Absil, P., and Harada, N. (1996). Effects of testosterone and its metabolites on aromatase-immunoreactive cells in the quail brain: Relationship with the activation of male reproduction behavior. *J. Steroid Biochem. Molec. Biol.* 56(1–6): 185–200.

Barter, R.A. and Klaassen, C.D. (1994). Reduction of thyroid hormone levels and alteration of thyroid function by four representative UDP-glucuronosyltransferase inducers in rats. *Toxicol. Appl. Pharmacol.* 128(1): 9–17.

Beato, M., Chavez, S., and Truss, M. (1996). Transcriptional regulation by steroid hormones. *Steroids* 61(4): 240–251.

Becher, G., Skaare, J.U., Polder, A., Sletten, B., Rossland, O.J., Hansen, H.K, and Ptashekas, J. (1995). PCDDs, PCDFs and PCBs in human milk from different parts of Norway and Lithuania. *J. Toxicol. Environ. Health* 46(2): 133–148.

Benecke, R., Keller, E., Vetter, B., and De Zeeuw, R.A. (1991). Plasma level monitoring of mitotane *(o,p'-*DDD) and its metabolite *(o,p'-*DDE) during long-term treatment of Cushing's disease with low doses. *European J. Clinical Pharmacol.* 41(3): 259–262.

Bishop, C.A., Lean, D.R.S, Brooks, R.J., Carey, J.H., and Ng, P. (1995). Chlorinated hydrocarbons in early life stages of the common snapping turtle *(Chelydra serpentina)* from a coastal wetland on Lake Ontario, Canada. *Environ. Toxicol. Chem.* 14(3): 421–426.

Bjerke, D.L., Sommer, R.J., Moore, R.W., and Peterson, R.E. (1994). Effects of *in utero* and lactational 2,3,7,8-tetrachlorodibenzo-p-dioxin exposure on responsiveness of the male rat reproductive system to testosterone stimulation in adulthood. *Toxicol. Appl. Pharmacol.* 127(2): 250–257.

Brann, D.W., Hendry, L.B., and Mahesh, V.B. (1994). Emerging diversities in the mechanism of action of steroid hormones. *Steroid Biochem. Molec. Biol.* 52(2): 113–133.

Brown, C., Sullivan, C., Bern, H., and Dickhoff, W. (1987). Occurrence of thyroid hormones in early developmental stages of teleost fish. *Am. Fish. Soc. Sympos.* 2: 144–150.

Brown, D.D., Wang, Z., Furlow, J.D., Kanamori, A., Schwartzman, R.A., Remo, B.F., and Pinder, A. (1996). The thyroid hormone-induced tail resorption program during *Xenopus laevis* metamorphosis. *Proc. Nat. Acad. Sci. USA* 93(5): 1924–1929.

Brown, R.E. (1993). Hormonal and experiential factors influencing parental behaviour in male rodents an integrative approach. *Behav. Proc.* 30(1): 1–27.

Brunetto, R., Leon, A., Burguera, J.L., and Burguera, M. (1993). Levels of DDT residues in human milk of Venezuelan women from various rural populations. *Sci. Total Environment.* 186(3): 203–207.

Buckbinder, L., and Brown, D.D. (1992). Thyroid hormone-induced gene expression changes in the developing frog limb. *J. Biol. Chem.* 267(36): 25786–25791.

Bull, J. J., Gutzke, W. H. N., and Crews, D. (1988). Sex reversal by estradiol in three reptilian orders. *Gen. Comp. Endocrinol.* 70(3): 425–428.

Bustos, S., Soto, J., and Tchernitchin, A. (1996). Estrogenic activity of p,p′-DDT. *Environ. Toxicol. Water Qual.* 11(3): 265–271.

Calikoglu, A.S., Gutierrez-Ospina, G., and D'Ercole, A.J. (1996). Congenital hypothyroidism delays the formation and retards the growth of the mouse primary somatic sensory cortex (S1). *Neuroscience Lett.* 213(2): 132–136.

Carson-Jurica, M.A., Schrader, W.T., and O'Malley, B.W. (1990). Steroid receptor family: Structure and function. *Endoc. Rev.* 11(2): 201–220.

Carter, C.S. (1989). The biology of social bonding in a monogamous mammal, In *Comparative Physiology, Vol. 9: Hormones, Brain and Behavior in Vertebrates, 2. Behavioural Activation in Males and Females: Social Interactions and Reproductive Endocrinology.* (J. Balthazart, ed.), pp. 154–164.

Chowen, J.A., Garcia-Segura, L.M., Gonzalez-Parra, S., and Argente, J. (1996). Sex steroid effects on the development and functioning of the growth hormone axis. *Cell. Molec. Neurobiol.* 16(3): 297–310.

Cooke, A.S. (1970). The effect of pp′-DDT on tadpoles of the common frog *(Rana temporaria). Environ. Pollut.* 1: 57–71.

Cooke, P.S., Zhao, Y.-D., and Hansen, L.G. (1996). Neonatal polychlorinated biphenyl treatment increases adult testis size and sperm production in the rat. *Toxicol. Appl. Pharmacol.* 136(1): 112–117.

Crews, D., Bergeron, J.M,. Bull, J.J., Flores, D., Tousignant, A., Skipper, J.K., and Wibbels, T. (1994). Temperature-dependent sex determination in reptiles: Proximate mechanisms, ultimate outcomes, and practical applications. *Develop. Genet.* 15(3): 297–312.

———, ———, McLachlan, J.A. (1995). The role of estrogen in turtle sex determination and the effect of PCBs. *Environ. Health Perspec.* 103(Suppl. 7): 73–77.

Darnerud, P.O., Morse, D., Klasson-Wehler, E., and Brouwer, A. (1996a).

Binding of a 3,3',4,4'-tetrachlorobiphenyl (CB-77) metabolite to fetal transthyretin and effects on fetal thyroid hormone levels in mice. *Toxicology* 106(1): 105–114.

———, Sinjari, T., and Jonsson, C.-J. (1996b). Foetal uptake of coplanar polychlorinated biphenyl (PCB) congeners in mice. *Pharmacol. Toxicol.* 78: 187–192.

Denver, R.J. (1988). Several hypothalamic peptides stimulate *in vitro* thyrotropin secretion by pituitaries of anuran amphibians. *Gen. Comp. Endocrinol.* 72: 383–393.

——— and Licht, P. (1989). Neuropeptide stimulation of thyrotropin secretion in the larval bullfrog: Evidence for a common neuroregulator of thyroid and interrenal activity during metamorphosis. *J. Exp. Zool.* 252: 101–104.

Dodd, M.H., and Dodd, J.M. (1976). The biology of metamorphosis. In *Physiology of the Amphibia, Vol. 3B.* (B.A. Lofts, ed.), pp. 467–599. Academic, New York.

Drent, M.L., and Van Der Veen, E.A. (1995). Endocrine aspects of obesity. *Neth. J. Med.* 47(3): 127–136.

Egeland, G., Sweeney, M., Fingerhut, M., Halperin, W., Wille, K. and Schnorr, T. (1992). Serum dioxin 2,3,7,8 Tetrachlorodibenzo-p-dioxin 2,3,7,8-Tcdd and total serum testosterone and gonadotropins in occupationally exposed men. *Am. J. Epidem.* 136(8): 1014.

———, ———, ———, ———, ———, and Halperin, W.E. (1994). Total serum testosterone and gonadotropins in workers exposed to dioxins. *Am. J. Epidem.* 139(3): 272–281.

Etkin, W., and Gilbert, L.I., eds. (1968). *Metamorphosis, a Problem in Developmental Biology.* Appleton-Century-Crofts, New York.

Finnson, K.W., and Eales, J.G. (1996). Identification of thyroid hormone conjugates produced by isolated hepatocytes and excreted in bile of rainbow trout, *Oncorhynchus mykiss. Gen. Comp. Endocrinol.* 101(2): 145–154.

Flouriot, G., Pakdel, F., Ducouret, B., and Valotaire, Y. (1995). Influence of xenobiotics on rainbow trout liver estrogen receptor and vitellogenin gene expression. *J. Molec. Endocrinol.* 15(2): 143–151.

Forrest, D., Erway, L.C., Ng, L., Altschuler, R., and Curran, T. (1996). Thyroid hormone receptor beta is essential for development of auditory function. *Nature Genetics* 13(3): 354–357.

Freedman, L.P. (1992). Anatomy of the steroid receptor zinc finger region. *Endocrine Reviews* 13(2): 129–145.

Gallien, L. (1974). Intersexuality. In *Physiology of the Amphibia. Vol. II.* (B. Lofts, ed.), pp. 523–549. Academic, New York.

Galton, V.A. (1990). Mechanisms underlying the acceleration of thyroid hormone-induced tadpole metamorphosis by corticosterone. *Endocrinology*

127: 2997–3002.

Giustina, A., and Wehrenberger, W.B. (1995). Influence of thyroid hormones on the regulation of growth hormone secretion. *Europ. J. Endocrinol.* 133(6): 646–653.

Goldman, D., and Yawetz, A. (1990). The interference of Aroclor 1254 with progesterone metabolism in guinea pig adrenal and testes microsomes. *J. Biochem. Toxicol.* 5(2): 99–108.

—— and —— (1991). Cytochrome P-450 mediated metabolism of progesterone by adrenal microsomes of PCB-treated and untreated barn owl *(Tyto alba)* and marsh turtle *(Mauremys caspica)* in comparison with the guinea-pig. *Comp. Biochem. Physiol. C: Pharmacol. Toxicol. & Endocrinol.* 99(1–2): 251–255.

—— and —— (1992). The interference of polychlorinated biphenyls (Aroclor 1254) with membrane regulation of the activities of cytochromes P-450C21 and P-45017a,lyase in guinea pig adrenal microsomes. *J. Steroid Biochem. Molec. Biol.* 42(1): 37–47.

Goldstein, R.E., Wasserman, D.H., Mcguinness, O.P., Lacy, D.B., Cherrington, A.D., and Abumbad, N.N. (1993). Effects of chronic elevation in plasma cortisol on hepatic carbohydrate metabolism. *Am. J. Physiol.* 264(1; Part 1): E119-E127.

Grandjean, P., Weihe, P., Needham, L.L., Burse, V.W., Patterson, D.G. Jr., Sampson, E.J., Jorgensen, P.J., and Vahter, M. (1995). Relation of a seafood diet to mercury, selenium, arsenic, and polychlorinated biphenyl and other organochlorine concentrations in human milk. *Environ. Research* 71(1): 29–38.

Guillette, L.J., Jr., Arnold, S.F., and McLachlan, J.A. (1996). Ecoestrogens and embryos: Is there a scientific basis for concern? *Animal Repro. Sci.* 42(1): 13–24.

——, Gross, T.S., Masson, G.R., Matter, J.M., Percival, H.F., and Woodward, A.R. (1994). Developmental abnormalities of the gonad and abnormal sex hormone concentrations in juvenile alligators from contaminated and control lakes in Florida. *Environ. Health Perspec.* 102(8): 680–688.

—— and Hose, J.E. (1995). Defining the role of pollutants in the disruption of reproduction in wildlife. *Environ. Health Perspec.* 103(Suppl. 4): 87–91.

Gupta, S., and Kanungo, M.S. (1996). Modulation of vitellogenin II gene by estradiol and progesterone in the Japanese quail. *Biochem. & Biophys. Res. Comm.* 222(1): 181–185.

Harper, M.-E., and Brand, M.D. (1995). Use of top-down elasticity analysis to identify sites of thyroid hormone-induced thermogenesis. *Proc. Soc. Exp. & Biol. Med.* 208(3): 228–237.

Hayes, T.B. (1995). Interdependence of corticosterone and thyroid hormones in larval growth and development in the western toad *(Bufo boreas)*. I. Thyroid hormone dependent and independent effects of corticosterone on growth and development. *J. Exp. Zool.* 271(2): 95–102.

———— (1997a). Steroid-mimicking environmental contaminants: Their potential role in amphibian declines. In *Herpetologia Bonnensis.* (Bhme, W., Bischoff, W., and Ziegler, T., eds.), pp.145–150. Bonn (SEH).

———— (1997b). Steroids as potential modifiers of thyroid hormone action in amphibian development. *Am. Zool.* 37: 185–195.

————, Chan, R., and Licht, P. (1993). Interactions of temperature and steroids in larval growth, development, and metamorphosis in a toad *(Bufo boreas). J. Exp. Zool.* 266(3): 206–215.

———— and Gill, T.N. (1995). Hormonal regulation of skin gland development in toad larvae *(Bufo boreas):* The role of the thyroid hormones and corticosterone. *Gen. Comp. Endocrinol.* 99: 161–168.

———— and Licht, P. (1992). Gonadal involvement in size sex dimorphism in the African bullfrog *(Pyxicephalus adspersus). J. Exp. Zool.* 264(2): 130–135.

———— and ———— (1995). Factors influencing testosterone metabolism by anuran larvae. *J. Exp. Zool.* 271(2): 112–119.

———— and Wu, T.-H. (1995a). Interdependence of corticosterone and thyroid hormones in larval growth and development in the western toad *(Bufo boreas).* II. Regulation of corticosterone and thyroid hormones. *J. Exp. Zool.* 271(2): 103–111.

———— and ———— (1995b). The role of corticosterone in anuran metamorphosis and its potential role in stress-induced metamorphosis. Proceedings of the 17th Conference of the European Society of Comparative Endocrinology. *Neth. J. Zool.* 45: 107–109

————, ————, and Gill, T.N. (1999). DDT-like effects as a result of corticosterone-treatment in an anuran amphibian: Is DDT a corticoid-mimic or a stressor? *Environ. Toxicol. Chem.* In press.

Hews, D.K., and Moore, M.C.(1995). Influence of androgens on differentiation of secondary sex characters in tree lizards, *Urosaurus ornatus. Gen. & Comp. Endocrinol.* 97(1): 86–102.

Ho, S.-M. (1991). Vitellogenesis. In *Vertebrate Endocrinology, Vol. 4. Fundamentals and Biomedical Implications, Part A. Reproduction.* (P.K.T. Pang and M.P. Schreibmen, eds.), pp. 91–126. Academic, San Diego.

Ho, K.K.Y., O'Sullivan, A.J., Weissberger, A.J, and Kelly, J.J. (1996). Sex steroid regulation of growth hormone secretion and action. *Hormone Research (Basel).* 45(1–2): 67–73.

Huisman, M., Eerenstein, S.E. J., Koopman-Esseboom, C., Brouwer, M., Fidler, V., Muskiet, F.A.J., Sauer, P.J.J., Boersma, E.R. (1995). Perinatal exposure to polychlorinated biphenyls and dioxins through dietary intake. *Chemosphere* 31(10): 4273–4287.

Hulbert, A.J., and Williams, C.A. (1988). Thyroid function in a lizard, a tortoise and a crocodile, compared with mammals. *Comp. Biochem. & Physiol A Comp. Physiol.* 90(1): 41–48.

Hunt, K., Wingfield, J.C., Astheimer, L.B., Buttemer, W.A., and Hahn, T.P. (1995). Temporal patterns of territorial behavior and circulating testosterone in the Lapland longspur and other Arctic passerines. *Am. Zool.* 35(3): 274–284.

Hyder, S.M., Stancel, G.M., Chiappetta, C., Murthy, L., Boettger-Tong, H.L., and Makela, S. (1996). Uterine expression of vascular endothelial growth factor is increased by estradiol and tamoxifen. *Cancer Res.* 56(17): 3954–3960.

Inui, Y., Yamano, K., and Miwa, S. (1995). The role of thyroid hormone in tissue development in metamorphosing flounder. Symposium on Application of Endocrinology to Pacific Rim Aquaculture, Bodega Bay, Calif., Sep. 7–11, 1994. *Aquaculture* 135(1–3): 87–98.

Janz, D.M., and Bellward, G.D. (1996). *In ovo* 2,3,7,8-tetrachlorodibenzo-p-dioxin exposure in three avian species. 2. Effects on estrogen receptor and plasma sex steroid hormones during the perinatal period. *Toxicol. Appl. Pharmacol.* 130(2): 292–300.

Janzen, F.J., and Paukstis, G.L. (1991). Environmental sex determination in reptiles ecology evolution and experimental design. *Q. Rev. of Biol.* 66(2): 149–180.

Johnson, D.C., Sen, M., and Dey, S.K. (1992). Differential effects of dichlorodiphenyltrichloroethane analogs, chlordecone, and 2,3,7,8-tetra-chlorodibenzo-p-dioxin on establishment of pregnancy in the hypophysectomized rat. *Proc. Soc. Exp. Biol. Med.* 199(1): 42–48.

Johnson, L., Wilker, C.E., Safe, S.H., Scott, B., Dean, D.D., and White, P.H. (1994). 2,3,7,8-Tetrachlorodibenzo-p-dioxin reduces the number, size, and organelle content of Leydig cells in adult rat testes. *Toxicology* 89 (1): 49–65.

Justice, M.J., and Logan, C.A. (1995). Effects of exogenous testosterone on male Northern Mockingbirds during the breeding season. *Wilson Bulletin* 107(3): 538–542.

Karin, M., Yang-Yen, H.-F., Chambard, J.-C., Deng, T., and Saatcioglu, F. (1993). Various modes of gene regulation by nuclear receptors for steroid and thyroid hormones. *Europ. J. Clinical Pharmacol.* 45(Suppl. 1): S9–S15.

Kanamadi, R.D., and Saidapur, S.K. (1993). Effect of testosterone on spermatogenesis Leydig cells and thumb pads of the frog *Rana cyanophlyctis* Schn. *J. Karnatak Univ. Sci.* 31: 157–162.

Kikuyama, S., Niki, K., Mayumi, M., Shibayama, R., Nishikawa, M., and Shitake, N. (1983). Studies on corticoid action on the toad tadpoles *in vitro*. *Gen. Comp. Endocrinol.* 52: 395–399.

————, Kawamura, K., Tanaka, S., and Yamamoto, K. (1993). Aspects of amphibian metamorphosis: Hormonal control. *Internat. Rev. Cytol.* 45: 105–148.

Klaushofer, K., Varga, F., Glantschnig, H., Fratzl-Zelman, N., Czerwenka, E., Leis, H.J., Koller, K., and Peterlik, M. (1995). The regulatory role of thyroid hormones in bone cell growth and differentiation. *J. Nutrition* 125 (Suppl. #7): 1996S–2003S.

Kelley, D.B. (1988). Sexually dimorphic behaviors. In *Annual Review of Neuroscience, Vol. 11.* XII. (W.M. Cowan, ed.), pp. 225–252. Annual Reviews, Palo Alto.

Kleeman, J.M., Moore, R.W., and Peterson, R.E. (1990). Inhibition of testicular steroidogenesis in 2,3,7,8-tetrachlorodibenzo-p-dioxin-treated rats: Evidence that the key lesion occurs prior to or during pregnenolone formation. *Toxicol. Appl. Pharmacol.* 106(1): 112–125.

Ko, J.C.H., Lock, T.F., Davis, J.L., and Smith, R.P. (1989). Spontaneous and oxytocin-induced uterine motility in cyclic and postpartum mares. *Theriogenology* 32(4): 643–652.

Kolle, R., Kierdorf, U., and Fischer, K. (1993). Effects of an antiandrogen treatment on morphological characters and physiological functions of male fallow deer (*Dama dama* L.). *J. Exp. Zool.* 267(3): 288–298.

Koopman-Esseboom, C., Morse, D.C., Weisglas-Kuperus, N., Lutkeschipholt, I.J., Van Der Paauw, C.G., Tuinstra, L. G.M.T., Brouwer, A., and Sauer, P.J.J. (1994). Effects of dioxins and polychlorinated biphenyls on thyroid hormone status of pregnant women and their infants. *Pediatric Research* 36(4): 468–473.

Kovacevic, R., Vojinovic-Miloradov, M., Teodorovic, I., and Andric, S. (1995). Effect of PCBs on androgen production by suspension of adult rat Leydig cells *in vitro. J. Steroid Biochem. Molec. Biol.* 52(6): 595–597.

Krug, E.C., Honn, K.V., Battista, J., and Nicoll, C. (1983). Corticosteroids in serum of *Rana catesbeiana* during development and metamorphosis. *Gen. Comp. Endocrinol.* 52: 232–241.

Lam, T. (1980). Thyroxine enhances larval development and survival in *Sarotherodon (Tilapia) mossambicus. Aquaculture* 21: 287–291.

Lans, M.C., Klasson-Wehler, E., Willemsen, M., Meussen, E., Safe, S., and Brouwer, A. (1993). Structure-dependent competitive interaction of hydroxy-polychlorobiphenyls dibenzo-p-dioxins and dibenzofurans with human transthyretin. *Chemico-Biological Interactions* 88(1): 7–21.

————, Spiertz, C., Brouwer, A., and Koeman, J.H.(1994). Different competition of thyroxine binding to transthyretin and thyroxine-binding globulin by hydroxy-PCBs, PCDDs and PCDFs. *Europ. J. Pharmacol. Environ. Toxicol. Pharmacol.* Sect. 3(2–3): 129–136.

Laugier, C., Courion, C., Pageaux, J.-F., Fanidi, A., Dumas, M.-Y., Sandoz, D., Nemoz, G., Prigent, A.-F., and Pacheco, H. (1988). Effect of estrogen on 3' 5' cyclic amp in quail oviduct possible involvement in estradiol-activated growth. *Endocrinology* 122(1): 158–164.

Lazier, C.B, Wiktorowicz, M., Dimattia, G.E., Gordon, D.A., Binder, R., Williams, D.L. (1994). Apolipoprotein (apo) B and apoII gene expression are both estrogen-responsive in chick embryo liver but only apoII is estrogen-responsive in kidney. *Molec. Cell. Endocrinol.* 106(1–2): 187–194.

Leatherland, J.F. (1985). Effects of 17β-estradiol and methyl testosterone on the activity of the thyroid gland in rainbow trout, *Salmo gairdneri* Richardson. *Gen. Comp. Endocrinol.* 60: 343–352.

Li, X., Weber, L.W.D., and Rozman, K.K. (1995). Toxicokinetics of 2,3,7, 8-tetrachlorodibenzo-*p*-dioxin in female Sprague-Dawley rats including placental and lactational transfer to fetuses and neonates. *Fund. Appl. Toxicol.* 27(1): 70–76.

Liu, C.-X., Takahashi, S., Murata, T., Hashimoto, K., Agatsuma, T., Matsukawa, S., and Highuchi, T. (1996). Changes in oxytocin receptor mRNA in the rat uterus measured by competitive reverse transcription-polymerase chain reaction. *J. Endocrinol.* 150(3): 479–486.

Lye, S.J. (1996). Initiation of parturition. *Animal Reproduction Sci.* 42(1–4): 495–503.

Maclatchy, D.L., Cyr, D.G., and Eales, J. (1986). Estradiol-17β depresses T_4 to T_3 conversion in rainbow trout. *Am. Zool.* 26: 24A.

Maiti, S., Doskow, J., Li, S., Nhim, R.P., Lindsey, J.S., and Wilkinson, M.F. (1996). The Pem homeobox gene: Androgen-dependent and independent promoters and tissue-specific alternative RNA splicing. *J. Biol. Chem.* 271(29): 17536–17546.

Maraud, R., Vergnaud, O., and Rashedi, M. (1990). New insights on the mechanism of testis differentiation from the morphogenesis of experimentally induced testes in genetically female chick embryos. *Am. J. Anat.* 188(4): 429–437.

Matt, K.S. (1990). Neuroendocrine and endocrine correlates of pair bonds and parental care in the seasonal reproductive cycle of the Siberian hamster phodopus-sungorus. (A. Epple, C.G. Scanes, and M.H. Stetson, eds.). *P. Clinic. Biol. Res.* 342: 648–652.

McNabb, F.M.A. (1995). Thyroid hormones, their activation, degradation and effects on metabolism. *J. Nutrition* 125(Suppl. 6): 1773S–1776S.

Mech, L.D., Phillips, M.K., Smith, D.W., and Kreeger, T.J. (1996). Denning behaviour of non-gravid Wolves, *Canis lupus. Canadian Field-Naturalist* 110(2): 343–345.

Mendonça, M.T., Chernetsky, S.D., Nester, K.E., and Gardner, G. L. Effects of gonadal sex steroids on sexual behavior in the big brown bat, *Eptesicus fuscus*, upon arousal from hibernation. *Horm. Behav.* 30(2): 153–161.

Miller, D.B., Earl, L., Gray, Andrews, J.E., Luebke, R.W., and Smialowicz, R.J. (1993). Repeated exposure to the polychlorinated biphenyl Aroclor 1254 elevates the basal serum levels of corticosterone but does not affect the stress-induced rise. *Toxicology* 81(3): 217–222.

Moore, M.C., Thompson, C.W., and Marler, C.A. (1991). Reciprocal changes in corticosterone and testosterone levels following acute and chronic handling stress in the tree lizard, *Urosaurus ornatus. Gen. Comp. Endocrinol.* 81(2): 217–226.

Moore, R.W., Jefcoate, C.R., and Peterson, R.E. (1991). 2,3,7,8-tetra-chlorodibenzo-p-dioxin inhibits steroidogenesis in the rat testis by inhibiting the mobilization of cholesterol to cytochrome $P450_{scc}$. *Toxicol. Appl. Pharmacol.* 109(1): 85–97.

Morley, J.E. (1995). Nutrition and the endocrine system. *Geriatric Nutrition: A Comprehensive Review, 2d ed.*, (J.E. Morley, Z. Glick, and L.Z. Rubenstein, eds.), pp. 265–269. Raven, New York.

Morse, D.C., Groen, D., Veerman, M., Van Amerongen, C.J., Koeter, H.B.W.M., Smits Van Prooije, A.E., Visser, T.J., Koeman, and J.H., Brouwer, A. (1993). Interference of polychlorinated biphenyls in hepatic and brain thyroid hormone metabolism in fetal and neonatal rats. *Toxicol. Appl. Pharmacol.* 122(1): 27–33.

———, Koeter, H.B.W.M., Van Prooijen, A.E.S., Brouwer, A. (1992). Interference of polychlorinated biphenyls in thyroid hormone metabolism: Possible neurotoxic consequences in fetal and neonatal rats. *Chemosphere* 25(1): 165–168.

———, Wehler, E.K., Wesseling, W., Koeman, J.H., and Brouwer, A. (1996). Alterations in rat brain thyroid hormone status following pre- and postnatal exposure to polychlorinated biphenyls (Aroclor 1254). *Toxicol. Appl. Pharmacol.* 136(2): 269–279.

Nacario, J. (1983). The effect of thyroxine on the larvae and fry of *Saratherodon niloticus* L. (*Tilapia nilotica*). *Aquaculture* 34: 78–83.

Nathanielsz, P.W., Guissani, D.A., Mecenas, C.A., Wu, W., Winter, J.A., Garcia-Villar, R., Baguma-Nibasheka, M., Honnebier, M.B.O.M., and McDonald, T.J. (1995). Regulation of the switch from myometrial contractures to contractions in late pregnancy: Studies in the pregnant sheep and monkey. *Reprod. Fertil. & Develop.* 7(3): 595–602.

Nitsche, E.M., Moquin, A., Adams, P.S., Guenette, R.S., Lakins, J.N., Sinnecker, G.H.G., Kruse, K., and Tenniswood, M.P. (1996). Differential display RT PCR of total RNA from human foreskin fibroblasts for investigation of androgen-dependent gene expression. *Am. J. Med. Genet.* 63(1): 231–238.

Nosek, J.A., Craven, S.R., Sullivan, J.R., Olson, J.R., and Peterson, R.E. (1992). Metabolism and disposition of 2,3,7,8-tetrachlorodibenzo-*p*-dioxin in ring-necked pheasant hens, chicks, and eggs. *J. Toxicol. Environ. Health* 35(3): 153–164.

Ogawa, S., Robbins, A., Kumar, N., Pfaff, D.W., Sundaram, K., and Bardin, C.W. (1996). Effects of testosterone and 7-alpha-methyl-19-nortestosterone (MENT) on sexual and aggressive behaviors in two inbred strains of male mice. *Horm. Behav.* 30(1): 74–84.

Ottaviani, E., and Franceschi, C. (1996). The neuroimmunology of stress from invertebrates to man. *Prog. Neurobiol. (Oxford)* 48(4–5): 421–440.

Palmer, B., and Palmer, S.K. (1995). Vitellogenin induction by xenobiotic estrogens in the red-eared turtle and african clawed frog. *Environ. Health Perspec.* 103(Suppl. 4): 19–25.

Pankhurst, N.W., and Carragher, J.F. (1995). Effect of exogenous hormones on reproductive behaviour in territorial males of a natural population of demoiselles, *Chromis dispilus* (Pisces: Pomacentridae). *Marine & Freshwater Research* 46(8): 1201–1209.

Pastor, D., Ruiz, X., Jover, L.,and Albaiges, J. (1996). The use of chorioallantoic membranes as predictors of egg organorine burden. *Environ. Toxicol. & Chem.* 15(2): 167–171.

Pawar, V.G., and Pancharatna, K. (1995). Estradiol-17-b induced oviductal growth in the skipper frog *Rana cyanophlyctis* (Schn). *J. Advanced Zool.* 16(2): 107–109.

Pierce, R.A., Mariencheck, W.I., Sandefur, S., Crouch, E.C., and Parks, W.C. (1995). Glucocorticoids upregulate tropoelastin expression during late stages of fetal lung development. *Am. J. Physiol.* 268(3; Part 1): L491–L500.

Poddar, R., Paul, S., Chaudhury, S., and Sarkar, P.K. (1996). Regulation of actin and tubulin gene expression by thyroid hormone during rat brain development. *Molecular Brain Research* 35(1): 111–118.

Quinsey, P.M., Donohue, D.C., and Ahokas, J.T. (1995). Persistence of organochlorines in breast milk of women in Victoria, Australia. *Food Chem. Toxicol.* 33(1): 49–56.

Rabelo, E.M.L., and Tata, J.R. (1993). Thyroid hormone potentiates estrogen activation of vitellogenin genes and autoinduction of estrogen receptor in adult *Xenopus* hepatocytes. *Molec. Cell. Endocrinol.* 96: 37–44.

———, Baker, B.S. and Tata, J.R. (1994). Interplay between thyroid hormone and estrogen in modulating expression of their receptor and vitellogenin genes during *Xenopus* metamorphosis. *Mechanisms Develop.* 45(1): 49–57.

Robertson, J.C., and Kelley, D.B. (1996). Thyroid hormone controls the onset of androgen sensitivity in the developing larynx of *Xenopus leavis. Dev. Biol.* 176: 108–123.

Rosner, W. (1990). The functions of corticosteroid-binding globulin and sex hormone-binding globulin: Recent advances. *Endoc. Rev.* 11(1): 80–92.

Schantz, S.L. (1996). Developmental neurotoxicity of PCBs in humans: What do we know and where do we go from here? *Neurotoxicol. Terat.* 18(3): 217–227.

———, Seo, B.-W., Moshtaghian, J., Peterson, R.E., and Moore, R.W. (1996). Effects of gestational and lactational exposure to TCDD or coplanar PCBs on spatial learning. *Neurotoxicol. Terat.* 18(3): 305–313.

Scheele, J., Teufel, M., and Niessen, K.-H. (1995). A comparison of the concentrations of certain chlorinated hydrocarbons and polychlorinated biphenyls in bone marrow and fat tissue of children and their concentrations in breast milk. *J. Environ. Path. Toxicol. & Oncology* 14(1): 11–14.

Schlaud, M., Seidler, A., Salje, A., Behrendt, W., Schwartz, F.W., Ende, M., Knoll, A., and Grugel, C. (1995). Organochlorine residues in human breast milk: Analysis through a sentinel practice network. *J. Epid. Comm. Health* 49(Suppl. 1): 17–21.

Schwabl, H. (1993). Yolk is a source of maternal testosterone for developing birds. *Proc. Nat. Acad. Sci., USA* 90(24): 11446–11450.

——— (1996). Maternal testosterone in the avian egg enhances postnatal growth. *Comp. Biochem. Physiol. A.* 114(3): 271–276.

Sekuli, M., Lovren, M., and Popovi, T. (1995). Estradiol-induced changes in TSH-like immunoreactivity of pituitary cells in female rats. *Experientia* 51: 335-338.

Seo, B.-W., Li, M.-H., Hansen, L.G., Moore, R.W., Peterson, R.E., and Schantz, S.L. (1995). Effects of gestational and lactational exposure to coplanar polychlorinated biphenyl (PCB) congeners or 2,3,7,8-tetrachlorodibenzo-p-dioxin (TCDD) on thyroid hormone concentrations in weanling rats. *Toxicol. Lett. (Shannon)* 78(3): 253–262.

Sheflin, L.G., Brooks, E.M., Keegan, B.P., and Spaulding, S.W. (1996). Increased epidermal growth factor expression produced by testosterone in the submaxillary gland of female mice is accompanied by changes in Poly-A tail length and periodicity. *Endocrinology* 137(5): 2085–2092.

Shi, Y-B. (1996). Biphasic intestinal development in amphibians: Embryogenesis and remodeling during metamorphosis. In *Current Topics in Developmental Biology, Vol. 32.* (R.A. Pedersen and G.P. Schatten, eds.), pp. 205–235. Academic, San Diego.

Silva, J.E. (1993). Hormonal control of thermogenesis and energy dissipation. *Trends Endocrinol. Metab.* 4(1): 25–32.

Simmen, F.A. (1987). Expression of the avian C-Erb B Egf receptor protooncogene during estrogen-promoted oviduct growth. *Biochim. Biophys. Acta.* 910(2): 182–188.

Sinjari, T., Klasson-Wehler, E., Oskarsson, A., and Darnerud, P.O. (1996). Milk transfer and neonatal uptake of coplanar polychlorinated biphenyl (PCB) congeners in mice. *Pharmacol. Toxicol.* 78: 181–186.

Snedeker, S.M. (1996). Hormonal and environmental factors affecting cell proliferation and neoplasia in the mammary gland. In *Progress in Clinical and Biological Research, Vol. 394. Cellular and Molecular Mechanisms of Hormonal Carcinogenesis: Environmental Influences.* (J. Huff, J. Boyd, and J.C. Barrett, eds.), pp. 211–253. Wiley-Liss, New York.

Soto, A.M., Chung, K.L., and Sonnenschein, C. (1994). The pesticides endosulfan, toxaphene, and dieldrin have estrogenic effects on human estrogen-sensitive cells. *Environ. Health Perspec.* 102(4): 380–383.

———, Sonnenschein, C., Chung, K.L., Fernandez, M.F., Olea, N., and Serrano, F.O. (1995). The E-SCREEN assay as a tool to identify estrogens: An update on estrogenic environmental pollutants. *Environ. Health Perspec.* 103(Suppl. 7): 113-122.

Spear, P.A., Bourbonnais, D.H., Norstrom, R.J., and Moon, T.W. (1990). Yolk retinoids (vitamin A) in eggs of the herring gull and correlations with polychlorinated dibenzo-p-dioxins and dibenzofurans. *Environ. Toxicol. Chem.* 9(8): 1053–1062.

Tagawa, M., De Jesus, E.G., and Hirano, T. (1994). The thyroid hormone monodeiodinase system during flounder metamorphosis. Symposium on Application of Endocrinology to Pacific Rim Aquaculture, Bodega Bay, Calif., Sep. 7-11, 1994. *Aquaculture* 135(1–3): 128–129.

Tata, J.R. (1994). Gene expression during post-embryonic development: Metamorphosis as a model. *Proc. Indian National Sci. Acad. Part B Biol. Sci.* 60(4): 287–301.

——— (1996). Amphibian metamorphosis: An exquisite model for hormonal regulation of postembryonic development in vertebrates. *Development Growth & Differentiation* 38(3): 223–231.

Thomas, A.,and Amico, J.A. (1996). Sequential estrogen and progesterone (P) followed by P withdrawal increases the level of oxytocin messenger ribonucleic acid in the hypothalamic paraventricular nucleus of the male rat. *Life Sci.* 58(19): 1615–1620.

Ulisse, S., and Tata, J.R. (1994). Thyroid hormone and glucocorticoid independently regulate the expression of estrogen receptor in male *Xenopus* liver cells. *Molec. Cell. Endocrinol.* 105: 45–53.

Van Birgelen, A.P.J.M., Smit, E.A., Kampen, I.M., Groeneveld, C.N., Fase, K.M., Van Der Kolk, J., Poiger, H., Van Den Berg, M., Koeman, J.H., and Brouwer, A. (1995). Subchronic effects of 2,3,7,8-TCDD or PCBs on thyroid hormone metabolism: Use in risk assessment. *Europ. J. Pharmacol. Environ. Toxicol. Pharmacol. Sect.* 5(1): 77–85.

Van den Hurk, R., Richter, C.J.J., and Janssen-Dommerholt, J. (1989). Effects of 17a-methyltestosterone and 11b-hydroxyandrostenedione on gonad differentiation in the African catfish, *Clarias gariepinus. Aquaculture* 83(1–2): 179–192.

Van Goozen, S.H.M., Cohen-Kettenis, P.T., Gooren, L.J.G., Frijda, N.H., and Van De Poll, N.E. (1995). Gender differences in behaviour: Activating effects of cross-sex hormones. *Psychoneuroendocrinol.* 20(4): 343–363.

————, Frijda, N., and Van De Poll, N. (1994). Anger and aggression in women: Influence of sports choice and testosterone administration. *Aggressive Behavior* 20(3): 213–222.

Varriale, B., and Serino, I. (1994). The androgen receptor mRNA is up-regulated by testosterone in both the Harderian gland and thumb pad of the frog, *Rana esculenta. J. Steroid Biochem. Molec. Biol.* 51(5–6): 259–265.

Vailes, L.D., Washburn, S.P., and Britt, J.H. (1992). Effects of various steroid milieus or physiological states on sexual behavior of Holstein cows. *J. Animal Sci.* 70(7): 2094–2103.

Vig, E., Barrett, T.J., and Vedeckis, W.V. (1994). Coordinate regulation of glucocorticoid receptor and c-jun mRNA levels: Evidence for cross-talk between two signaling pathways at the transcriptional level. *Molec. Endocrinol.* 8(10): 1336–1346.

Waliszewski, S.M., Pardio Sedas, V.T., Chantiri, P. J.N., Infanzon, R.R.M., and Rivera, J. (1996). Organochlorine pesticide residues in human breast milk from tropical areas in Mexico. *Bull. Environ. Contam. Toxicol.* 57(1): 22–28.

Walker, M.K., Cook, P.M., Batterman, A.R., Butterworth, B.C., Berini, C., Libal, J. J., Hufnagle, L.C., and Peterson, R.E. (1996). Translocation of 2,3,7,8-tetrachlorodibenzo-p-dioxin from adult female lake trout *(Salvelinus namaycush)* to oocytes: Effects on early life state development and sac fry survival. *Can. J. Fish Aquat. Sciences* 51(6): 1410–1419.

Wallace, H., Pate, A., and Bishop, J.O. (1995). Effects of perinatal thyroid hormone deprivation on the growth and behaviour of newborn mice. *J. Endocrinol.* 145(2): 251–262.

Wang, W.L., Thomsen, J.S., Porter, W., Moore, M., and Safe, S. (1996). Effect of transient expression of the oestrogen receptor on constitutive and inducible CYP1A1 in Hs578T human breast cancer cells. *British J. Cancer* 73(3): 316–322.

Whitworth, J.A., Brown, M.A., Kelly, J.J., and Williamson, P.M. (1995). Mechanisms of cortisol-induced hypertension in humans. *Steroids* 60(1): 76–80.

Wingfield, J. C., O'Reilly, K.M., and Astheimer, L.B. (1995). Modulation of the adrenocortical responses to acute stress in Arctic birds: A possible ecological basis. *Am. Zool.* 35(3): 285–294.

Witt, D.M., Carter, C.S., and Insel, T.R. (1991). Oxytocin receptor binding in female prairie voles: Endogenous and exogenous estradiol stimulation. *J. Neuroendocrinol.* 3(2): 155–162.

Young, L.J., Nag, P.K., and Crews, D. (1995). Species differences in behavioral and neural sensitivity to estrogen in whiptail lizards: Correlation with hormone receptor messenger ribonucleic acid expression. *Neuroendocrinology* 61(6): 680–686.

Zhou, F., and Thompson, E.B. (1996). Role of c-jun induction in the glucocorticoid-evoked apoptotic pathway in human leukemic lymphoblasts. *Molec. Endocrinol.* 10(3): 306–316.

Chapter 3

BIOACCUMULATION, STORAGE, AND MOBILIZATION OF ENDOCRINE-ALTERING CONTAMINANTS

Charles F. Facemire

Ivy Tech State College
590 Ivy Tech Drive
Madison, IN 47250

Introduction

A major tenet of toxicology holds that the "dose" of a contaminant determines the compound's toxicity. However, it is not actually the dose that is important but the concentration of the compound at the site of action. As Rozman and Klaassen (1996) state, "the concentration of the compound at the site of action is proportional to the dose, but the same dose of two or more chemicals may lead to vastly different concentrations in a particular target organ of toxicity." This concept particularly deserves attention in the study of endocrine-disrupting contaminants, given the numerous studies recently showing an apparent lack of dose-response behavior of chemicals. This chapter will review the major factors affecting the concentration of endocrine-altering contaminants at the site of action: namely, bioaccumulation, storage, and mobilization of compounds.

What is the relationship between exposure to a pollutant and the resulting concentration of the pollutant in the tissues of an organism? It is well known that most if not all biota accumulate a wide variety of chemical residues from the media in which they live and from the food and liquid that they ingest. For the purposes of the discussion that follows, I have chosen, with others (e.g., Moriarty, 1983; Barber et al., 1991), to label the accumulation of a pollutant from all possible sources as bioaccumulation rather than bioconcentration. The latter term is generally used to refer to the uptake of a pollutant by an organism directly from the medium in which it lives, resulting in a concentration of the pollutant in the organism greater than that occurring in the environment (LeBlanc, 1995).

Tissue concentrations of some pollutants increase as they are transferred from one organism to another in a food chain. Although some authors (e.g., Clarkson, 1995) also refer to this phenomenon as bioaccumulation, it is generally termed biomagnification. However, all of these phenomena are ultimately dependent on the availability of the pollutant to the organism. Bioavailability of a pollutant is governed by several factors; but, as noted by Anderson et al. (1987), ultimately the question of bioavailability of any toxicant can be answered only by observation of the organisms so exposed (but see Huckins et al., 1990; Prest et al., 1995; Lebo et al., 1996).

The literature contains numerous examples of sublethal concentrations of pollutants that disrupt the endocrine system of virtually every species investigated. While some of these pollutants are short-lived hydrophilic pollutants (i.e., an organophosphate that changes the activity of a steroid-producing enzyme), most endocrine-altering contaminants are lipophilic compounds stored in fatty tissues throughout the body, and these lipophilic endocrine-disruptors are the focus of this chapter. The purpose of this chapter is not to present an exhaustive review of this literature but to explore, insofar as is possible, how and where these pollutants are accumulated, stored, and subsequently mobilized. Consequently, the references I have cited are intended primarily to provide an avenue for entry into the literature.

Bioavailability

By definition, a pollutant is a contaminant (something released into the environment due to man's activities) that elicits a biological effect (Moriarty, 1983). I shall try to adhere to this distinction, although I have been unable to call to mind a contaminant that does not in some way affect at least one species, and so, in practice, I tend to use the terms synonymously. In any case, if a contaminant is to become a pollutant, it must be assimilated into the cells or intercellular interstices of a living organism.

The availability of a contaminant for uptake by a living organism is influenced by several factors, including the organic carbon content, redox potential (Eh), and pH of the medium (i.e., soil or water), as well as the water solubility of the compound (Kenaga and Goring, 1980; Adams, 1987; Jenne and Zachara, 1987; Rodgers et al., 1987). However, Adams (1987) suggested that bioavailability is primarily a function of: (1) feeding habits (i.e., where an organism forages and what is ingested) and (2) the route of exposure (e.g., ingestion, inhalation, dermal contact). Although the first of these may not be of major importance for aquatic organisms for which dietary intake appears, at least in some cases, to be a relatively minor route of exposure (Adams, 1987), both must be considered when referring to terrestrial organisms. For example, Roelke et al. (1990) observed that only those Florida panthers *(Felis concolor coryi)* that fed almost exclusively on raccoons were at risk due to methylmercury exposure.

Bioavailability of a given contaminant is also dependent on the persistence of the contaminant in the environment (Table 3-1). Many organisms can tolerate high concentrations of contaminants for a short period of time, whereas chronic exposure to a much lower concentration may result in serious injury. Organochlorine pesticides and other halogenated hydrocarbons such as polychlorinated biphenyls (PCBs) generally are the most persistent contaminants in the environment. For example, it has been estimated that mirex residues may be biologically available in some systems for up to 600 years (Eisler, 1985). As a direct result, residues of these highly persistent pollutants are ubiquitous in the environment. Thus, they are still found in plant and animal (including human) tissues, not only in countries where such organochlorine pesticides are still used (Kashyap et al., 1994; Gladen and Rogan, 1995; Mukherjee and Gopal, 1996) but also in the United States (Youngs et al., 1994; Facemire et al., 1995; Kidwell et al., 1995) and other countries (Hirai and Tomokuni, 1993) where they have been prohibited for several years.

Organophosphate, carbamate, and synthetic pyrethroid pesticides are generally less persistent in the environment, respectively, as they undergo rapid

Table 3-1. Estimated environmental persistence of selected groups of man-made organic compounds[a,b]

Group	Estimated half-life
Lead-, arsenic-, copper-, and mercury-based pesticides	10–30 years
Organochlorine insecticides and polychlorinated biphenyls	2–>4 years
Bipyridylium herbicides	>2 years
Triazine herbicides	0.5–2 years
Substituted urea herbicides and picloram	4–12 months
Benzoic acid, dinitroaniline, and amide herbicides	3–12 months
Phenoxy and toluidine herbicides	1–8 months
Carbamates and aliphatic acid herbicides	0.5–6 months[b]
Synthetic pyrethroid insecticides	2–13 weeks[b]
Organophosphorus insecticides	1–12 weeks[b]
Fungicides	1–4 weeks[b]

[a]After Witkowski et al. (1987).

[b]Sources in addition to those cited in the original: Ahmad et al. (1979), Bennett et al. (1986), Eisler (1992), Frederick et al. (1994), Liu and Hsiang (1994), Odenkirchen and Eisler (1988), and Smith et al. (1995).

transformation to different chemical forms. However, the transformation (breakdown) products may be more persistent and more toxic than the parent compound. For example, half-lives of phoratoxon sulfone and terbufoxon sulfone, oxidation products of phorate and terbufos, are about 2.5 times greater than the half-lives of their respective parent products (Ahmad et al., 1979).

The degradation rate or half-life of a contaminant may be influenced by a number of factors (Pritchard, 1987; Gambrel and Patrick, 1988; Facemire, 1991; Szeto and Price, 1991; Smith et al., 1995); however, organic matter appears to be the most important factor in soil. Szeto and Price (1991) found that mean organochlorine pesticide residues present in Fraser Valley, British Columbia, muck, soils in 1989 were still approximately 7% (aldrin) to 40% (lindane) of those reported from the same farms in 1971.

In general, it has been assumed that chemicals sorbed to soil particles are not bioavailable and thus not subject to microbial degradation. However, this does not seem to be true in all cases. Rodgers et al. (1987) illustrated that, although sorption to particulates altered the bioavailability of chemicals to aquatic organisms, the direction and degree of alteration were not easily predictable, and Pritchard (1987) reported greatly increased biodegradation of fenthion, an organophosphate insecticide, with the addition of sediments to seawater.

Bioaccumulation, Bioconcentration, and Biomagnification

Bioaccumulation of a pollutant is affected by numerous factors, both extrinsic and intrinsic to an organism (Table 3-2).

Organisms in both terrestrial (Leita et al., 1991; McLachlan, 1996) and aquatic (Ekelund et al., 1990; Ahel et al., 1993) food chains accumulate certain pollutants to the extent that they may become toxic (Leita et al., 1991; Roelke et al., 1991; Clarkson, 1995). However, most investigations have focused on the aquatic food chain. There are at least two reasons for this emphasis. First, as Clarkson (1995) has noted, we have little control over contaminants entering the aquatic system. Second, aquatic food chains are generally longer and more complex than terrestrial ones. This usually results in pollutant concentrations in top predators (whether aquatic or terrestrial) many thousands (or millions; e.g., Colburn, 1993) of times greater than the concentration in the water or sediment. Thus, whereas the pollutant concentration in water may be below the limit of detection of currently used analytical methods, serious injury or death may result in organisms at the top of the food chain (e.g., Facemire et al., 1995; White and Hoffman, 1995; Henny et al., 1996).

Clarkson (1995) stated that three special properties are required for significant bioaccumulation (i.e., biomagnification) to occur. The pollutant must: (1) have a high octanol-water coefficient (P) (i.e., be lipophilic), (2) be chemically and metabolically stable in water and in organisms in the food chain,

Table 3-2. Factors influencing pollutant bioaccumulation in an aquatic system[a]

Factors affecting the amount of bioaccumulation

Type of Factor	Dietary uptake	Uptake from water	Clearance

Extrinsic	Nature of chemical
	Water temperature
	Dietary pollutant
	  concentration
	Nature of diet
		Nature of chemical
		Water temperature
		Water quality parameters
		Water concentration
		  of pollutant
		Oxygen concentration
		  in water
			Nature of chemical
			Water temperature
			Water quality parameters
			Presence of other
			  pollutants in water
			Concentration of
			  pollutant in tissues

Intrinsic	Body size
	Metabolic rate
	Growth rate
	Assimilation coefficient
	  of pollutant
	Capacity to degrade
	  pollutant
	Satiation volume
	Gastrointestinal tract
		Body size
		Metabolic rate
		Growth rate
		Assimilation coefficient
		  of pollutant
		Oxygen assimilation
		  efficiency
			Body size
			Metabolic rate (?)
			Adiposity
			Capacity to degrade
			  pollutant

[a]After Roberts et al. (1979), Table 1.

respectively, and (3) have a relatively low toxicity to organisms at the low end of the food chain to allow accumulation by top predators. The most notable examples of the few chemical compounds that meet these requirements are PCBs, dioxins, organochlorine pesticides, and methylmercury (Mason et al., 1986; Roelke et al., 1991; Clarkson, 1995); but based on data presented by Ekeland et al. (1990), it would appear that nonylphenols should be included in this list.

However, a positive correlation between pollutant concentration and trophic level may not necessarily be an indication of biomagnification. LeBlanc (1995) illustrated that the concentration of certain lipophilic pollutants increased with trophic level, not due to biomagnification (which was only noted to occur with compounds having a bioconcentration factor[1] (BCF>114, 000 or a log P > 6.3) but rather due to two other factors: (1) a positive, but apparently nonlinear, correlation between lipid content and trophic level and (2) a negative correlation between chemical elimination efficiency (depuration rate) and trophic level. It is also apparent that different species assimilate and metabolize the same pollutant differently, resulting in the retention of different amounts of any given pollutant (Moriarty, 1983). However, Gobas et al. (1993) clearly demonstrated that biomagnification can and does occur, at least in goldfish and human infants, by simple diffusion of lipophilic pollutants across the mucosa of the gastrointestinal tract.

Someone once said, "All models are wrong, but some are useful." Several of the latter that address the accumulation of pollutants by various organisms (mainly fish) have been developed during the last 2 or 3 decades. A thorough review of many earlier models, all of which address bioconcentration by aquatic organisms, has been provided by Kenaga and Goring (1980). They concluded that water solubility and P (the octanol-water partition coefficient) of a chemical compound were the two factors that could be used to predict the bioconcentration potential of a compound in fish (and the sorption of the compound by soil) when experimental values were not available. More recent models addressing accumulation by both aquatic and terrestrial organisms have been proposed by a number of authors including Roberts et al. (1979), DiToro et al. (1987), Lake et al. (1987), Walker (1987), and McLachlan (1996).

Models for aquatic and terrestrial species differ primarily because aquatic species (as well as some soil-dwelling organisms, bacteria, and lichens) accumulate pollutants from, and lose some portion of their pollutant load into, the medium in which they live. For example, Roberts et al. (1979) examined the usefulness of a model based on the assumption that the rate of pollutant uptake from water and food is directly linked to the food and oxygen requirements of the fish. Stated in terms of the pollutant body burden (P) in grams

$$\frac{dP}{dt} = A_w\, C_w\, V_r + A_f\, C_f\, V_f - kclF(x)\, P\, w(\delta+1)$$

where C_w and C_f represent pollutant concentrations in water and food, respectively; V_r is the respiration volume (grams of water passing the gill per day); and V_f is the estimated daily food requirement (g). The terms A_w and A_f represent the net amount of pollutant from water and food actually assimilated, and the expression $kclF(w)\, P\, w(\delta+1)$ represents the clearance rate of the pollutant. In this expression, kcl is the clearance rate constant, $F(x)$ other control factors that might influence clearance (Table 3-2), P the body burden, and $w(\delta+1)$ the dilution factor (change in pollutant concentration due to change in weight), respectively. The clearance rate of methylmercury was not temperature-dependent, and the rates for cis- and $trans$-chlordane were dependent on body size. Thus, the authors concluded that clearance rate was the only variable seemingly not always controlled by the metabolic rate.

Roberts et al. (1979) also noted that : (1) dietary net assimilation coefficients (A_f) of persistent, highly lipophilic compounds are much greater than those for more hydrophilic ones; (2) for some chlorinated hydrocarbons, A_f increases and clearance rates $(kclF(x)\, P\, w(\delta+1))$ decrease with increasing amount of adipose; and (3) the assimilation coefficient for persistent chlorinated hydrocarbons from water (A_w) is generally greater than 0.75. Although not stored in fat, methylmercury exhibited the highest A_f (0.63–0.95) of any pollutant tested; DDT ($A_f = 0.89$) ranked second. This model worked well in predicting the accumulation patterns of chlorinated hydrocarbons and methyl mercury in laboratory studies and, according to the authors, was helpful in explaining much of the variation noted in field studies.

DiToro et al. (1987), with many others (see Kenaga and Goring, 1980), suggested that the potential for biological uptake of an organic contaminant is a function of the fugacity (thermodynamic activity) of the contaminant in what they referred to as the "exposure environment." They proposed that the addition of a sediment to a fixed volume of water which contains a fixed quantity of contaminant will result in a decrease of chemical activity with a concomitant decrease in potential for biological uptake. Thus, assuming a state of sorption equilibrium, chemical activity in both the sorbed and solution phases should be equal. Consequently, the potential for bioconcentration, in a thermodynamic sense, from either sediment or water also will be equal. The actual source is dependent on the relative kinetics of the processes involved, the physical properties of the system, and the habits of the organism involved. And, as no suitable kinetic model that addresses all these factors is currently available, they (DiToro et al., 1987) proposed that the maximum concentration of a lipophilic compound in an organism can be predicted by

$$\underset{\text{org}}{C^{max}} = \frac{P_{l,oc}\, f_l\, C_{sed}}{f_{oc}}$$

where $P_{l,oc}$ equals the ratio of the activity coefficient, γ, of the contaminant for organic carbon (oc) and lipid (l) phases, respectively; f_l and f_{oc} represent the mass fraction of lipid and organic carbon in the organism and sediment, respectively; and C_{sed} is the sediment contaminant concentration. The advantage of using this model, as noted by the authors, is that one can estimate the maximum concentration of any given nonpolar contaminant in an organism without knowing how the contaminant is distributed between sediment and water, the BCF, or the mode of uptake.

Schrap and Opperhuizen (1990) confirmed that bioconcentration is significantly reduced when fish are exposed in a sediment suspension and that the reduction is positively correlated (i.e., the reduction is greater) with hydrophobicity of the compound. However, they also noted that sediment concentrations in the intestine of the fish itself also reduces total body burden of the compound.

Terrestrial organisms, however, are more limited. As Walker (1987) noted, they have neither the problems nor benefits associated with uptake and loss of pollutants by diffusion from or to the ambient medium. In these species, pollutant uptake is primarily by ingestion and subsequent gastrointestinal absorption (Gobas et al., 1993; Jan et al., 1996), and loss occurs mainly due to metabolism of the pollutant into a form that may be easily excreted.[2] Walker (1987) proposed that if an animal ingests a pollutant at a constant rate, a steady-state concentration will eventually result and that

$$\text{Rate of uptake } (R_U) = \frac{C_o \times A_i}{N} \text{ mg absorbed kg body weight}^{-1} \text{ day}^{-1}$$

where C_o is the pollutant concentration in food (parts per million [ppm] by weight), Ai is the fraction of C_o that is absorbed, and N is the number of days required to ingest an amount of food equal to its own body weight.

Conversely, the

$$\text{Rate of loss} = \frac{0.5 \ Ci}{t50} = \text{mg lost kg body weight}^{-1} \text{ day}^{-1}$$

where C_i is the initial concentration (ppm) in the whole animal at the steady-state and $t50$ is the initial half-life (in days). Given that, by definition, the steady-state concentration is obtained when uptake and loss are equal, i.e., when

$$\frac{C_o \times A_i}{N} = \frac{0.5 \ C_i}{t50}$$

by rearrangement of terms, it is then evident that the

$$\text{Bioaccumulation factor (BAF)} = \frac{2 \times A_i \times t50}{N}$$

But whereas this model considers steady-state pollutant concentrations in terrestrial species, McLachlan (1996) questions whether a steady-state concentration is ever reached in human populations due to the existence of a positive correlation between tissue pollutant concentrations and age[3] (to be discussed later).

Storage in Adipose Tissue

Prior to discussing the storage of lipophilic pollutants in fatty tissues of humans, other animals, and plants, I have chosen to present a brief review of the literature pertaining to fatty acids and the synthesis of triacylglycerols: the form of lipid stored in adipose tissues. There are several excellent sources (e.g., Lehninger, 1975; Snyder, 1977) that provide in-depth discussions of fatty acid and triacylglycerol synthesis and metabolism, and the reader should refer to these or other similar sources for information additional to that presented hereafter.

Most fatty acids in plant or animal tissue occur as esters of the alcohol glycerol, with only traces occurring in free (unesterified) form in cells and tissues (Lehninger, 1975). The most abundant forms have an even number of carbon atoms and vary in length from 14 to 22 carbon atoms. Of this group, those with 16 or 18 carbon atoms are dominant. The carbon chain may be saturated (each carbon with the maximum possible number of covalently bonded hydrogen atoms) or unsaturated (one or more hydrogens missing with resultant double bonds between carbons). Fatty acids are often identified using a shorthand notation (Table 3-3) that identifies the number of carbon atoms and the position of the double bond(s), if any. Thus, stearic acid, a saturated 18-carbon fatty acid, is designated 18:0, whereas oleic acid, which has 18 carbon atoms with a double bond between carbons 9 and 10, is designated as $18:1\omega9$.

Palmitic (16:0), stearic (18:0), oleic ($18:1\omega9$), and linoleic ($18:2\omega6$) acids are dominant in both plant and animal tissues (White et al., 1973; Lehninger, 1975; Table 3-4), but unsaturated fatty acids are predominant, especially in higher plants and animals that live in colder climates (Lehninger, 1975).

As previously mentioned, fatty acids are usually found in plant and animal tissues as glycerol esters; i.e., as a constituent of triacylglycerol molecules that function as depot or storage lipids (Lehninger, 1975). Triacylglycerols are synthesized, in vertebrates, mainly by liver and fat cells (adipocytes). Adipose tissue, the major storage site for triacylglycerols, constitutes about 18% of the body weight of a normally fed human (Shapiro, 1977) and thus provides a major energy store, as complete oxidation of a triacylglycerol yields 2.5 times as many molecules of ATP as does an equal amount (by weight) of glycogen (Lehninger, 1975).

As triacylglycerols in the blood are either in the form of chylomicrons (which result from the intestinal absorption of fat) or lipoproteins (formed in the liver), it is to be expected that, when fatty acids are involved, "one is what one eats." Marckmann et al. (1995) found that daily fish consumption was positively correlated with adipose tissue docosahexaenoic acid ($22:6\omega4$, 7, 10, 13, 16, 19) content. Positive correlations between five common dietary fatty

Table 3-3. Melting points of some naturally occurring fatty acids[a]

Symbol	Structure	Common name	m.p. (°C)
	Saturated fatty acids		
14:0	$CH_3(CH_2)_{12}COOH$	Myristic	53.9
16:0	$CH_3(CH_2)_{14}COOH$	Palmitic	63.1
18:0	$CH_3(CH_2)_{16}COOH$	Stearic	69.6
20:0	$CH_3(CH_2)_{18}COOH$	Arachidic	76.5
	Unsaturated fatty acids		
$16:1\omega7$	$CH_3(CH_2)_5CH=CH(CH_2)_7COOH$	Palmitoleic	-0.5
$18:1\omega9$	$CH_3(CH_2)_7CH=CH(CH_2)_7COOH$	Oleic	13.4
$18:2\omega6$	$CH_3(CH_2)_4CH=CHCH_2CH=CH(CH_2)_7COOH$	Linoleic	-5.0
$18:3\omega3$	$CH_3CH_2CH=CHCH_2CH=CHCH_2CH=CH(CH_2)_7COOH$	Linolenic	-11.0
$20:4\omega6$	$CH_3(CH_2)_4(CH=CHCH_2)_3CH=CH(CH_2)_3COOH$	Arachidonic	-49.5

[a] After Lehninger (1975), Table 11.2.

Table 3-4. Fatty acid composition of depot, liver, and milk fat from various sources[a]

		Depot fat				Liver fat	Milk fat	
Acid	Symbol	Human	Cow	Pig	Sheep	Cow	Human[b]	Cow
Butyric	4:0							9
Caproic	6:0							3
Caprylic	8:0							2
Capric	10:0							4
Lauric	12:0							3
Myristic	14:0	3	7	1	2	3		11
Palmitic	16:0	23	29	28	25	35	25	23
Stearic	18:0	6	21	10	26	5	7	9
Other saturated fatty acids[c]								11
Palmitoleic	16:1ω7	5				10	3	4
Oleic	18:1ω9	50	41	58	42	36	35	26
Linoleic	18:2ω6	10	2	3	5	8	12	3
Other unsaturated fatty acids		3	5					

a After White et al. (1973), Table 4.4.

b From Martin et al. (1993).

c Includes 10:0, 12:0, 14:0, 15:0, 17:0, 20:0, 22:0, and 24:0.

acids and adipose tissue concentrations of the same compounds in rabbits were reported by Lin et al. (1993). Similar phenomena have been noted in human subjects (Felton et al., 1994; but see Rodbell, 1960). Thus, the ultimate source for depot fat is blood-borne triacylglycerols. However, hydrolysis (accomplished in part by lipoprotein lipase) and resynthesis of a triacylglycerol molecule is essential for incorporation of these fatty acids into adipose tissue (Lehninger, 1975; Shapiro, 1977).

However, all adipocytes are apparently not the same. In an *in vitro* study of adipocytes from three different abdominal sites (subcutaneous epigastric, greater omentum, intestinal mesentery) of severely obese men and women, Edens et al. (1993) found that triacyl- and diacylglycerol synthesis rates differed significantly between sites. Whereas total acylglycerol synthesis was similar in intra-abdominal tissues from both sexes, total lipid synthesis was 69% higher in subcutaneous tissue from women than in adipocytes from the same location in men. Triacylglycerol, which comprises a higher percentage of total newly synthesized lipid in female subcutaneous tissue than in men (Edens et al., 1990), synthesis in the other two sites was much lower: 33% and

54% for omental and mesenteric tissues, respectively. In men, triacylglycerol synthesis in adipose tissue from all three sites was less than that observed in female subcutaneous tissue. Diacylglycerol synthesis was greatest (and lipolysis least) in intra-abdominal tissues from both males and females.

In humans, the synthesis of triacylglycerol from fatty acids is regulated by adipsin/acylation stimulating protein (ASP), which is synthesized by maturing adipocytes, via the enzyme diacylglycerol acyltransferase (Cianflone et al., 1994). Triacylglycerols in depot fats of plant and animal tissues usually contain two or more different fatty acids, but little is known regarding the mechanisms responsible for specifying the identity and relative position of the different fatty acids (Lehninger, 1977). Uptake is maximal under conditions of high caloric uptake and minimal when the animal is starved (Shapiro, 1977). However, site-specific differences in the fatty acid composition of triacylglyc-erols (triglycerides) have been noted (Bhattacharyya et al., 1987; Malcom et al., 1989; Phinney et al., 1994).

Phinney et al. (1994) found that concentrations of saturated fatty acids were significantly greater ($P < 0.01$ to $P < 0002$) in abdominal adipose tissue than in adipose tissue from either inner or outer thigh. Conversely, polyunsaturates were greater ($P < 0.06$) in outer thigh than in abdominal tissue. Some differences were also noted between inner and outer thigh areas. As the melting point of saturated and unsaturated fatty acids differs greatly (Table 3-3), the mean melting points of the accumulated fatty acids also varied significantly ($P < 0.03$ to $P < 0.008$) by site (abdomen = 22.3°C, inner thigh = 20.6°C, and outer thigh = 19.7°C). Bhattacharyya et al. (1987) and Malcom et al. (1989) found similar differences when comparing fatty acid composition of human adipose tissue from perirenal, abdominal, and buttock areas. Saturates were greater and monounsaturates less in perirenal than in buttock adipose tissues; however, polyunsaturates were similar at all sites. These data tend to indicate that fatty acid accumulation and composition may be metabolically controlled and struc-tured to the specific conditions, including temperature, of the site. Thus, as noted by Phinney et al. (1994), although it is partly true that "one is what one eats, " it may be more accurate to state that "one is what one saves from what one eats."

It is interesting to note that Bhattacharyya et al. (1987) also found that the percentage of palmitoleic acid was consistently lower ($P < 0.001$) and that of stearic acid consistently greater ($P < 0.001$) in black males than in white males. In general, the percentages of total saturates and polyunsaturates decreased and monounsaturates increased with age in both perirenal and buttock adipose tissue in blacks, but only for the perirenal site for whites. The cause and impli-cations of these differences, particularly in the present context, are unknown.

The selective retention of saturated fatty acids in abdominal and visceral adipose tissue may be associated with increased health risk as several studies (e.g., Sellers et al., 1992; Bengtsson et al., 1993; Shapira et al., 1993) have found a positive correlation between central obesity (high waist:hip circumference ratio) and incidence of breast cancer, total mortality, and mortality due to myocardial infarction. Human males are apparently at greater risk than human females due to the accumulation of significantly greater amounts of adipose tissue around the abdomen and viscera in relation to total body fat (Lemieux et al., 1993). The possible role that lipophilic pollutants stored in abdominal and visceral adipose tissue may play will be discussed hereafter.

Although it is generally supposed that obesity is the result of eating too much, mounting evidence indicates that the storage of fat is genetically controlled and is regulated by the brain via a protein that is capable of sensing the amount of fat present in the body (Bennett, 1995; Lindpainter, 1995). Thus, at least to a certain extent, we may not have much choice in regulating body fat deposits.

Storage of Lipophilic Contaminants

As with fatty acids, one is also what one eats when speaking of contaminants. Studies by Skerfving (1988), Roelke et al. (1991b), Hovinga et al. (1993), Asplund et al. (1994), and Schantz et al. (1994), for example, have shown that the amount and type of food ingested are good predictors of lipophilic contaminant concentrations in humans and other animals. However, even though variability of bioaccumulation of a given contaminant in a predator seems to be tied to the variability of the contaminant concentration in the prey (Mendenjian and Carpenter, 1993), contaminants are selectively stored in adipose tissue. That is, contaminant concentrations in adipose tissue do not necessarily match relative concentrations in the diet.

In a study cited by Stickel (1973), although 30% of technical grade DDT fed to hens was o,p'-DDT, only 10% of the total DDT stored was o,p'-DDT; the remaining 90% was p,p'-DDT. The same study also illustrated that some organochlorine pesticides have a greater propensity for storage than others as, in hens, bioaccumulation of heptachlor epoxide ≥ dieldrin > endrin > DDT > lindane. Of perhaps greater importance in the present context, DDE, dieldrin, and heptachlor epoxide residues were still present after 26 weeks of uncontaminated diet.

Other studies cited by Stickel (1973) showed that chemicals fed to mammals in combination were stored at different relative concentrations than when fed separately. Thus, it appears that the presence of one chemical may either enhance or inhibit the storage of another.[4]

Most contaminants contained in solution in ingested fatty acids are carried by serum lipoproteins and are readily assimilated into resynthesized triacylglycerols for storage in adipose tissue (but see Gobas et al., 1993). However, dioxin (2,3,7,8-tetrachlorodibenzo-p-dioxin, TCDD), is transported primarily by chylomicrons in the lymphatic system (Lakshmanan et al., 1986). In blood, dioxin is predominantly associated with low-density lipoproteins (55%) as opposed to very-low-density lipoproteins (17%) or high-density lipoproteins (27%) (Marinovich et al., 1983; Henderson and Patterson, 1988). Regardless of the carrier molecule involved, dioxin leaves the circulatory system rapidly. Lakshmanan et al. (1986) reported that the initial half-life (loss of 67% of the compound) of TCDD in the blood compartment is less than 1 min and that the remaining amount has a half-life of approximately 30 min. Most of the compound is ultimately stored in adipose and liver tissue (Ryan et al., 1985; Poiger and Schlatter, 1986). The tissue distributions of coplanar PCBs and polybrominated biphenyls (PBBs) appears to be similar to TCDD (Clevenger et al., 1989; Wehler et al., 1989); however, several investigators (e.g., Abraham et al., 1988, Poiger et al., 1989; Leung et al., 1990) have noted that tissue distribution of TCDD is dose-dependent (but see Rose et al., 1976).

Residue Concentrations

As discussed earlier in this chapter, Walker (1987) constructed a model that predicts the steady-state pollutant concentration in a terrestrial species. However, McLachlan (1996) suggested that a steady-state concentration is never reached in human populations due to the existence of increasing tissue concentrations with age. Several studies (e.g., Mori et al., 1983; Ansari et al., 1986; Jan and Tratnik, 1988; Mes, 1990; Gómez-Catalán et al., 1991, 1995; Duarte-Davidson et al., 1994; Falandysz et al., 1994; Burgaz et al., 1995; Waliszewski et al., 1995) have shown that tissue concentrations of chlorinated hydrocarbons, including PCBs, increase with age of the contaminated individual and indicate that contamination of the human population by these chemicals is apparently ubiquitous.

Focardi et al. (1986), Williams et al. (1988), Burgaz et al. (1995), and Gómez-Catalán et al. (1995) showed significantly greater concentrations of some of these chemicals (generally all organochlorines other than DDT and associated metabolites (but see Williams et al., 1988) in adipose tissue of females than in that of males. Others (Mori et al., 1983; Ansari et al., 1986; Mes et al., 1990; Falandysz et al., 1994; Waliszewski et al., 1995) reported greater mean residue concentrations of some pollutants (notably total DDT, p,p'-DDE, and some PCB congeners) in males than in females. However, the differences noted in either case were not always significant.

Clark and Krynitsky (1983) found that DDE concentrations (ppm gram lipid[-1]) were significantly (28%) greater in brown than in white adipose tissue of hibernating bats. They believed that this difference was due to a difference in fatty acid composition of the two types of adipose tissue, although this was not determined. Laukola (1980) reported differences in fatty acid composition of brown and white fats (in other species), whereas Smalley and Dryer (1967) did not (in bats). If a difference in fatty acid composition of brown and white fats does exist, the difference noted in DDE concentrations would be indicative of a close association of this pollutant with one or more particular fatty acids.

Effects of Pollutants on Lipid Metabolism

Pollutants may also have an adverse effect on lipid metabolism and, thus, may indirectly influence their own sotrage and longevity in the body. Kahn (1990), in an excellent review of the literature, reported abnormal fatty acid metabolism in rats that resulted in dermatitis of the skin and tail as a result of exposure to DDT. This condition seems to be related to a decreased ability of liver microsomes to desaturate palmitate and 8,11,14-ercosatrienoate to arachidonate. He also noted that DDT enhances hepatic metabolism of estrogens, especially in avian species. Exposure to other pollutants produces a variety of effects on body lipids, including the following.

Dioxin affects hepatic metabolism of lipids and lipoproteins, increases the concentration of unsaturated fatty acids and triacylglycerols (in rat blood and liver), and causes fat necrosis, serous fat atrophy, and hypercholesterolemia (a condition in which greater than normal amounts of cholesterol are present in the blood) in humans and other species (Kahn, 1990). Exposure to dioxin also causes an apparent increase in lipid content of interscapular brown adipose tissue (IBAT) followed a few days later by increasing lipid depletion concomitant with the accumulation of glycogen in IBAT (in rats) (Rozman et al., 1986). These authors also noted a total depletion of both lipids and glycogen, as well as several swollen or ruptured mitochondria, and that some 14 days postexposure, IBAT in exposed animals ceased to be functional.

Mirex, an organochlorine insecticide, causes hypercholesterolemia, hyperlipidemia (an excess of lipids in the plasma), and fatty liver (lipid accumulation in the liver), increases concentrations of triacylglycerols in (rat) blood and liver, and apparently affects the hepatic composition of cholesterol, cholesterol esters, and other hepatic lipids in rats. Decreased choline phosphoglyceride concentrations concomitant with increased concentrations of triacylglycerols, monoacylglycerols, cholesterol esters, and ethanolamine-, serine-, and inositol-phosphoglycerides accompanied by high concentrations of 16:1, 18:1 and 20:4 fatty acids with a reduction in saturated fatty acids have been observed. Mirex and lindane, an organochlorine insecticide, cause a significant increase of arachodonic acid in the fatty acid fraction of lung and kidney phosphatityl ethanolamine in rats (Kahn, 1990).

Exposure to fenvalerate, a synthetic pyrethroid insecticide, results in hypercholesterolemia in rats (Kahn, 1990). Kahn (1990) also noted that the synthesis of cholesterol and fatty acids in rat liver homogenates is altered by the phenoxyacetic acids 2,4-D and 2,4,5-T, and that many chemicals (e.g., mirex, TCDD, DDT, PCBs, carbaryl, and the cyclodienes) cause lipid mobilization, which is addressed later in this chapter.

Biotransformation

In animals, most lipophilic contaminants are metabolized, resulting in products that are less hydrophobic than the parent compound. Such metabolites may either be excreted directly or conjugated with an easily excretable group such as an amino acid, sugar, or sulfate (Harwood et al., 1994). However, some of these metabolites may be more toxic, more soluble, or exhibit greater storage potential or persistency than the parent compound.

For example, Stickel (1973) cited a study by J. G. Cummings and coworkers that clearly illustrated the metabolism in chickens of stored DDT into DDE long after presentation of treated food had ceased. Although DDT has been shown to be mildly estrogenic (Arnold et al., 1996), DDE is a potent antiandrogen (Kelce et al., 1995).

Although most biotransformation takes place in the liver, the mammalian gonad (both testes and ovaries) is also capable of metabolizing a number of exogenous chemicals. Both cytochrome P-450 and arylhydrocarbon hydroxylase are present in testicular tissue, and a variety of mixed-function oxidases and cytochrome systems are found in the ovary (Thomas, 1991). Consequently, several enzymes in pathways leading to steroidogenesis may be affected by the metabolites of a number of agricultural and industrial contaminants, including dibromochloropropane (DCP), phthalate acid esters, epichlorohydrin, tri-o-cresyl phosphate, n-hexane, 2-methoxyethanol, and acrylamide (Thomas, 1991).

Xenobiotic Lipids

Xenobiotic Fatty Acids

Some compounds or their metabolites may resemble naturally occurring lipids to the extent that they act as substrates for enzymes involved in lipid biosynthesis (Caldwell and Marsh, 1983; Harwood et al., 1994). These authors note that the simplest class of xenobiotic lipids are those resulting from elongation of a xenobiotic carboxylic acid to form a xenobiotic fatty acid. Although there are a few examples in the literature of some herbicides being elongated by metabolic processes in plants, similar data are generally lacking for animal species. However, Harwood et al. (1994) reported that mammalian (rat) tissues can convert cyclopropanecarboxylic acid, a metabolite of the miticide cycloprate, into 10-, 12-, or 14-carbon xenobiotic fatty acids, apparently via the action of fatty acid synthetase.

Although there has been no report of toxicity resulting from xenobiotic fatty acids (Harwood et al., 1994), one cannot rule out possible effects on the endocrine system, especially in light of the fact that estrogen biosynthesis occurs in human adipose tissue (Zhao et al., 1995).

Fatty Acid Esters of Xenobiotic Alcohols

Xenobiotic alcohols may form fatty acid esters that subsequently may be incorporated into triacylglycerols (Harwood et al., 1994). A number of halogenated pollutants, including 2-chloro-, 2-bromo-, and trichloroethanol; pentachlorophenol; 3-chloro-1,2-propanediol; chlorambucil; and metabolites of DDT, have been shown to form fatty acid esters in animal tissue, apparently due to the action of an acyl-CoA acyltransferase. Palmitate (16-carbon) esters are most frequently found; however, Harwood et al. (1994) reported that neither chain length nor degree of unsaturation appeared to have much bearing on esterification of the xenobiotic alcohol.

Xenobiotic fatty acid esters may elicit a toxic response, as most studies suggest that the properties of the xenobiotic ester are similar to those of the parent compound. For example, fatty acid ethyl esters have been shown to disrupt plasma membranes (Hungund et al., 1988 [cited in Harwood et al., 1994]). Once the xenobiotic ester is conjugated, it generally becomes more lipophilic and is likely incorporated into a molecule of triacylglycerol for storage and concentration in adipose tissue. Thus, results are usually observed only after hydrolysis of the triacylglycerol and subsequent liberation and mobilization of the ester.

Neutral Acylglycerols

The incorporation of xenobiotic acids into analogues of triacyl- or diacylglycerol apparently occurs much more frequently than the two situations discussed above (Harwood et al., 1994). This group of compounds, which includes the 2-substituted propanoic acids (e.g., "profen" nonsteroidal anti-inflammatory drugs and "fop" herbicides), may also form xenobiotic cholesterol esters. Little is known regarding the mechanism involved in the synthesis of xenobiotic acylglycerols, but there are several enzymes, including diacyl acyltransferase and glycerol 3-phosphate acyltransferase, which may drive the necessary reactions. Although there appears to be no evidence that xenobiotic acylglycerols are toxic, there is evidence that at least some may be able to mimic second messenger activities of their natural counterpart (Harwood et al., 1994).

One could perhaps consider the storage (in adipose tissue) of a xenobiotic compound, regardless of chemical form, a protective event. For example, Lassiter and Hallam (1990) found that toxic response to a pollutant was inversely proportional to the amount of body fat and concluded that, in similarly exposed organisms, the "fattest survives the longest." However, as noted by Harwood et al. (1994), storage increases the persistence and concentration of the compound that, when released due to the action of a lipolytic hormone, may result in serious consequences for the host in times of high energetics, such as reproduction or migration.

Mobilization

As noted by Shapiro (1977), fat reserves are stored in adipocytes as large droplets consisting mainly of triacylglycerols. Hydrolysis releases free fatty acids, glycerol, and any lipophilic pollutants (e.g., see Van Velzen et al., 1972) present into the blood stream.[5] The rate of release is dependent on several factors, including the nutritional state of the individual (Van Velzen et al., 1972; Shapiro, 1977; Ross and Rissanen, 1994). However, Shapiro (1977) notes that many of the fatty acids resulting from hydrolysis may be reesterified and restored rather than released into the blood stream.

Triacylglycerol hydrolysis is catalyzed by a "hormone-sensitive" lipase, which is activated by a number of "lipolytic hormones" including, but not necessarily limited to, epinephrine, norepinephrine, adrenocorticotropic hormone, thyroid-stimulating hormone, and glucagon (Shapiro, 1977). On the other hand, insulin causes a decrease in fatty acid release, resulting in part from an increase in fatty acid reesterification in the adipocytes. Insulin has also been noted to inhibit epinepherine-induced lipolysis (Shapiro, 1977).

Reduced energy intake, especially in combination with increased activity, results in lipid mobilization. Ross and Rissanen (1994) noted significant reductions in both visceral and subcutaneous adipose tissue of human females as a result of reduced dietary uptake and either resistance or aerobic exercise. However, the magnitude of reduction differed with tissue location. That is, there was a preferential reduction of subcutaneous adipose tissue from the abdomen when compared with the lower body, and reduction of visceral adipose tissue was significantly greater in the upper abdominal region (15 cm above L4) than that noted in the area of the lower abdomen (5 cm below L4). As noted by Schaefer et al. (1983), reduction of adipose tissue during dietary restriction is almost entirely due to the mobilization of triacylglycerols rather than cholesterol and tocopherol (but see Andersen et al., 1995).

Lipid mobilization is also induced by pollutant exposure (Khan, 1990; Matsumura, 1995). Hyperlipidemia has been reported in several species as a result of exposure to various pollutants, including mirex, carbaryl, and PCBs. In addition to hyperlipidemia, Khan (1990) reported a complete loss of perirenal fat pads, 60% to 70% reduction of epididymal fat stores, and nearly complete depletion of subcutaneous fat in guinea pigs 15 days after injection of 2 µg kg⁻¹ i.p. TCDD. Increased blood concentrations of cholesterol, triglycerides, free fatty acids, and phospholipids were observed within 3 days postinjection. Perhaps of greatest interest, at least in the present context, was that the release of fatty acids after injection was apparently selective. That is, 18:2 fatty acids were dominant, and 16:0, 18:0, and 18:1 fatty acids low in serum lipids after treatment. In adipocytes of treated animals, a slight increase in 16:0 and 16:1 concomitant with a decrease in 18:2 and 17:0 fatty acids was also noted (Kahn, 1990).

Ross and Rissanen (1994) noted that the incidence of blood lipid disorders, coronary heart disease, and noninsulin-dependent diabetes are greater in those individuals who preferentially store fat in the abdominal area. Thus, they suggested that a regimen leading to reduction of these particular fat stores should reduce health risks associated with upper-body obesity. However, when lipids are mobilized, lipophilic pollutants are transported throughout the body as well. In at least one case, pollutant mobilization resulted in death of the organism.

In a study by Van Velzen et al. (1972) designed to show the effects of DDT mobilization due to restricted diet, 20 brown-headed cowbirds *(Molothrus ater)* were fed 100 ppm of DDT for 13 days, full rations of untreated food for 2 days, followed by a restricted ration (43% of normal). Seven of the 20 birds died within 4 days after being placed on the restricted diet, whereas none of the control group (similar treatment except for diet restriction) died. In a separate pilot study, 21 of 30 birds died within 6 days of being placed on restricted rations. Six of the remaining nine birds died during a second period of restricted rations some 4 months later. Clark and Krynitsky (1983) reported body burdens of DDE in two species of bats sufficient to cause death if mobilized and transferred to the brain, which, they believed, could likely occur during arousal from hibernation.

Other effects of pollutant mobilization are more subtle. Noting that premature births and spontaneous abortions were correlated with high levels of PCBs in plasma of human females, Rao and Banerji (1988) sampled amniotic fluid from 26 women during normal delivery. They found PCBs in all 26 samples with concentrations ranging from 0.001 ppm to 1.162 ppm (mean = 0.131±0.26 S.D.). Given the small size of an embryo in comparison to the amount of amniotic fluid, they believed that, if the concentrations noted were representative of the general population, many fetuses are at substantial risk. Using rats as a model, Khanna et al. (1991) found that lindane from maternal fat depots readily crossed the placenta of pregnant females and was deposited in fetal tissues during gestation.

Lead has long been known to cause reproductive impairment in virtually all mammalian species. Maternal exposure to this toxic metal has resulted in numerous manifestations, including reduced litter size, retardation of fetal development, and reduced survival of offspring. Unlike those pollutants previously discussed, lead is not lipophilic, but, like uranium, is stored in the bone. During conditions of bone demineralization, usually prompted by low blood calcium, lead is released from the bone and mobilized into the blood. Al-Saleh et al. (1995) reported that, although lead concentrations in umbilical cord blood were lower than those noted in maternal blood, concentrations were still sufficient (7–10 µg/dl) in cord blood of 7% of 126 newborns examined to cause concern, as the Center for Disease Control has identified central nervous system impairment in children due to blood lead concentrations as low as 10 µg/dl.

As in mammalian adults, infants are what they eat. Maternal diet and mobilized fatty acids are the sources for essential fatty acids of their offspring (Martin et al., 1993). The metabolism of adipose tissue, which is directed toward mobilization rather than accretion during lactation, is also a major source of pollutants. Several studies (Conway et al., 1985; Skerfving, 1988; Galetin-Smith et al., 1990; Dewailly et al., 1991; Krauthacker, 1991; Sitarska et al., 1991; Georgii et al., 1995) have reported the presence of elevated concentrations of pollutants in human and other mammalian breast milk. In female mammals, the major route of excretion for lipophilic pollutants and methylmercury is via milk secretion. Although breast feeding may serve to reduce the pollutant body burden of a lactating female, the offspring are at greater risk. For example, unlike lead, concentrations of methylmercury in blood of infants are greater than concentrations in maternal blood (Nelson et al., 1971; Suzuki et al., 1971). Given that the excretion of methylmercury from infants is probably slower than for adults and that the central nervous system of an infant is likely more susceptible to this toxin than that of an adult (Clarkson et al., 1985), it is not difficult to see why this may be the case. However, more recent studies (Gladen and Rogan, 1995; Neville and Walsh, 1995) have indicated that milk secretion is impaired by exposure to many pollutants, including DDT and dioxin. This may reduce the risk to newborns somewhat.

As previously mentioned, it is apparent in humans that those individuals who store fat primarily in the abdominal area are more likely to experience blood lipid disorders, coronary heart disease, and noninsulin-dependent diabetes. Abdominal obesity has also been correlated with an increased incidence of breast cancer (Sellers et al., 1992; Shapira et al., 1993).

As abdominal adipose tissue is mobilized preferentially when additional energy reserves are required, it seems likely that the concomitant mobilization of lipid-soluble pollutants could be a major factor in the proliferation of breast cancer. Wolff et al. (1993) believed that exposure to DDE was strongly associated with increased risk for breast cancer (but see Key and Reeves, 1994).

In any case, there are still several unanswered questions regarding the accumulation and storage of lipophilic pollutants. For example, are they stored preferentially? That is, is one more likely to find a given pollutant in saturated or unsaturated fats, or in monounsaturated as opposed to polyunsaturated fats? Is there a connection between increased rates of breast cancer and the presence of estrogenic or antiandrogenic pollutants in abdominal adipose tissue? Are adipose cells stimulated to produce estrogens by xenobiotics, and could these estrogens be active in the proliferation of breast cancer? As is usually the case in an exercise such as this, I seem to have raised more questions than I have answered. But I hope that I have provided a stimulus for further research that will enlarge our understanding, not only of how our bodies assimilate, store, or metabolize pollutants, but also how we may be able to counteract the effects of these pollutants.

Footnotes

1 Bioconcentration factor = steady-state concentration of the pollutant in organism tissue (dry weight) ÷ pollutant concentration in water or other medium.
2 There are some exceptions to this general rule. Methylmercury, for example, is sequestered in the hair, fur, or feathers of terrestrial organisms, and this appears to be a major route of excretion for this nonlipophilic pollutant.
3 This same phenomenon has been noted in largemouth bass *(Micropterus salmoides)* contaminated with methylmercury (e.g., see Lange et al., unpublished manuscript; Brim, 1994).
4 There is also some evidence that the presence of one pollutant may alter the toxicity of another. For example, several studies have demonstrated that selenium may ameliorate the toxic effects of methylmercury (Eisler, 1985, 1987). Several authors have reported synergistic effects from exposure to two or more pollutants (see Facemire, 1991).
5 The reader should remember that some of these free fatty acids and glycerol subunits may be modified pollutants (see the section Xenobiotic Lipids, pp. 66).

References

Abraham, K., Krowke, R., and Neubert, D. (1988). Pharmacokinetics and biological activity of 2,3,7,8-tetrachlorodibenzo-*p*-dioxin. 1. Dose-dependent tissue distribution and induction of hepatic ethoxyresorufin o-deethylase in rats following a single injection. *Arch. Toxicol.* 62: 359–368.

Adams, W.J. (1987). Bioavailability of neutral lipophilic organic chemicals contained on sediments: A review. In *Fate and Effects of Sediment-Bound Chemicals in Aquatic Systems* (K.L. Dickson, A.W. Maki, and W.A. Brungs, eds.), pp. 219–244. Pergamon, New York.

Ahel, M., McEvoy, J., and Giger, W. (1993). Bioaccumulation of the lipophilic metabolites of nonionic surfactants in freshwater organisms. *Environ. Pollut.* 79: 243–248.

Ahmad, N., Walgenbach, D.D., and Sutter, G.R. (1979). Comparative disappearance of fonofos, phorate, and terbufos soil residues under similar South Dakota field conditions. *Bull. Environ. Contam. Toxicol.* 23: 423–429.

Al-Saleh, I., Khalil, M.A., and Taylor, A. (1995). Lead, erythrocyte protoporphyrin, and hematological parameters in normal maternal and umbilical cord blood from subjects of the Riyadh region, Saudi Arabia. *Arch. Environ. Health* 50: 66–73.

Andersen, R.E., Wadden, T.A., Bartlett, S.J., Vogt, R.A., and Weinstock, R.S. (1995). Relation of weight loss to changes in serum lipids and lipoproteins in obese women. *Am. J. Clin. Nutr.* 62: 350–357.

Anderson, J., Birge, W., Gentile, J., Lake, J., Rodgers, J. Jr., and Swartz, R. (1987). Biological effects, bioaccumulation, and ecotoxicology of sediment-associated chemicals. In *Fate and Effects of Sediment-Bound Chemicals in Aquatic Systems* (K. L. Dickson, A.W. Maki, and W.A. Brungs, eds.), pp. 267–296. Pergamon, New York.

Ansari, G.A.S., James, G.P., Hu, L.A., and Reynolds, E.S. (1986). Organochlorine residues in adipose tissue of resident of the Texas Gulf Coast. *Bull. Environ. Contam. Toxicol.* 36: 311–316.

Arnold, S.F., Robinson, M.K., Notides, A.C., Guillette, L.J., Jr., and McLachlan, J.A. (1996). A yeast estrogen screen for examining the relative exposure of cells to natural and xenoestrogens. *Environ. Health Perspect.* 104: 544–548.

Asplund, L., Svensson, B-G., Nilsson, A., Eriksson, U., Jansson, B., Jensen, S., Wideqvist, U., and Skerfving, S. (1994). Polychlorinated biphenyls, 1, 1, 1-trichloro-2, 2-bis(*p*-chlorophenyl)ethane (*p,p'*-DDT) and 1, 1-dichloro-2, 2-bis(*p*-chlorophenyl)-ethylene (*p,p'*-DDE) in human plasma related to fish consumption. *Arch. Envrion. Health* 49: 477–486.

Barber, M.C., Suávez, L.A., and Lassiter, R.R. (1991). Modeling bioaccumulation of organic pollutants in fish with an application to PCBs in Lake Ontario salmonids. *Can. J. Fish. Aquat. Sci.* 48: 318–337.

Bengtsson, C., Bjorkelund, C., Lapidus, L., and Lissner, L. (1993). Associations of serum lipid concentrations and obesity with mortality in women: 20-year follow up of participants in prospective population study in Gothenburg, Sweden. *Brit. Med. J.* 307: 1385–1388.

Bennett, W.I. (1995). Beyond overeating. *N. Engl. J. Med.* 332: 673–674.

Bennett, R.S., Klaas, E.E., Coats, J.R., and Kolbe, E.J. (1986). Fenvalerate concentrations in the vegetation, insects, and small mammals of an old-field ecosystem. *Bull. Environ. Contam. Toxicol.* 36: 785–792.

Bhattacharyya, A.K., Malcom, G.T., Guzman, M.A., Kokatnur, M.G., Oalmann, M.C., and Strong, J.P. (1987). Differences in adipose tissue fatty acid composition between black and white men in New Orleans. *Am. J. Clin. Nutr.* 46: 41–46.

Brim, M. (1994). Mercury in largemouth bass of the Lake Woodruff National Wildlife Refuge. Publ. No. PCFO-EC 94-05, U.S. Fish and Wildlife Service, Panama City, FL, 14 pp.

Burgaz, S., Afkham, B.L., and Karakaya, A.E. (1995). Organochlorine pesticide contaminants in human adipose tissue collected in Tebriz (Iran). *Bull. Environ. Contam. Toxicol.* 54: 546–553.

Caldwell, J., and Marsh, M.V. (1983). Interrelationships between xenobiotic metabolism and lipid biosynthesis. *Biochem. Pharmacol.* 32: 1667–1671.

Cianflone, K., Roncari, D.A.K., Maslowska, M., Baldo, A., Forden, J., and Sniderman, A.D. (1994). Adipsin/acylation stimulating protein system in human adipocytes: Regulation of triacylglycerol synthesis. *Biochemistry* 33: 9489–9495.

Clark, D.R., Jr., and Krynitsky, A.J. (1983). DDE in brown and white fat of hibernating bats. *Environ. Pollut.* (Series A) 31: 287–299.

Clarkson, T.W. (1995). Environmental contaminants in the food chain. *Am. J. Clin. Nutr.* 61: 682–686.

———, Nordberg, G.F., and Sager, P.R. (1985). Reproduction and developmental toxicity of metals. *Scand. J. Work Environ. Health* 11: 145–154.

Clevenger, M.A., Roberts, S.M., Lattin, D.L., Harbison, R.D., and James, R.C. (1989). The pharmacokinetics of 2, 2′, 5, 5′-tetrachlorobiphenyl and 3, 3′, 4, 4′-tetrachlorobiphenyl and its relationship to toxicity. *Toxicol. Appl. Pharmacol.* 100: 315–327.

Colburn, T. (1993). Animal/health connection. *In* Proceedings of the U.S. Environmental Protection Agency's National Technical Workshop "PCBs in Fish Tissue," EPA/823-R-93-003." U.S. Environmental Protection Agency, Washington, DC.

Conway, I.S., Perriman, W.S., and Whitney, S. (1985). Organochlorine pesticide residue levels in human milk in Western Australia, 1979–1980. *Arch. Environ. Health* 40: 102–108.

Dewailly, É. Weber, J-P., Gingras, S., and Laliberté, C. (1991). Coplanar PCBs in human milk in the Province of Québec, Canada: Are they more toxic than dioxin for breast fed infants? *Bull. Environ. Contam. Toxicol.* 47: 491–498.

Duarte-Davidson, R., Wilson, S.C., and Jones, K.C. (1994). PCBs and other organochlorines in human tissue samples from the Welsh population: I-Adipose. *Environ. Pollut.* 84: 69–77.

Edens, N.K., Fried, S.K., Kral, J.G., Hirsch, J., and Leibel, R.L. (1993). In vitro lipid synthesis in human adipose tissue from three abdominal sites. *Am. J. Physiol.* 265 (*Endocrinol. Metab.* 28): E374–E379.

———, Leibel, R.L., and Hirsch, J. (1990). Lipolytic effects on diacylglycerol accumulation in human adipose tissue *in vitro*. *J. Lipid Res.* 31: 1351–1359.

Eisler, R. (1985). Mirex hazards to fish, wildlife, and invertebrates: A synoptic review. Biol. Rep. 85(1.1). U.S. Fish and Wildlife Service, Washington, DC, 42 pp.

——— (1985). Selenium hazards to fish, wildlife, and invertebrates: A synoptic review. Biol. Rep. 85(1.5). U.S. Fish and Wildlife Service, Washington, DC, 57 pp.

——— (1987). Mercury hazards to fish, wildlife, and invertebrates: A synoptic review. Biol. Rep. 85(1.10). U.S. Fish and Wildlife Service, Washington, DC, 90 pp.

——— (1992). Fenvalerate hazards to fish, wildlife, and invertebrates: A synoptic review. Biol. Rep. 2. U.S. Fish and Wildlife Service, Washington, DC, 43 pp.

Ekelund, R., Bergmann, Å., Granmo, Å., and Berggren, M. (1990). Bioaccumulation of 4-nonylphenol in marine animals: A re-evaluation. Environ. Pollut. 64: 107–120.

Facemire, C.F. (1991). The impact of agricultural chemicals on wetland habitats and associated biota with special reference to migratory birds: A selected and annotated bibliography. Agric. Exp. Sta. Publ. No. 708, South Dakota State University, Brookings, 65 pp.

———, Gross, T.S., and Guillette, L.J., Jr. (1995). Reproductive impairment in the Florida panther: Nature or nurture? Environ. Health Perspect. 103(Suppl. 4): 79–86.

Falandysz, J., Kannan, K., Tanabe, S., and Tatsukawa, R. (1994). Concentrations and 2, 3, 7, 8-Tetrachlorodibenzo-*p*-dioxin toxic equivalents of non-*ortho* coplanar PCBs in adipose tissue of Poles. Bull. Environ. Contam. Toxicol. 53: 267–273.

Felton, C.V., Crook, D., Davies, M.J., and Oliver, M.F. (1994). Dietary polyunsaturated fatty acids and composition of human aortic plaques. Lancet 344: 1195–1196.

Focardi, S., Fossi, C., Leonzio, C., and Romei, R. (1986). PCB congeners, hexachlorobenzene, and organochlorine insecticides in human fat in Italy. Bull. Environ. Contam. Toxicol. 36: 644–650.

Frederick, E.K., Bischoff, M., Throssell, C.S., and Turco, R.F. (1994). Degradation of chloroneb, triadimefon, and vinclozolin in soil, thatch, and grass clippings. Bull. Environ. Contam. Toxicol. 53: 536–542.

Galetin-Smith, R., Pavkov, S., and Roncevic, N. (1990). DDT and PCBs in human milk: Implication for breast feeding infants. Bull. Environ. Contam. Toxicol. 45: 811–818.

Gambrell, R.P., and Patrick, W.H., Jr. (1988). The influence of redox potentials on the environmental chemistry of contaminants in soils and sediments. In The Ecology and Management of Wetlands, Vol. 1: Ecology of Wetlands (D.D.

Hook et al., eds.), pp. 319–333. Timber, Portland.

Georgii, S., Bachour, G., Elmafda, I., and Brunn, H. (1995). PCB congeners in human milk in Germany from 1984/85 and 1990/91. *Bull. Environ. Contam. Toxicol.* 54: 541–545.

Gladen, B.C., and Rogan, W.J. (1995). DDE and shortened duration of lactation in a Northern Mexico town. *Am. J. Public Health* 85: 504–508.

Gobas, F.A.P.C., McCorquodale, J.R., and Haffner, G.D. (1993). Intestinal absorption and biomagnification of organochlorines. *Environ. Toxicol. Chem.* 12: 567–576.

Gómez-Catalán, J., Lezaun, M., To-Figueras, J., and Corbella, J. (1995). Organochlorine residues in the adipose tissue of the population of Navarra (Spain). *Bull. Environ. Contam. Toxicol.* 54: 534–540.

———, Sabroso, M., To-Figueras, J., Planas, J., and Corbella, J. (1991). PCB residues in the adipose tissue of the population of Barcelona (Spain). *Bull. Environ. Contam. Toxicol.* 47: 504–507.

Harwood, J.L., Cryer, A., Gurr, M.I., and Dodds, P. (1994). Medical and agricultural aspects of lipids. In *The Lipid Handbook, 2d ed.* (F. D. Gunstone, J. L. Harwood, and F. B. Padley, eds.), pp. 665–707. Chapman and Hall, London.

Henderson, L.O., and Patterson, D.G., Jr. (1988). Distribution of 2, 3, 7, 8-tetrachlorodibenzo-*p*-dioxin in human whole blood and its association with, and extractability from lipoproteins. *Bull. Environ. Contam. Toxicol.* 40: 604–611.

Henny, C. J., Grove, R.A., and Hedstrom, O.R. (1996). A field evaluation of mink and river otter on the Lower Columbia River and the influence of environmental contaminants. *Final Report to the Lower Columbia River Bi-State Water Quality Program*, National Biological Service Forest and Rangeland Ecosystem Science Center, Northwest Research Station, Corvallis, OR.

Hirai, Y., and Tomokuni, K. (1993). Levels of chlordane, oxychlordane, and nonachlor on human skin and in human blood. *Bull. Environ. Contam. Toxicol.* 50: 316–324.

Hovinga, M. E., Sowers, M., and Humphrey, H.E.B. (1993). Environmental exposure and lifestyle predictors of lead, cadmium, PCB and DDT levels in Great Lakes fish eaters. *Arch. Environ. Health* 48: 98–104.

Huckins, J.N., Tubergen, M.W., and Manuweera, G.K. (1990). Semipermeable membrane devices containing model lipid: A new approach to monitoring the bioavailability of lipophilic contaminants and estimating their bioconcentration potential. *Chemosphere* 20: 533–552.

Jan, J., and Tratnik, M. (1988). Polychlorinated biphenyls in residents around the River Krupa, Slovenia, Yugoslavia. *Bull. Environ. Contam. Toxicol.* 41:

809–814.

———, Logar, B., and Jan, J. (1996). Tissue distribution of co-planar and non-planar tetra- and hexa-chlorobiphenyl isomers in guinea pigs after oral ingestion. *Bull. Environ. Contam. Toxicol.* 56: 375–380.

Jenne, E.A., and Zachara, J.M. (1987). Factors influencing the sorption of metals. In *Fate and Effects of Sediment-Bound Chemicals in Aquatic Systems* (K.L. Dickson, A.W. Maki, and W.A. Brungs, eds.), pp. 83–98. Pergamon, New York.

Kashyap, R., Iyer, L.R., and Singh, M.M. (1994). Evaluation of dietary intake of dichloro-diphenyl-trichloroethane (DDT) and benzene hexachloride (BHC) in India. *Arch. Environ. Health* 49: 63–66.

Kelce, W.R., Monosson, E., Gamcsik, M.P., Laws, S.C., and Gray, L.E.J. (1995). Persistent DDT metabolite *p,p'*-DDE is a potent androgen receptor antagonist. *Nature* 375: 581–585.

Kenaga, E. E., and Goring, C.A.I. (1980). Relationship between water solubility, soil sorption, octanol-water partitioning, and concentration of chemicals in biota. In *Aquatic Toxicology, ASTM STP 707* (J.G. Eaton, P.R. Parrish, and A.C. Hendricks, eds.), pp. 78–115. American Society for Testing and Materials, Philadelphia.

Key, T., and Reeves, G. (1994). Organochlorines in the environment and breast cancer. *Brit. Med. J.* 308: 1520–1521.

Khan, M.A.Q. (1990). Biochemical effects of pesticides on mammals. *Chem. Plant Protect.* 6: 109–171.

Kidwell, J.M., Phillips, L.J., and Birchard, G.F. (1995). Comparative analysis of contaminant levels in bottom feeding and predatory fish using the National Contaminant Biomonitoring Program data. *Bull. Environ. Contam. Toxicol.* 54: 919–923.

Krauthacker, B. (1991). Levels of organochlorine pesticides and polychlorinated biphenyls (PCBs) in human milk and serum collected from lactating mothers in the Northern Adriatic Area of Yugoslavia. *Bull. Environ. Contam. Toxicol.* 46: 797–802.

Lake, J. L., Rubinstein, N., and Pavignano, S. (1987). Predicting bioaccumulation: Development of a simple model for use as a screening tool for regulating ocean disposal of wastes. In *Fate and Effects of Sediment-Bound Chemicals in Aquatic Systems* (K.L. Dickson, A.W. Maki, and W.A. Brungs, eds.), pp. 151–166. Pergamon, New York.

Lakshmanan, M.R., Campbell, B.S., Chirtel, S.J., Ekarohita, N., and Ezekiel, M. (1986). Studies on the mechanism of absorption and distribution of 2,3,7,8-tetrachlorodibenzo-*p*-dioxin in the rat. *J. Pharmacol. Exp. Therap.* 239: 1059–1064.

Lange, T., Royals, H., and Connor, L. (unpublished manuscript). Influence of water chemistry on mercury concentration in largemouth bass from Florida Lakes. Florida Game and Fresh Water Fish Commission, Eustis Fisheries Laboratory, Eustis, FL.

Lassiter, R.R., and Hallam, T.G. (1990). Survival of the fattest: Implications for acute effects of lipophilic chemicals on aquatic populations. *Environ. Toxicol. Chem.* 9: 585–595.

Laukola, S. (1980). Seasonal changes in the fatty acid spectrum of the hedge-hog's white and brown adipose tissue. *Ann. Zool. Fennici* 17: 191–201.

LeBlanc, G.A. (1995). Trophic-level differences in the bioconcentration of chemicals: Implications in assessing environmental biomagnification. *Environ. Sci. Technol.* 29: 154–160.

Lebo, J. A., Zajicek, J.L., Orazio, C.E., Petty, J.D., Huckins, J.N., and Douglas, E.H. (1996). Use of the semipermeable membrane device (SPMD) to sample polycyclic aromatic hydrocarbon pollution in a lotic system. *Polycyclic Aromatic Compounds* 8: 53–65.

Lehninger, A.L. (1975). *Biochemistry: The Molecular Basis of Cell Structure and Function, 2d ed.* Worth, New York.

Leita, L., Enne, G., De Nobili, M., Baldini, M., and Sequi, P. (1991). Heavy metal accumulation in lamb and sheep bred in smelting and mining areas of S. W. Sardinia (Italy). *Bull. Environ. Contam. Toxicol.* 46: 887–893.

Lemieux, S., Prud'homme, D., Bouchard, C., Tremblay, A., and Després, J-P. (1993). Sex differences in the relation of visceral adipose tissue accumulation to total body fatness. *Am. J. Clin. Nutr.* 58: 463–467.

Leung, H.-W., Poland, A.P., Paustenbach, D.J., and Anderson, M.E. (1990). Dose dependent pharmacokinetics of [125I]-2-iodo-3, 7, 8-trichlorodibenzo-*p*-dioxin in mice: Analysis with a physiological modeling approach. *Toxicol. Appl. Pharmacol.* 103: 411–419.

Lin, D.S., Conner, W.E., and Spenler, C.W. (1993). Are dietary saturated, monounsaturated, and polyunsaturated fatty acids deposited to the same extent in adipose tissue of rabbits? *Am. J. Clin. Nutr.* 58: 174–179.

Lindpainter, K. (1995). Finding an obesity gene: A tale of mice and man. *N. Engl. J. Med.* 332: 679–681.

Liu, L. X., and Hsiang, T. (1994). Bioassays for benomyl adsorption and persis-tence in soil. *Soil Biol. Biochem.* 26: 317–324.

Malcom, G.T., Bhattacharyya, A.K., Velez-Duran, M., Guzman, M.A., Oalmann, M.C., and Strong, J.P. (1989). Fatty acid composition of adipose tissue in humans: Differences between subcutaneous sites. *Am. J. Clin. Nutr.* 50: 288–291.

Marckmann, P., Lassen, A., Haraldsdóttir, J., and Sandström, B. (1995). Biomarkers of habitual fish intake in adipose tissue. *Am. J. Clin. Nutr.* 62: 956–959.

Marinovich, M., Sirtori, C.R., Galli, C.L., and Paoletti, R. (1983). The binding of 2, 3, 7, 8-tetrachlorodibenzodioxin to plasma lipoproteins may delay toxicity in experimental hyperlipidemia. *Chem.-Biol. Interact.* 45: 393–399.

Martin, J-C., Bougnoux, P., Fignon, A., Theret, V., Antoine, J-M., Lamisse, F., and Couet, C. (1993). Dependence of human milk essential fatty acids on adipose stores during lactation. *Am. J. Clin. Nutr.* 58: 653–659.

Mason, C.F., Ford, T.C., and Last, N.I. (1986). Organochlorine residues in British Otters. *Bull. Environ. Contam. Toxicol.* 36: 656–661.

Matsumura, F. (1995). Mechanism of action of dioxin-type chemicals, pesticides, and other xenobiotics affecting nutritional indexes. *Am. J. Clin. Nutr.* 61(Suppl.): 695S–701S.

McLachlan, M.S. (1996). Bioaccumulation of hydrophobic chemicals in agricultural food chains. *Environ. Sci. Technol.* 30: 252–259.

Mendenjian, C.P., and Carpenter, S.R. (1993). Accumulation of PCBs by lake trout (*Salvelinus namaycush*): An individual-based model approach. *Can. J. Fish. Aquat. Sci.* 50: 97–109.

Mes, J. (1990). Trends in the levels of some chlorinated hydrocarbon residues in adipose tissue of Canadians. *Environ. Pollut.* 65: 269–278.

Mori, Y., Kikuta, M., Okinaga, E., and Okura, T. (1983). Levels of PCBs and organochlorine pesticides in human adipose tissue collected in Ehime Prefecture. *Bull. Environ. Contam. Toxicol.* 30: 74–79.

Moriarty, F. (1983). *Ecotoxicology: The Study of Pollutants in Ecosystems.* Academic, London. 233 pp.

Mukherjee, I., and Gopal, M. (1996). Insecticide residues in baby food, animal feed, and vegetables by gas liquid chromatography. *Bull. Environ. Contam. Toxicol.* 56: 381–388.

Nelson, N., Byerly, T.C., Kolbye, A.C., Kurland, L.T., Shapiro, R.E., Shibko, S.I., Stickel, W.H., Thompson, J.E., Van Den Berg, L.A., and Weissler, A. (1971). Hazards of mercury: Special report to the Secretary's Pesticide Advisory Committee, Department of Health, Education, and Welfare, Nov. 1970. *Environ. Res.* 4: 1–69.

Neville, M.C., and Walsh, C.T. (1995). Effects of xenobiotics on milk secretion and composition. *Am. J. Clin. Nutr.* 61(Suppl.): 687S–694S.

Papadopoulou-Mourkidou, E. (1991). Postharvest-applied agrochemicals and their residues in fresh fruits and vegetables. *J. Assoc. Off. Anal. Chem.* 74: 745–765.

Phinney, S.D., Stern, J.S., Burke, K. E., Tang, A.B., Miller, G., and Holman, R.T. (1994). Human subcutaneous adipose tissue shows site-specific differences in fatty acid composition. *Am. J. Clin. Nutr.* 60: 725–729.

Poiger, H., and Schlatter, C. (1986). Pharmacokinetics of 2, 3, 7, 8-TCDD in man. *Chemosphere 15: 1489–1494.*

————, Pluess, N., and Buser, H.R. (1989). The metabolism of selected PCDFs in the rat. *Chemosphere* 18: 259–264.

Prest, H.F., Huckins, J.N., Petty, J.D., Herve, S., Paasivirta, J., and Heinonen, P. (1995). A survey of recent results in passive sampling of water and air by semipermeable membrane devices. *Mar. Pollut. Bull.* 31: 4–12.

Pritchard, P. H. (1987). Assessing the biodegradation of sediment associated materials. In *Fate and Effects of Sediment-Bound Chemicals in Aquatic Systems* (K. L. Dickson, A.W. Maki, and W.A. Brungs, eds.), pp. 109–135. Pergamon, New York.

Rao, C.V., and Banerji, A.S. (1988). Polychlorinated biphenyls in human amniotic fluid. *Bull. Environ. Contam. Toxicol.* 41: 798–801.

Roberts, J.R., DeFrietas, A.S.W., and Gidney, M.A.J. (1977). Control factors on uptake and clearance of xenobiotic chemicals by fish. In *Animals as Monitors of Environmental Pollutants* (F. Peter, ed.), pp. 3–12. National Academy of Sciences, Washington, DC.

Rodbell, M. (1960). The removal and metabolism of chylomicrons by adipose tissue *in vitro. J. Biol. Chem.* 235: 1613–1620.

Roelke, M.E., Schultz, D.P., Facemire, C.F., Sundlof, S.F., and Royals, H.E. (1991b). Mercury contamination in Florida Panthers. Florida Game and Fresh Water Fish Commission, Gainesville, FL, 48 pp.

————, ————, ————, and ———— (1991a). Mercury contamination in the free-ranging endangered Florida Panther *(Felis concolor coryi). Am. Assoc. Zoo. Vet. Annu. Proc.* 1991: 277–283.

Rodgers, J.H., Jr., Dickson, K.L., Saleh, F.Y., and Staples, C.A. (1987). Bioavailability of sediment-bound chemicals to aquatic organisms: Some theory, evidence and research needs. In *Fate and Effects of Sediment-Bound Chemicals in Aquatic Systems* (K.L. Dickson, A.W. Maki, and W.A. Brungs, eds.), pp. 245–266. Pergamon, New York.

Rose, J.Q., Ramsey, J.C., Wentzler, T.H., Hummel, R.A., and Gehring, P.J. (1976). The fate of 2, 3, 7, 8-tetrachlorodibenzo-*p*-dioxin following single and repeated oral doses to the rat. *Toxicol. Appl. Pharmacol.* 36: 209–226.

Ross, R., and Rissanen, J. (1994). Mobilization of visceral and subcutaneous adipose tissue in response to energy restriction and exercise. *Am. J. Clin. Nutr.* 60: 695–703.

Rozman, K.K., and Klassen, C.D. (1996). Absorption, Distribution, and Excretion of Toxicants. In *Casarett and Doull's Toxicology: The Basic Science of Poisons, 5th ed.*, p. 50. McGraw-Hill, New York.

————, Pereira, D., and Iatropoulos, M.J. (1986). Histopathology of interscapular brown adipose tissue, thyroid, and pancreas in 2, 3, 7, 8-tetrachlorodibenzo-*p*-dioxin (TCDD)-treated rats. *Toxicol. Appl. Pharmacol.* 82: 551–559.

Ryan, J.J., Schecter, A., Lizotte, R., Sun, W.-F., and Miller, L. (1985). Tissue distribution of dioxins and furans in humans from the general population. *Chemosphere* 14: 929–932.

Schaefer, E.J., Woo, R., Kibata, M., Bjornsen, L., and Schreibman, P.H. (1983). Mobilization of triglyceride but not cholesterol or tocopherol from human adipocytes during weight reduction. *Am. J. Clin. Nutr.* 37: 749–754.

Schantz, S.L., Jacobson, J.L., Humphrey, H.E.B., Jacobson, S.W., Welch, R., and Gasior, D. (1994). Determinants of polychlorinated biphenyls (PCBs) in the sera of mothers and children from Michigan farms with PCB-contaminated silos. *Arch. Environ. Health* 49: 452–458.

Schapira, D.V., Kumar, N.B., and Lyman, G.H. (1993). Variation in body fat distribution and breast cancer risk in the families of patients with breast cancer and control families. *Cancer* 71: 2764–2768.

Schrap, S.M., and Opperhuizen, A. (1990). Relationship between bioavailability and hydrophobicity: Reduction of the uptake of organic chemicals by fish due to the sorption on particles. *Environ. Toxicol. Chem.* 9: 715–724.

Sellers, T.A., Kushi, L.H., Potter, J.D., Kaye, S.A., Nelson, C.L., McGovern, P.G., and Folsom, A.R. (1992). Effect of family history, body-fat distribution, and reproductive factors on the risk of postmenopausal breast cancer. *N. Engl. J. Med.* 326: 1323–1329.

Shapiro, B. (1977). Adipose tissue. In *Lipid Metabolism in Mammals 1* (F. Snyder, ed.), pp. 287–316. Plenum, New York.

Sitarska, E., Klucinski, W., Winnicka, A., and Ludwicki, J. (1991). Residues of organochlorine pesticides in milk gland secretion of cows in perinatal period. *Bull. Environ. Contam. Toxicol.* 47: 817–821.

Skerfving, S. (1988). Mercury in women exposed to methylmercury through fish consumption, and in their newborn babies breast milk. *Bull. Environ. Contam. Toxicol.* 41: 475–482.

Smalley, R.L., and Dryer, R.L. (1967). Brown fat in hibernation. In *Mammalian Hibernation, Vol. 3* (K.C. Fisher, A.R. Dawe, C.P. Lyman, E. Schonbaum, and F.E. South, Jr., eds.), pp. 325–345. American Elsevier, New York.

Smith, S., Jr., Willis, G.H., and Cooper, C.M. (1995). Cyfluthrin persistence in soil as affected by moisture, organic matter, and redox potential. *Bull. Environ. Contam. Toxicol.* 55: 142–148.

Stickel, L.F. (1973). Pesticide residues in birds and mammals. In *Environmental Pollution by Pesticides* (C.A. Edwards, ed.), pp. 234–312. Plenum, London.

Suzuki, T., Miyama, T., and Katsunuma, H. (1971). Comparison of mercury contents in maternal blood, umbilical cord blood, and placental tissues. *Bull. Environ. Contam. Toxicol.* 5: 502–508.

Szeto, S.Y., and Price, P.M. (1991). Persistence of pesticide residues in mineral and organic soils in the Fraser Valley of British Columbia. *J. Agric. Food Chem.* 39: 1679–1684.

Thomas, J.A. (1991). Toxic responses of the reproductive system. In *Casarett and Doull's Toxicology: The Basic Science of Poisons, 4th ed.* (M.O. Amdur, J. Doull, and C.D. Klaassen, eds.), pp. 484–520. Pergamon, New York.

Van Velzen, A. C., Stiles, W. R., and Stickel, L. F. (1972). Lethal mobilization of DDT by cowbirds. *J. Wildl. Manage.* 36: 733–739.

Waliszewski, S.M., Pardio Sedas, V.T., Infanzon, R.M., and Rivera, J. (1995). Determination of organochlorine pesticide residues in human adipose tissue: 1992 study in Mexico. *Bull. Environ. Contam. Toxicol.* 55: 43–49.

Walker, C.H. (1987). Kinetic models for predicting bioaccumulation of pollutants in ecosystems. *Environ. Pollut.* 44: 227–240.

Wehler, E.K., Bergman, A., Brandt, I., Darnerud, P.O., Wachtmeister, C.A. (1989). 3, 3', 4, 4'-tetrachlorobiphenyl excretion and tissue retention of hydroxylated metabolites in the mouse. *Drug Metabol. Dispos.* 17: 441–448.

White, A.W., Handler, P., and Smith, E.L. (1973). *Principles of Biochemistry, 5th ed.* McGraw-Hill, New York.

White, D.H., and Hoffman, D.J. (1995). Effects of polychlorinated dibenzo-*p*-dioxins and dibenzofurans on nesting wood ducks *(Aix sponsa)*. *Environ. Health Perspect.* 103(Suppl. 4): 79–86.

Williams, D.T., LeBel, G.L., and Junkins, E. (1988). Organohalogen residues in human adipose autopsy samples from six Ontario municipalities. *J. Assoc. Off. Anal. Chem.* 71: 410–414.

Witkowski, P.J., Smith, J.A., Fusillo, T.V., and Chiou, C.T. (1987). A review of surface-water sediment fractions and their interactions with persistent manmade organic compounds. Circular 993. U.S. Geological Survey, Denver, 39 pp.

Wolff, M.S., Toniolo, P.G., Lee, E.W., Rivera, M., and Dubin, N. (1993). Blood levels of organochlorine residues and risk of breast cancer. *J. Nat. Cancer Inst.* 85: 648–652.

Youngs, W.D., Gutenmann, W.H., Josephson, D.C., Miller, M.D., and Lisk, D.J. (1994). Residues of *p,p'*-DDE in lake trout in Little Moose Lake in New York State. *Chemosphere* 29: 405–406.

Zhao, Y., Nichols, J.E., Bulun, S.E., Mendelson, C.R., and Simpson, E.R. (1995). Aromatase P450 gene expression in human adipose tissue: Role of a Jak/STAT pathway in regulation of the adipose-specific promoter. *J. Biol. Chem.* 27: 16449–16457.

Chapter 4

CONTAMINANT INTERACTIONS WITH STEROID RECEPTORS: EVIDENCE FOR RECEPTOR BINDING

Andrew A. Rooney and Louis J. Guillette, Jr.

Department of Zoology
University of Florida
Gainesville, FL 32611

Introduction

Steroid receptors are important determinants of endocrine disrupter consequences. As the most frequently proposed mechanism of endocrine-disrupting contaminant (EDC) action, steroid receptors are not only targets of natural steroids but are also commonly sites of nonsteroidal compound action. In fact, for many years, steroid receptor binding has been used as evidence of the endocrine-disrupting potential of a compound. Functional activation of steroid receptors has also been studied for years. Reporter assays such as the chloramphenicol acetyltransferase (CAT) assay have been used with both estrogen and androgen receptors (ERs and ARs) to test steroid receptor activation. In this review, significant factors from generalized steroid receptor interactions to species differences in steroid receptor sequences are explored in an effort to more fully discern the significance of EDC activation of steroid receptors. Interaction or "cross-talk" among growth factors, thyroid hormones, and natural hormones is considered in light of EDC ligands. Although this review concentrates on steroid receptors, the aryl hydrocarbon (Ah) receptor is discussed briefly due to its similarities to steroid receptors. Factors affecting the result of EDC/steroid receptor binding are then presented, beginning with a discussion of agonistic and antagonistic responses. This is followed by an exploration of the techniques used to assess receptor binding and contributions of accessory proteins such as binding proteins and receptor-associated proteins. The importance of receptor changes is then examined with discussions of receptor translocation, receptor stabilization/dissociation, receptor dimers/heterodimers, and ligand-induced receptor conformational changes. Ontogenetic and phylogenetic receptor differences are then explored, including age, season, tissue, and species-related differences. Finally EDC synergy and implications of environmental exposure are presented.

Typical Steroid/Receptor Interactions

Steroid receptor reactions begin shortly after steroid-binding protein interactions end. Circulating steroids dissociate from plasma-binding proteins and, being hydrophobic, penetrate the target cell's membrane by simple diffusion (Fig. 4-1). Once inside the cell, steroids bind to cytoplasmic or nuclear receptors and the steroid-receptor complex undergoes a conformational change called activation or transformation. The principal location of steroid receptor binding depends on the receptor type: The majority of estrogen receptors (ERs) and androgen receptors (ARs) are concentrated in the nucleus; corticosteroid receptors (CRs) are localized within the cytoplasm; and the location of progesteroid receptors (PRs) is currently unclear (Norris, 1997). Each receptor in its ligand-free state associates with a specific group of heat shock proteins that stabilize the receptor and prevent binding of the receptor to specific transcription enhancing steroid-responsive elements (SREs) of the DNA. The translocation of any cytoplasmic steroid receptors into the nucleus—including any cytoplasmic ERs or ARs—is stimulated by ligand binding to the receptor. Binding of the steroid to its receptor also causes the receptor to dissociate from heat shock proteins during the conformational change that leads to dimerization of the steroid receptor complex and allows binding to SREs. Finally, as a dimer, the steroid receptor complex binds to a specific SRE, which leads to hormone-specific activation of genes (Kalimi and Dave, 1995).

Steroid/Receptor Interactions—Other Factors

General Receptor Properties

The steroid receptor superfamily is comprised of receptors for steroids, thyroid hormones, vitamin D3, retinoic acid, and several "orphan" receptors for which the ligand has yet to be identified (Evans, 1988; Green, 1990; Landers and Spelsberg, 1992). This review concentrates on steroid receptors, although the

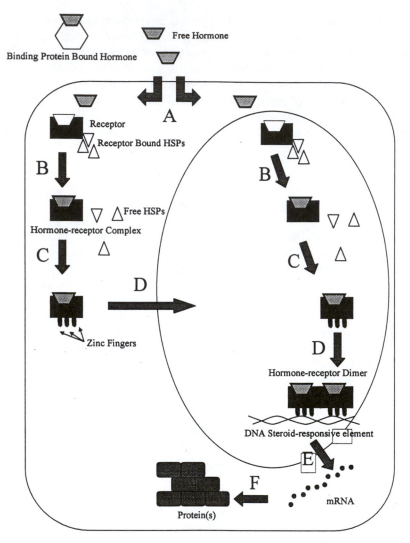

Figure 4-1.
Most steroid hormone molecules outside of cells are attached to binding proteins. Hormones separate from the binding proteins prior to entering target cells (A) and bind to steroid receptors within either the cytoplasm or the nucleus. Binding of hormone to receptors (B) causes heat shock proteins to disassociate from the steroid receptor. Hormone binding also causes conformational changes in the hormone receptor complex (C), including zinc fingers that allow binding to the DNA. The hormone receptor complex then forms a dimer (D) either from within the nucleus or while entering the nucleus and binds to steroid response elements on the DNA. The DNA-bound steroid receptor dimer stimulates production of mRNA (E), which then exits the nucleus. Finally, the newly formed mRNA is translated into proteins (F), which lead to the response generated by the steroid hormone.

function of thyroid hormone receptors and orphan receptors are of interest when examining the interactions between endocrine-disrupting contaminants (EDCs) and receptors. Evaluating the vulnerability of steroid receptors to EDCs is an ongoing process, complicated by five major factors: (1) age/developmental stage, (2) expression, (3) tissue specificity, (4) receptor subtypes, and (5) species differences. Evidence for age-dependent receptors is found in mice, which have a developmentally restricted nuclear receptor that binds to the estrogen response element. This orphan receptor is found only between day 6.5 and 7.5 postcoitum and is hypothesized to participate in overlapping gene networks with the estrogen receptor during the formation of the chorion (Pettersson et al., 1996). Ligands that bind the new receptor are unknown, but developmentally restricted receptors introduce the possibility of developmentally restricted EDC action on those receptors.

The second receptor property affecting sensitivity to EDCs is variation in the number or expression of steroid receptors in a given tissue. Using the ER as an example, the number of estrogen receptors in fetal vaginal tissue of the guinea pig is quite high—approximately 7,000 fmol/mg DNA. ER number decreases significantly after birth and then increases slightly at 2 weeks of age (Nguyen et al., 1986). Similar changes take place in AR, ER, and PR during reproductively significant times or in response to hormonal changes (Byers et al., 1997; Fisher et al., 1997; Kashon et al., 1996; Spencer and Bazer, 1995). For example, maximal ER and PR expression in ewes is found immediately following mating (Spencer and Bazer, 1995). Steroid receptors also exhibit age specific fluctuations such as the changes in ER expression in male rats and marmosets associated with development and function of efferent ducts and Leydig cells (Fisher et al., 1997). Steroid receptors are also often induced by the ligand itself, as in the case of E_2 induction of ER in mammals, amphibians and fish (Green and Chambon, 1986; Mommsen and Lazier, 1986; Pakdel et al., 1991; Riegel et al., 1987).

The third receptor property affecting sensitivity to EDCs is tissue specificity. Simply stated, some endocrine tissues have steroid receptors and others don't. The expression of steroid receptors also differs on a more subtle level between different tissue types. For example, thyroid hormone receptors differ by 3-fold among the following tissues: brain, heart, lung, and liver (Bernal and Pekonen, 1984). Additional tissue-related differences in steroid or EDC response may relate to fundamental molecular differences in the way different cells interpret the ligand signal as a consequence of distinct transcription factor pools or the promoter context in which the SREs reside (Katzenellenbogen, 1996). For example, for transcriptional activation of the human ER, some promoters require activity of both the TAF1 and TAF2 transactivation regions of the ER, whereas other promoters can accomplish ER-dependent transcription with only TAF1 or TAF2 (Tzukerman et al., 1994). Furthermore, the activities of TAF1 and TAF2 are cell-type-dependent (Tora et al., 1989).

There is also evidence of the fourth receptor factor: the existence of several different steroid receptor subtypes within a single species. Multiple forms of estrogen and thyroid receptors have been identified (Evans, 1988; Landers and Spelsberg, 1992). For example, a new estrogen receptor (ER-β) has been characterized and shown to coexist with the classic ER (ER-α) in rats, mice, and humans (Katzenellenbogen and Korach, 1997; Kuiper and Gustafsson, 1997; Mosselman et al., 1996; Pennisi, 1997; Pettersson et al., 1997). There is also evidence that different receptor types are differentially distributed and differentially active. Early distribution differences hypothesized at a cellular level have been confirmed up to the tissue level, with ER-β and ER-α displaying distinct tissue association patterns (Giambiagi et al., 1984; Golding and Korach, 1988; Kuiper et al., 1997; Kuiper and Gustafsson, 1997; Paech et al., 1997). Additionally, ER-β binding characteristics are similar to ER-α, but transcriptional activation is distinct for estrogens as well as antiestrogens (Kuiper et al., 1997; Paech et al., 1997).

Finally, the fifth receptor property affecting EDC action is species differences in steroid receptors. As steroid receptors are proteins produced via transcription/translation of DNA, evolutionary divergence among species creates structural divergence among receptors. The significance of the high species differences in steroid receptor DNA sequences relative to the highly conserved steroid sequence is unknown; however, species differences in EDC action are suggested by differences at this molecular level (see Chapter 1 for further discussion).

Membrane-Bound Steroid Receptors

The speed of steroid action through cytosolic and nuclear receptors typical of steroid receptors is much slower than the typical second messenger system, and yet, researchers have been aware of numerous fast-acting steroid hormone responses for many years (Wehling, 1997). This "faster" response of some aldosterone (Wehling et al., 1993), androgen (Sachs and Leipheimer, 1988), estrogen (Pietras and Szego, 1979), glucocorticoid (Moore and Orchinik, 1991), and progesterone (Eisen et al., 1997) reactions has been shown to be mediated through membrane-bound steroid receptors. Although evidence for membrane-bound steroid receptors has been accumulating for some time, they were not widely accepted until the 1990s. In 1991, a membrane-bound corticosterone receptor was isolated from newt brain (Orchinik et al., 1991), contributing to wider acceptance of membrane-bound steroid receptors. The differential action of steroid receptor agonists or antagonists in different tissues may be related to the type or location of the receptor (Wehling, 1995). However, the similarities between nuclear and membrane-bound glucocorticoid receptors suggests that binding may be identical for both receptors, negating the significance of possessing different receptor types. Such similarity in binding is countered by the failure of [³H]tamoxifen aziridine to bind to cytosolic or membrane fractions of mouse uterine cells and suggests that if membrane ER exist, their structural conformations differ from that of nuclear ER (Ignar-Trowbridge et al., 1991). The significance of membrane-bound receptors requires further study and debate, particularly in light of EDCs (Watson et al., 1995). Furthermore, the possibility of EDC action indirectly affecting steroid action by altering properties of the cell membrane are suggested by membrane changes induced by EDCs such as polycyclic aromatic hydrocarbons (PAHs; Pallardy et al., 1992).

Receptor Interactions and Cross-Talk with Growth Factors

Cross-talk among various ligands and their corresponding receptors presents additional complications to the already intricate relationship between a receptor and its ligand. For example, treatment of mice with epidermal growth factor (EGF) results in many of the changes associated with estrogen treatment, including ER translocation, increased DNA synthesis, phosphatidylinositol turnover, and uterine growth. Although the mechanisms are unknown, EGF appears to partially mediate the effects of estrogens on the reproductive tract. In fact, some physiologic actions of several peptide growth factors are dependent on the ER. There is *in vitro* evidence of positive stimulation of the ER, or ER agonism, by various growth factors. EGF, transforming growth factor-α (TGFα), and insulin-like growth factor I (IGF-I) all enhance transcription from a consensus estrogen response element in ER-positive BG-1 human ovarian adenocarcinoma cells. Furthermore, ICI 164,384, a specific ER blocker, or antagonist, inhibits the response to EGF, TGFα, and IGF-I (Ignar-Trowbridge et al., 1996). Growth factor cross-talk has also been reported for progesterone and EGF on progestin response element-dependent transcription in T47D breast cancer cells (Krusekopf et al., 1991). The AR also can be stimulated by several growth factors, including IGF-I and EGF. DU-145 cells, transfected with an AR expression vector, were equally stimulated by IGF-I or the synthetic androgen methyltrienolone with all androgen-inducible promoters tested. Keratinocyte growth factor (KGF) and EGF were also effective in stimulating AR-dependent transcription when an artificial promoter with two androgen-responsive elements was utilized (Culig et al., 1994). As in ER-growth factor interactions, the ability of growth factors to stimulate AR-dependent transcription was blocked by a known receptor antagonist.

Additional evidence indicates that the growth factors are capable of ligand-dependent activation of a SRE using a mechanism unique from that of the normal ligand. Combinations of EGF, TGFα, or IGF-I acted on BG-1 transcription in an additive manner; however, combinations of E_2 with either TGFα or IGF-I induced a greater than additive, or synergistic response (Ignar-Trowbridge et al., 1996). Synergy among ligands has become a recent topic in EDC evaluation, and is discussed separately later. Although the mechanism of peptide growth factor cross-talk with steroid receptors is unknown, the occurrence of synergy and growth factor stimulation in mutant cell lines that lack normal ligand binding suggests that at least some growth factors function by a mechanism that is distinct from normal ligand binding (Ignar-Trowbridge et al., 1996). The implication of growth factor cross-talk on EDC effects is 2-fold: (1) EDC action may function indirectly through growth factor modifications, and (2) EDC action may operate through a mechanism similar to that of growth factors, such that EDC interaction is receptor-dependent and yet distinct from the normal ligand-receptor interplay.

Receptor Interactions and Cross-Talk with Thyroid Hormones

There is also strong evidence for interaction among E_2, ER, thyroid hormones, and thyroid hormone receptors (Dellovade et al., 1995). For example, in the African clawed frog, *Xenopus laevis*, triiodothyronine (T_3) up-regulates thyroid hormone receptor β, but not α, and enhances autoinduction of ER by E_2. Furthermore, administration of T_3 rapidly potentiates the activation of vitellogenin (VTG) genes by E_2 in cultured *Xenopus* hepatocytes (Rabelo and Tata, 1993). However, in embryonic chick hepatocytes, T_3 reduced the level of nuclear ERs and inhibited activation of VTG (Elbrecht and Lazier, 1985). Direct interaction between TR and ER has also been suggested by studies of a rat E_2 sensitive pituitary tumor cell line. T_3 or E_2 alone stimulated proliferation of these cells; however, simultaneous administration of both hormones resulted in mutual inhibition (Zhou-Li et al., 1993). Dellovade et al. (1995) suggest that TR and ER bind to parts of the same SRE and compete with each other. Under this hypothesis, the result of TR and ER interactions would depend on the presence of ligands and the affinity for SREs. Interactions at the level of SREs are supported by several studies demonstrating TR interaction with the estrogen response element (e.g., Glass et al., 1988). Although the relationship between TR and ER is unclear at this time, the likelihood of a relationship is strong and complicates EDC interactions with either receptor.

Receptor Interactions and Cross-Talk with Other Hormones

Another illustration of the problem of cross-talk is the ability of steroids to bind to the steroid receptor other than their recognized receptor. For example, in the fish hepatocyte vitellogenin (VTG) assay, E_2, T, 11-ketotestosterone (11-KT), and P_4 all stimulated the production of VTG through an ER-mediated mechanism (Pelissero et al., 1993). The concentration of T, 11-KT, and P_4 necessary for similar levels of VTG production was 1,000 times higher than E_2, but recognition of multiple steroids is clearly a possibility for steroid receptors. The physiological relevance of this cross-receptor binding through the ER is questionable in two respects: (1) The concentration of these hormones necessary to bind ER and promote VTG synthesis is far beyond physiological levels, and (2) T activity may be through E_2 following aromatization. The activity of these compounds through the ER is supported by two other lines of evidence: (1) the effects of 11-KT and P_4 are most likely to be through direct ER binding because of the increased difficulty in transforming these steroids into E_2, and (2) the known ER antagonist tamoxifen blocks the VTG stimulation of E_2, T, 11-KT, and P_4 (Pelissero et al., 1993). Steroids may also affect the action of other steroids by altering receptor levels. For example, it is well known that E_2 increases both ER and PR in mammalian and avian reproductive tracts, and progesterone decreases both receptors. A slightly different steroid/receptor relationship appears to function in the turtle, *Chrysemys picta*, in which progesterone has no effect on ER expression (Giannoukos and Callard, 1996). The AR is more permissive for natural steroids than the ER, although estrogens bind the AR with a greatly reduced affinity when compared to androgens (Wilson and French, 1976). This decreased selectivity of the AR suggests that EDCs should bind the AR with a greater frequency than the ER. Although a number of EDCs, particularly breakdown products of the fungicide vinclozolin (Kelce et al., 1997), have been shown to bind the AR, the vast majority of EDC receptor binding has been demonstrated with the ER. At this time it is unclear whether the current bias is the result of an experimental approach or a physiological phenomenon.

Ah Receptors

As discussed above, steroid receptors are a group of related receptors that function through ligand-dependent transcription. Although the aryl hydrocarbon (Ah) receptor is not a member of the steroid receptor superfamily, activation of the Ah receptor follows a similar mechanism (Cuthill et al., 1988; Poland and Knutson, 1982). Although the natural ligand of the Ah receptor is unknown, many of the toxic effects of 2,3,7,8-tetrachlorodibenzo-*p*-dioxin (TCDD) and structurally related halogenated aryl hydrocarbons (HAHs), including polychlorinated biphenyls (PCBs), appear to be mediated via the Ah receptor. For example, HAH ligand binding is followed by interaction of the receptor-ligand complex with a specific DNA sequence (in this case the dioxin response element—DRE) that induces transcription of structural genes encoding mRNA for CYP1A1 enzyme activity and the expression of additional unidentified genes, the products of which are hypothesized to mediate HAH toxicity (Kerkvliet and Burleson, 1994). Although the majority of HAHs interact with the Ah receptor, some PCBs have been shown to directly bind to the ER (Korach et al., 1987; Nesaretnam et al., 1996). The similarities between the Ah and steroid receptors suggest that EDC-receptor interactions of both types should be studied as aspects of an overall related phenomenon.

Endocrine-Disrupting Contaminants/Receptor Effects

Positive EDC/Receptor Interactions—Agonism

The level of interaction of an EDC with a steroid receptor is variable, and the extent to which the contaminant blocks or mimics normal steroids will determine the overall extent of endocrine modification induced by that contaminant. Indeed, classification of EDCs is done in reference to a variety of endpoints from receptor binding to the effects on organ culture. For example, compounds such as diethylstilbestrol (DES) and 1,1,1-trichloro-2-(*p*-chlorophenyl)-2(*o*-chlorophenyl)ethane (*o,p'*-DDT), a component of the technical grade insecticide DDT, are both classified as estrogenic because exposure to these compounds causes effects that mimic the response of tissues and cells to estrogen. In this case estrogenic is synonymous with estrogen receptor agonist. Steroid receptor agonists act by achieving the normal ligand-activated transcription with an endogenous or novel ligand. The *o,p'*-isomer of DDT will be used to further illustrate the classification and characterization of agonistic EDCs. Early studies relied on organ culture and biological response systems. For example, estrogenic effects of *o,p'*-DDT were indicated by *o,p'*-DDT–induced estrogen-dependent responses in rodents and birds, such as increased uterine glycogen content and increased uterine weight (Bitman et al., 1968). The limited information provided by these early bioassays was later supplemented with evidence of EDCs functioning at multiple levels of normal steroid receptor interactions (see arrows B–F in Fig. 4-1). The ability of EDCs such as *o,p'*-DDT to compete with $[^3H]E_2$ for binding to the ER is perhaps the best evidence of direct interaction between EDCs and the ER. Examples of *o,p'*-DDT–binding inhibition include studies on rat uterine ER (Nelson, 1974), rat testicular ER (Bulger et al., 1978), and alligator ER (Vonier et al., 1996). Binding studies are considered crucial to demonstrating EDC action because true agonistic activity must involve receptor binding. Substances that appear estrogenic may in fact be antiandrogenic, be goiterogenic (i.e., modify thyroid hormones), modify growth factors, alter transcription, or influence receptor expression. For example, the phenotypic alterations attributed to DDT isomers are consistent with a ER agonist; however, some isomers such as *p,p'*-DDE have been demonstrated to antagonistically bind the AR, not antagonistically bind the ER (Kelce et al., 1995). Therefore receptor interaction is critical in evaluating activity of an EDC.

Predicting EDC interaction *in vivo* requires a knowledge of the mechanisms of that interaction. Estrogenic effects of *o,p'*-DDT were indicated by later, more mechanistic studies demonstrating estrogen-like effects such as *o,p'*-DDT induced ER translocation from the cytoplasm to uterine nuclei in rats (Robison and Stancel, 1982; see arrow D in Fig. 4-1). Continuing studies by Robison and his lab demonstrated that *o,p'*-DDT stimulated the *in vitro* production of an estrogen-inducible protein and that *o,p'*-DDT could support the growth of an estrogen-responsive cell line (Robison et al., 1984; 1985). The final class of evidence for EDC agonistic response is *in vivo* assays for tissue, hormonal, or even behavioral changes. This can be as simple as positive response in a rodent bioassay for vaginal cornification or uterine weight gain. These bioassays are similar or in some cases the same tests that were originally used to detect endocrine activity of compounds. However, just as the original tests were not specific, new ER bioassays may be complicated by nonestrogenic substances that are capable of activating estrogen-inducible genes. For example, growth factors appear to stimulate ligand-dependent receptor-mediated transcription in a pathway that is distinct from the normal ligand (see above). Evidence of *in vivo* hormonal modification includes wildlife data of increased gonadal E_2 synthesis by alligators from a polluted lake that may be linked to DDT metabolites (Guillette et al., 1994) and experimental evidence of accelerated pubertal (or first) ovulation induced by *o,p'*-DDT in rats (Welch et al., 1969). Behavioral evidence of the estrogenicity of *o,p'*-DDT includes wildlife evidence in birds (Fry, 1995) and experimental evidence in mice (vom Saal et al., 1995). Both classes of *in vivo* tests indicate an effect independent of mechanism, and thus must be coupled with mechanistic assays of EDC action for a clear indication of agonistic activity.

Full agonist activity can only be concluded on the basis of several tests. The logical first test of EDC activity is a competitive binding study to demonstrate EDC interaction with a given receptor. A transcriptional or mitotic assay is then necessary to demonstrate EDC effectiveness in stimulating the receptor interactions that follow receptor binding and precede the production of mRNA. The transcription and mitotic endpoints used by Robison (Robison et al., 1995a, 1995b) are part of many well-established tests for agonistic activity, especially for estrogenicity. For an EDC to promote transcription, or mitotic division, it is believed that it must perform all or nearly all of the functions of the ligand, i.e., binding to the receptor, promoting a conformational change in the receptor, promoting receptor dimerization and binding to SREs, and then promoting transcription at the SRE. Examples of current transcriptional tests include the yeast estrogen screen (YES) consisting of a yeast line transfected with human estrogen receptor and two estrogen response elements linked to the *lacZ* gene (Arnold et al., 1996; Tran et al., 1996). Androgenic detection systems include an androgen-dependent transcription system created by cotransfection of monkey kidney CV1 cells with the pCMVhAR expression vector and mouse mammary tumor virus luciferase reporter vector. In both of the above transcription systems, the transcription product is an enzyme that simplifies detection such as the *lacZ* gene that produces β-galactosidase, which in turn can be measured in absorbance at 420 nm. The natural ligand—estrogen in the case of the YES system and androgen in the case of the monkey CV1 transfected with human AR—causes a given level of transcription, allowing EDCs to be judged relative to the activity of the natural ligand. The natural ligand is also a benchmark in mitotic assays. Mitotic tests rely on cell lines and culture conditions that require ligand for full mitotic activity. EDCs can then be tested for mitotic activity relative to the natural ligand. For example, the E-SCREEN test of Soto et al. is a mitotic test for agonist activity in the ER (Soto et al., 1995). Soto et al. (1995) rely on two critical properties of human cells and human serum for the mitotic test in their E-SCREEN: (1) A human serum-borne molecule specifically inhibits the proliferation of human estrogen-sensitive cells, and (2) estrogens cancel this inhibitory effect, allowing proliferation of estrogen-sensitive cells. Therefore, compounds such as the natural estrogen E_2, the synthetic estrogen DES, the mycoestrogen zearalenone, and the pesticide DDT are judged to be estrogenic because all of the above compounds induce proliferation in the E-SCREEN to various degrees. Results of the E-SCREEN are expressed as relative proliferative potency (RPP) or the ratio of the lowest dose of the test compound needed for maximal cell yield and the minimum concentration of E_2 to achieve maximum proliferation (e.g., RPP of E_2 = 100%, RPP of DES = 1,000%, RPP of zearalenone = 1%, and RPP of DDT = 0.0001%). The third step in EDC examination is to perform an appropriate *in vivo* assay to evaluate the effectiveness of a given EDC within an intact animal. *In vivo* assays are important due to unforeseen effects of natural conditions such as the effects of binding proteins and or antagonistic receptor effects in the presence of natural ligand (see below).

Negative EDC/Receptor Interactions—Antagonism

Steroid antagonists act by inhibiting normal ligand-activated transcription; put simply, antagonists inhibit the effect of agonists (Nimrod and Benson, 1996). The actual inhibition leading to steroid antagonism can take place at several places. First, the antagonist can block ligand binding through direct receptor binding and, therefore, competition with the natural ligand. For example the synthetic antiandrogen hydroxyflutamide binds the AR with moderate affinity and promotes nuclear translocation. Hydroxyflutamide, however, fails to promote binding of the AR-hydroxyflutamide complex to the SRE, and therefore inhibits androgen-mediated transcription (Wong et al., 1995). The fungicide vinclozolin appears to inhibit AR-mediated transcription in a similar manner to hydroxyflutamide, although vinclozolin is 2- to 3-fold less effective than hydroxyflutamide as an antiandrogen (Wong et al., 1995). The most active antiandrogen resulting from vinclozolin exposure is the metabolite 3′,5′- dichloro-2-hydroxy-2-methylbut-3-enanilide (M2), with 2-[[(3,5- dichlorophenyl)-carbamoyl]oxy]-2-methyl-3-butenoic acid (M1) and the native vinclozolin being less effective antiandrogens. Similarly *p,p*′-DDE, a metabolite of DDT, is a potent antiandrogen based on AR binding and inhibition of normal AR-dependent gene expression (Kelce et al., 1995).

Binding is not evidence of agonism or antagonism taken by itself because antagonists are often partial but inefficient agonists. The end result of EDC agonism or antagonism becomes a relative point on a continuum that depends on the test conditions such as the presence of ligand. For example, Wong et al. (1995) showed that M2 and hydroxyflutamide were androgenic as well as antiandrogenic in the same system. M2 and hydroxyflutamide are antiandrogenic in the presence of androgen and androgenic in the absence of androgen. Promoter context can also control agonist/antagonist behavior of a compound. For example, cell differences in the transactivation regions TAF1 and TAF2 of the human ER allows tamoxifen to establish partial agonist activity in some cells while functioning as a antagonist in others (see Tissue Specificity below for a full discussion; Tzukerman et al., 1994).

Demonstrating antagonism *in vivo* is critical because the effects suggested by binding or transcription studies are not always evident in an intact animal. For example, further examination of *p,p'*-DDE and vinclozolin demonstrated antiandrogenic properties *in vivo* including suppression of testosterone-induced seminal vesicle weight gain, reduction of testosterone induced mRNA (Kelce et al., 1997), and a possible wildlife effect of *p,p'*-DDE on reduced penis size in an alligator population in Florida (Guillette et al., 1996). On the other hand, the progesterone receptor (PR) has many known pharmaceutical antagonists (e.g., RU 486; Allan et al., 1992), and yet EDC binding has only been shown to the alligator PR (Vonier et al., 1996). Both antiandrogenic metabolites of vinclozolin (i.e., M1and M2) bind to the progesterone receptor *in vitro*; however, the effects of M1 and M2 in this system are unknown because no agonistic or antagonistic effect of vinclozolin treatment was found *in vivo* (Laws et al., 1996). Further exploration of antagonistic effects of EDCs suggests that the underlying mechanisms of agonism, partial agonism, or antagonism are all related to alterations in receptor conformation (see below).

Factors Affecting Endocrine-Disrupting Contaminants/Receptor Interactions

Receptor Binding

The agonistic or antagonistic activity of a compound is generally judged relative to the natural ligand. For example, estrogenicity is almost always judged relative to E_2, although DES has also been used for comparison. The ability to bind the AR is often judged in relation to DHT or a synthetic agonist in a competitive binding assay (i.e., [3H]R1881; Wong et al., 1995). Competitive PR binding is also judged with a synthetic agonist (i.e., [3H]R5020; Vonier et al., 1996). Receptor binding itself is not a clear concept when applied to the endocrine disruption literature. Receptor binding in a competitive-binding or inhibition-binding assay is judged in three major ways, all of which depend on the concentration of the competing compound necessary to displace 50% of bound, labeled standard relative to the concentration of unlabeled standard required for 50% displacement. The first measure, relative binding affinity–RBA, relies on the concentration of standard required for 50% displacement set to 100% (e.g., Soto et al., 1995). The second measure, the concentration of unlabeled inhibitor yielding half-maximal specific binding relative to standard–C50, is a relative measure with the concentration of standard required for 50% displacement set to 1.0 (e.g., Korach et al., 1987). The third measure, the inhibitor concentrations necessary for 50% inhibition–IC_{50}, is a specific measure in (μM concentration of the inhibitor (e.g., Vonier et al., 1996). In addition, some data have been presented relative to a third compound such as [³H]17β-estradiol displacement of alkyl phenols relative to tetrahydronaphthol-2 (Mueller and Kim, 1978).

In addition to the relationship between the three measures of binding inhibition, there are other important factors to consider when examining binding inhibition in the literature. EDC inhibitors typically require 1,000 or more times the concentration of the natural ligand to displace labeled standard (Arnold et al., 1996). Concentrations of inhibitors required to displace 50% may be clearly pharmacological, and yet only the IC_{50} method clearly displays these concentrations. Solubility limits can be reached prior to 50% binding inhibition. If 50% binding inhibition is not possible, then the compound is often judged to be nonsignificantly reactive with the receptor. Some binding inhibition data are from actual data points, and some are from extrapolation (e.g., Hammond et al., 1979). This calls the extrapolated points into question, and the points are generally used as weak evidence or relative evidence without conclusion of full binding. Another factor to consider is that, until recently, binding inhibition of compounds was generally tested individually, although exposure is more likely to be a multiple-compound phenomenon. Interactions and the effects of mixtures are discussed in the synergism section below.

The relationship between a compound's structure and its ability to bind to steroid receptors has been explored for years. For example, Mueller and Kim (1978) demonstrated that the effectiveness of ER binding of alkyl phenols was related to the size and hydrophobic character of the alkyl group. In contrast to the logical relationship between structure and ER binding of alkyl phenols, Hammond et al. (1979) concluded that the caged ring structure of chlordecone was not predictive of its ER binding ability. Recent attempts at explaining steroid receptor binding with structural models include Wong et al.'s (1995) demonstration of the effectiveness of the vinclozolin metabolite M2 in interacting with the AR and its similarity to hydroxyflutamide. The conformational restriction of compounds also has been demonstrated for ER binding; extensive structural correlation between hydroxy PCB compounds and the ability of these compounds to interact with the ER has been found (Korach et al., 1987). However, the diversity of compounds that bind steroid receptors suggest that binding cannot be predicted by structural bases alone (Soto et al., 1995).

Tables 4-1 to 4-3 are an attempt at an inclusive collection of contaminants that have been shown to interact with steroid receptors. Much of this information was adapted from the 1996 review by Guillette et al., with the following exceptions and stipulations. Compounds were included only if there was experimental evidence for direct steroid receptor interactions (i.e., binding inhibition). Significant receptor binding was determined by original authors because of the difficulty in establishing significance of EDC receptor interaction. For example, a potency of 1,000 to 2,000 times less than E_2 was experimentally determined for the phytoestrogens biochanin-A, coumestrol, daidzein, equol, formononetin, and genistein in a VTG transcription assay (Pelissero et al., 1993). The results were judged to indicate estrogenic or weakly estrogenic activity for these compounds, although the authors indicate that the biological relevance is unclear in light of cross-reactive steroids such as androgens that were equally effective in this ER transcription system (Pelissero et al., 1993). The tables include a single example of binding evidence for each tissue type that served as a receptor source, as well as each species that evidence was obtained from. When available, a single example of the category of receptor binding (i.e., estrogenic) is also included for each known effect of a compound (e.g., there are two examples of the effect of M2 because there is evidence that M2 is antiandrogenic as well as androgenic). Citations denoting the effect of a compound were preferentially chosen from steroid responsive cell culture evidence over steroid responsive tissue evidence because of the increased likelihood of indirect interactions in steroid responsive tissues. Finally, cross-taxa or wildlife examples were also included when available.

Binding Proteins and Test Conditions

The experimental conditions under which a compound is tested for receptor interactions are critical, with the presence or absence of serum proteins becoming vital for EDC potency. As discussed in the binding protein section of Chap. 1, this is linked to the interaction of natural estrogens, phytoestrogens, and synthetic estrogens to serum proteins. Binding proteins affect the likelihood that each compound will actually encounter the steroid receptor. Natural steroids such as E_2 may be bound to plasma proteins such that less than 1% of circulating E_2 is free to diffuse into cells and exert a biological action (vom Saal et al., 1995). Synthetic estrogens or EDCs generally do not bind to binding proteins, or bind with very low affinity (vom Saal et al., 1992). Therefore, EDCs may have a greater effect relative to total circulating concentrations than natural steroids due to the increased availability of EDCs within the cell (Arnold et al., 1996). Binding inhibition studies and transcriptional activation studies that do not have serum proteins in the experimental media may greatly underestimate the *in vivo* effects of EDCs relative to natural steroids. Several authors have attempted to address the effects of low relative interaction between EDCs and serum proteins. For example, the competitive binding

Table 4-1. Food contaminants, fungicides, herbicides, and pesticides that bind to steroid receptors [a]

Compound	Source/use	Mode of action		Receptor system	Reference
		Interaction	Effect		
Diethylstilbestrol	Food contaminant	ER binding		Sheep uterine ER	[Shutt, 1972]
		ER binding		Rat uterine ER	[Kelce, 1995]
		ER binding		Human mammary ER	[Whitliff, 1975]
		ER binding		Mouse uterine ER	[Okey, 1978]
		ER binding		Mouse mammary ER	[Okey, 1978]
		ER binding		Trout ER	[Pakdel, 1990]
		E responsive cells	Estrogenic	YES (human ER)	[Arnold, 1996]
		AR binding		Rat prostate AR	[Kelce, 1995]
Butylated hydroxyanisole	Food additive	ER binding		Trout hepatic ER	[Jobling, 1995]
		E responsive cells	Estrogenic	Cell culture (ZR-75)	[Jobling, 1995]
Vinclozolin	Fungicide	AR binding		Rat Epididymides AR	[Kelce, 1994]
		AR binding		Human AR in m. kid. COS	[Wong, 1995]
		A responsive cells	Antiandrogenic	Human AR in m. kid. CV1	[Wong, 1995]
2-[[(3,5-dichlorophenyl)-carbamoyl]oxy]-2-methyl-3-butenoic acid	Fungicide (M1; Vinclozolin)	AR binding		Rat Epididymides AR	[Kelce, 1994]
		AR binding		Human AR in m. kid. COS	[Wong, 1995]
		A responsive cells	Antiandrogenic	Human AR in m. kid. CV1	[Wong, 1995]
		PR binding		Rat uterine PR	[Laws, 1996]
3',5'-dichloro-2-hydroxy-2-methylbut-3-enanilide	Fungicide (M2; Vinclozolin)	AR binding		Rat Epididymides AR	[Kelce, 1994]
		AR binding		Human AR in m. kid. COS	[Wong, 1995]
		A responsive cells	Antiandrogenic	Human AR in m. kid. CV1	[Wong, 1995]
		A responsive cells	Androgenic	Human AR in m. kid. CV1	[Wong, 1995]
		PR binding		Rat uterine PR	[Laws, 1996]
Alachlor	Herbicide	ER binding		Alligator oviductal ER	[Vonier, 1996]
		E responsive cells	Estrogenic	YES (human ER)	[Klotz, 1996]

Table 4-1. *Continued*

Compound	Source/use	Mode of action		Receptor system	Reference
		Interaction	Effect		
Atrazine	Herbicide (triazine)	ER binding		Alligator oviductal-ER	(Vonier, 1996)
		ER binding		Human ER	(Tran, 1996)
		E responsive cells	Antiestrogenic	Yeast system (human ER)	(Tran, 1996)
		PR binding		Alligator oviductal-PR	(Vonier, 1996)
		AR binding		Rat pituitary AR	(Kniewald, 1979)
Cyanazine	Herbicide (triazine)	ER binding		Alligator oviductal-ER	(Vonier, 1996)
		ER binding		Human ER	(Tran, 1996)
		E responsive cells	Antiestrogenic	Yeast system (human ER)	(Tran, 1996)
		PR binding		Alligator oviductal-PR	(Vonier, 1996)
Desisopropyl	Herbicide (triazine)	ER binding		Human ER	(Tran, 1996)
		E responsive cells	Antiestrogenic	Yeast system (human ER)	(Tran, 1996)
Simazine	Herbicide (triazine)	ER binding		Human ER	(Tran, 1996)
		E responsive cells	Antiestrogenic	Yeast system (human ER)	(Tran, 1996)
Prometryne	Herbicide	AR binding		Rat pituitary AR	(Kniewald, 1979)
o,p'-DDD	Insecticide (DDT)	ER binding		Rat uterine ER	(Nelson, 1974)
		ER binding		Alligator oviductal ER	(Vonier, 1996)
		E responsive cells	Estrogenic	YES (human ER)	(Klotz, 1996)
p,p'-DDD	Insecticide (DDT)	ER binding		Recombinant human ER	(Klotz, 1996)
		E responsive cells	Estrogenic	YES (human ER)	(Klotz, 1996)
		AR binding		Rat prostate AR	(Kelce, 1995)
p,p'-DDE	Insecticide (DDT)	AR binding		Rat prostate AR	(Kelce, 1995)
		A responsive cells	Antiandrogenic	Human AR in monkey kidney CV1 cells	(Kelce, 1995)
		A responsive cells	Androgenic	YAS (human AR)	(Gaido, 1997)
		E responsive cells	Estrogenic	E-SCREEN (MCF-7)	(Soto, 1995)

Table 4-1. *Continued*

Compound	Source/use	Mode of action		Receptor system	Reference
		Interaction	Effect		
o,p'-DDE	Insecticide (DDT)	ER binding		Rat uterine ER	[Nelson, 1974]
		ER binding		Alligator oviductal ER	[Vonier, 1996]
		E responsive cells	Estrogenic	E-SCREEN (MCF-7)	[Soto, 1995]
o,p'-DDT	Insecticide (DDT)	ER binding		Alligator oviductal ER	[Vonier, 1996]
		ER binding		Rat uterine ER	[Nelson, 1974]
		ER binding		Rat testicular ER	[Bulger, 1978]
		ER binding		Recombinant human ER	[Klotz, 1996]
		ER binding		Breast cancer ER (MCF-7)	[vom Saal, 1995]
o,p'-DDT	Insecticide (DDT)	E responsive cells	Estrogenic	E-SCREEN (MCF-7)	[Soto, 1995]
		E responsive cells	Estrogenic	Trout hepatocyte VTG	[Sumpter, 1995]
		AR binding		Rat prostate AR	[Kelce, 1995]
		A responsive cells	Antiandrogenic	Human AR in monkey kidney CV1 cells	[Kelce, 1995]
p,p'-DDT	Insecticide (DDT)	AR binding		Rat prostate AR	[Kelce, 1995]
DDOH	Insecticide (DDT)	ER binding		Alligator oviductal ER	[Vonier, 1996]
		PR binding		Alligator oviductal-PR	[Vonier, 1996]
2,4-Dichlorophenol	Insecticide	ER binding		Trout hepatic ER	[Jobling, 1995]
Chlordecone (Kepone)	Insecticide	ER binding		Rat oviductal ER	[Hammond, 1979]
		ER binding		Chicken oviductal ER	[Palmiter, 1978]
		ER binding		Alligator oviductal ER	[Vonier, 1996]
		E responsive cells	Estrogenic	E-SCREEN (MCF-7)	[Soto, 1995]
		PR binding		Alligator oviductal PR	[Vonier, 1996]
		AR binding		Rat prostate AR	[Kelce, 1995]
Dicofol	Insecticide	ER binding		Alligator oviductal ER	[Vonier, 1996]
		PR binding		Alligator oviductal-PR	[Vonier, 1996]

Table 4-1. *Continued*

Compound	Source/use	Mode of action		Receptor system	Reference
		Interaction	Effect		
Endosolfan sulfate	Insecticide	PR binding		Alligator oviductal PR	(Vonier, 1996)
β-Endosolfan	Insecticide	ER binding		Breast cancer ER (MCF-7)	(Soto, 1995)
		E responsive cells	Estrogenic	E-SCREEN (MCF-7)	(Soto, 1995)
Methoxychlor (tech. grade)	Insecticide	ER binding		Rat uterine ER	(Nelson, 1974)
		ER binding		Breast cancer ER (MCF-7)	(vom Saal, 1995)
		E responsive cells	Estrogenic	YES (human ER)	(Gaido, 1997)
2,2 bis (p-Hydroxyphenyl-1,1, 1-trichloroehane (HPTE)	Insecticide (Methoxychlor)	ER binding		Rat testicular ER	(Bulger, 1978)
cis-Nonachlor	Insecticide (Chlordane)	ER binding		Alligator oviductal-ER	(Vonier, 1996)
		E responsive cells	Estrogenic	YES (human ER)	(Klotz, 1996)
trans-Nonachlor	Insecticide (Chlordane)	ER binding		Alligator oviductal-ER	(Vonier, 1996)
		E responsive cells	Estrogenic	YES (human ER)	(Klotz, 1996)
Toxaphene	Insecticide	ER binding		Breast cancer ER (MCF-7)	(Soto, 1995)
		E responsive cells	Estrogenic	E-SCREEN (MCF-7)	(Soto, 1995)

[a] Compounds were only included if there was competitive binding evidence of significant steroid receptor interactions. Significant receptor binding was determined by original authors, as explained in the text. Binding evidence was cited for each tissue type that served as a receptor source, as well as each species that evidence was obtained from. When available, a single example of the category of receptor binding was also included for each known effect of a compound. Citations denoting the effect of a compound were preferentially chosen from steroid responsive cell culture evidence over steroid responsive tissue evidence. Finally, cross-taxa or wildlife examples were also included.

Table 4-2. Industrial chemicals that bind to steroid receptors [a]

Compound	Source/use	Mode of action Interaction	Mode of action Effect	Receptor system	Reference
Tamoxifen	Drug	ER binding		Cell culture–uterine cells	[Webb, 1995]
		ER binding		Calf uterine ER	[Kiang 1978]
		ER binding		Mouse uterine ER	[Kohno, 1994]
		E responsive cells	Estrogenic	Cell culture–(Ishikawa)	[Webb, 1995]
		E responsive tissue	Estrogenic	Fetal guinea pig uterus–*in vitro*	[Pasqualini, 1987]
		ER binding		Breast cancer ER (MCF-7)	[Nesaretnam, 1996]
		E responsive tissue	Antiestrogenic	Fetal guinea pig uterus–*in vitro*	[Pasqualini, 1987]
		E responsive cells	Antiestrogenic	Cell culture–(MCF-7)	[Webb, 1995]
Aroclor 1242 (PCB mixture)	Industrial chemical	ER binding		Alligator oviductal ER	[Vonier, 1996]
p-sec-Amyl phenol	Industrial chemical	ER binding		Rat uterine ER	[Mueller, 1978]
p-Isoamyl phenol	Industrial chemical	ER binding		Rat uterine ER	[Mueller, 1978]
p-tert-Amyl phenol	Industrial chemical	ER binding		Rat uterine ER	[Mueller, 1978]
o-sec-amyl phenol	Industrial chemical	ER binding		Rat uterine ER	[Mueller, 1978]
bensophenone	Industrial chemical	ER binding		Trout hepatic ER	[Jobling, 1995]
Bisphenol-A	Industrial chemical	ER binding		Rat uterine ER	[Krishnan, 1993]
		E responsive cells	Estrogenic	Cell culture (MCF-7)	[Krishnan, 1993]
		E responsive cells	Estrogenic	Trout hepatocyte VTG	[Sumpter, 1995]
bis(2-Ethylyexyl) phthalate	Industrial chemical	ER binding		Trout hepatic ER	[Jobling, 1995]
bis(2-Ethylyexyl) adipate	Industrial chemical	ER binding		Trout hepatic ER	[Jobling, 1995]
6-Bromonaphthol-2	Industrial chemical	ER binding		Rat uterine ER	[Mueller, 1978]
		E responsive cells	Estrogenic	E-SCREEN (MCF-7)	[Soto, 1995]
4-sec-Butylphenol	Industrial chemical	ER binding		Rat uterine ER	[Mueller, 1978]
		E responsive cells	Estrogenic	E-SCREEN (MCF-7)	[Soto, 1995]

Table 4-2. *Continued*

Compound	Source/use	Mode of action — Interaction	Mode of action — Effect	Receptor system	Reference
n-butyl-benzene	Industrial chemical	ER binding		Trout hepatic ER	(Jobling, 1995)
di-n-Butylphthalate	Industrial chemical	ER binding E responsive cells	Estrogenic	Trout hepatic ER Cell culture (ZR-75)	(Jobling, 1995) (Jobling, 1995)
Butylbenzyl phthalate	Industrial chemical	ER binding E responsive cells		Trout hepatic ER Cell culture (ZR-75)	(Jobling, 1995) (Jobling, 1995)
4,4'-Dihydroxy 2' chloro biphenyl	Industrial chemical	ER binding		Mouse uterine ER	(Korach, 1987)
4,4'-Dihydroxybiphenyl	Industrial chemical	ER binding E responsive cells	Estrogenic	Rat uterine ER E-SCREEN (MCF-7)	(Mueller, 1978) (Soto, 1995)
4-Hydroxy, 2',4',6'-trichloro biphenyl	Industrial chemical	ER binding E responsive cells E responsive tissue	Estrogenic Estrogenic	Mouse uterine ER E-SCREEN (MCF-7) Turtle gonadal sex	(Korach, 1987) (Soto, 1995) (Bergeron, 1994)
4-Hydroxy, 2',3',4',5'-tetrachloro biphenyl	Industrial chemical	ER binding E responsive cells E responsive tissue	Estrogenic Estrogenic	Mouse uterine ER E-SCREEN (MCF-7) Turtle gonadal sex	(Korach, 1987) (Soto, 1995) (Bergeron, 1994)
1-Naphthol	Industrial chemical	ER binding		Rat uterine ER	(Mueller, 1978)
2-Naphthol	Industrial chemical	ER binding		Rat uterine ER	(Mueller, 1978)
4-nitrotoluene	Industrial chemical	ER binding		Trout hepatic ER	(Jobling, 1995)
p-Nonylphenol	Industrial chemical	ER binding ER binding ER binding		Breast cancer ER (MCF-7) Trout hepatic ER Mouse ER	(Soto, 1995) (White, 1994)
p-Nonylphenol	Industrial chemical	E responsive cells E responsive cells E responsive cells	Estrogenic Estrogenic Estrogenic	CEFs with mouse ER Trout hepatocyte VTG E-SCREEN (MCF-7)	(White, 1994) (White, 1994) (Soto, 1995)

Table 4-2. *Continued*

Compound	Source/use	Mode of action		Receptor system	Reference
		Interaction	Effect		
Nonylphenol	Industrial chemical	ER binding		Trout hepatic ER	[Sumpter, 1995]
		ER binding		Roach hepatic ER	[Sumpter, 1995]
		E responsive cells	Estrogenic	Trout hepatocyte VTG	[Sumpter, 1995]
		E responsive cells	Estrogenic	E-SCREEN (MCF-7)	[Soto, 1995]
4-Nonylphenoxy carboxylic acid	Industrial chemical	ER binding		Trout hepatic ER	[White, 1994]
		ER binding		Mouse ER	[White, 1994]
		E responsive cells	Estrogenic	CEFs with mouse ER	[White, 1994]
		E responsive cells	Estrogenic	Trout hepatocyte VTG	[White, 1994]
		E responsive cells	Estrogenic	Cell culture (MCF-7)	[White, 1994]
4-*tert*-Octylphenol	Industrial chemical	ER binding		Trout hepatic ER	[White, 1994]
		ER binding		Mouse ER	[White, 1994]
		E responsive cells	Estrogenic	CEFs with mouse ER	[White, 1994]
		E responsive cells	Estrogenic	Trout hepatocyte VTG	[White, 1994]
		E responsive cells	Estrogenic	Cell culture (MCF-7)	[White, 1994]
α-Phenylcresol	Industrial chemical	ER binding		Rat uterine ER	[Mueller, 1978]
Tetrahydronaphthol-2	Industrial chemical	ER binding		Rat uterine ER	[Mueller, 1978]
3,4,3',4'-tetrachloro biphenyl	Industrial chemical	ER binding		Breast cancer ER (MCF-7)	[Nesaretnam, 1996]
		E responsive cells	Estrogenic	Cell culture (MCF-7)	[Nesaretnam, 1996]

[a] Compounds were only included if there was competitive binding evidence of significant steroid receptor interactions. Significant receptor binding was determined by original authors, as explained in the text. Binding evidence was cited for each tissue type that served as a receptor source, as well as each species that evidence was obtained from. When available, a single example of the category of receptor binding was also included for each known effect of a compound. Citations denoting the effect of a compound were preferentially chosen from steroid responsive cell culture evidence over steroid responsive tissue evidence. Finally, cross-taxa or wildlife examples were also included.

assay of vom Saal et al. (1995)—the relative binding affinity-serum modified access (RBA-SMA) assay—was developed to evaluate EDCs in serum-free and 100% serum media. The YES transcriptional activation assay of Arnold et al. (1996) can also be used to test EDC transcriptional activation of the ER in the presence and absence of serum proteins. The YES assay demonstrated that charcoal-stripped human serum, sex hormone–binding globulin, or α-fetoprotein all reduced the estrogenic response of E_2 and the phytoestrogen coumestrol by 75% (Arnold et al., 1996). The activity of genistein, another phytoestrogen, was reduced by serum proteins, although the reduction was smaller than that of E_2. In contrast, the transcriptional activity of synthetic estrogens, DES, o,p'-DDT, p,p'-DDD, octyl phenol, and kepone, was only minimally reduced by the addition of serum proteins (Arnold et al., 1996).

In competition assays such as binding competition and competitive yeast reporter assays, the concentration of natural ligand is also a critical factor. These assays are generally performed with the inhibitor or test substance competing with the minimal concentration of standard to produce the desired effect. For example, in a competitive ER yeast reporter assay, 20 nM of E_2 induced maximal reporter activity. When tested against 20 nM of E_2, the commonly used herbicides atrazine, desisopropyl, cyanazine, and simazine (all chloro-S-triazine–derived compounds) failed to inhibit transcription. However, all four triazine herbicides decreased transcriptional activity in a dose-dependent manner in the presence of a submaximal dose of E_2 (0.5 nM) (Tran et al., 1996).

The presence of binding proteins may explain the discrepancy between *in vivo* and *in vitro* studies. For example, in fetal guinea pigs, tamoxifen is an ER agonist *in vivo* (Nguyen et al., 1986) but an ER antagonist *in vitro* (Pasqualini et al., 1987), again suggesting the importance of experimental conditions. The differences between *in vivo* and *in vitro* results suggest a role for serum proteins because embryos have a high concentration of binding proteins, including the developmentally restricted α-fetoprotein. However, the differences between *in vivo* and *in vitro* results clearly demonstrates the importance of test conditions when evaluating EDC interaction with steroid receptors.

Receptor-Associated Proteins

The role of receptor-associated proteins, such as the heat shock proteins, is currently an important topic in regulation of endogenous and environmental hormone actions. For example, recent evidence indicates that ER-associated proteins enhance human ER-DNA interactions and that maximal interaction of human ER with vitellogenin SRE requires at least two ER-associated proteins (Landel et al., 1995). Similar accessory proteins have been shown to enhance AR and glucocorticoid receptor (GR) binding to SREs (Kupfer et al., 1993). Furthermore, Landel et al. (1995) suggest that the mechanism by

Table 4-3. Mycotoxins and phytoestrogens that bind to steroid receptors [a]

Compound	Source/use	Mode of action		Receptor system	Reference
		Interaction	Effect		
Angolensin	Phytoestrogen	ER binding		Sheep uterine ER	(Shutt, 1972)
Apigenin	Phytoestrogen	ER binding		Human uterine ER in COS-7 cells	(Miksicek, 1993)
		ER binding		Breast cancer ER (MCF-7)	Collins, sub
		E responsive cells	Estrogenic	Cell culture (MCF-7)	(Miksicek, 1993)
		E responsive cells	Antiestrogenic	Cell culture (MCF-7)	Collins, sub
Biochanin-A	Phytoestrogen	ER binding		Rat hepatic ER	(Rosenblum, 1993)
		ER binding		Recombinant human ER in COS-7 cells	(Micksick, 1994)
		ER binding		Rat uterine ER	(Verdeal, 1980)
		E responsive cells	Estrogenic	Fish hepatocyte VTG	(Pelissero, 1993)
		E responsive cells	Estrogenic	Cell culture (MCF-7)	Collins, sub
		E responsive cells	Antiestrogenic	Cell culture (MCF-7)	Collins, sub
Chrysin	Phytoestrogen	ER binding		Breast cancer ER (MCF-7)	Collins, sub
		E responsive cells	Estrogenic	Cell culture (MCF-7)	Collins, sub
		E responsive cells	Antiestrogenic	Cell culture (MCF-7)	Collins, sub
Coumestrol	Phytoestrogen	ER binding		Rat uterine ER	(Verdeal, 1980)
		ER binding		Sheep uterine ER	(Shutt, 1972)
		ER binding		Rabbit uterine ER	(Shemesh, 1972)
		ER binding		Calf uterine ER	(Lee, 1977)
		ER binding		Breast cancer ER (MCF-7)	(Martin, 1978)
		E responsive cells	Estrogenic	Cell culture (MCF-7)	(Martin, 1978)
		E responsive cells	Estrogenic	Fish hepatocyte VTG	(Pelissero, 1993)
		E responsive cells	Antiestrogenic	Cell culture (MCF-7)	Collins, sub
Daidzein	Phytoestrogen	ER binding		Sheep uterine ER	(Shutt, 1972)
		ER binding		Recombinant human ER in COS-7 cells	(Miksicek, 1994)

Table 4-3. *Continued*

Compound	Source/use	Mode of action		Receptor system	Reference
		Interaction	Effect		
Daidzein	Phytoestrogen	ER binding		Rat uterine ER	(Verdeal, 1980)
		ER binding		Rat hepatic ER	(Rosenblum, 1993)
		E responsive cells	Estrogenic	Cell culture (Ishikawa-I)	(Markiewicz, 1993)
		E responsive cells	Estrogenic	Fish hepatocyte VTG	(Pelissero, 1993)
o-Desmethylangolensin	Phytoestrogen	ER binding		Sheep uterine ER	(Shutt, 1972)
4,4'-Dihydroxychalcone	Phytoestrogen	ER binding		Human uterine ER in COS-7 cells	(Miksicek, 1993)
		E responsive cells	Estrogenic	Cell culture (MCF-7)	(Miksicek, 1993)
4,7'-Dihydroxyflavanone	Phytoestrogen	ER binding		Human uterine ER in COS-7 cells	(Miksicek, 1993)
		E responsive cells	Estrogenic	Cell culture (MCF-7)	(Miksicek, 1993)
Equol	Phytoestrogen	ER binding		Sheep uterine ER	(Shutt, 1972)
		ER binding		Rat uterine ER	(Thompson, 1984)
		ER binding		Breast cancer ER (MCF-7)	(vom Saal, 1995)
		E responsive tissue	Estrogenic	Rat uterine growth/ER	(Tang, 1980)
		E responsive tissue	Antiestrogenic	Rat uterine growth/ER	(Tang, 1980)
		E responsive cells	Estrogenic	Cell culture (Ishikawa-I)	(Markiewicz, 1993)
		E responsive cells	Estrogenic	Fish hepatocyte VTG	(Pelissero, 1993)
Genistein	Phytoestrogen	ER binding		Rat uterine ER	(Verdeal, 1980)
		ER binding		Rat hepatic ER	(Rosenblum, 1993)
		ER binding		Sheep uterine ER	(Shutt, 1972)
		ER binding		Rabbit uterine ER	(Shemesh, 1972)
		ER binding		Human uterine ER in COS-7 cells	(Miksicek, 1993)
		ER binding		Breast cancer ER (MCF-7)	(Martin, 1978)
		ER binding		Trout hepatic ER	(Sumpter, 1995)
		ER binding		Roach hepatic ER	(Sumpter, 1995)

Table 4-3. *Continued*

Compound	Source/use	Mode of action		Receptor system	Reference
		Interaction	Effect		
Genistein	Phytoestrogen	E responsive cells	Estrogenic	Cell culture (MCF-7)	[Martin, 1978]
		E responsive cells	Estrogenic	Fish hepatocyte VTG	[Pelissero, 1993]
		E responsive cells	Antiestrogenic	Cell culture (MCF-7)	Collins, sub
Isoliquiritigenin	Phytoestrogen	ER binding		Human uterine ER in COS-7 cells	[Miksicek, 1993]
Kaempferide	Phytoestrogen	ER binding		Breast cancer ER (MCF-7)	Collins, sub
		E responsive cells	Estrogenic	Cell culture (MCF-7)	Collins, sub
		E responsive cells	Antiestrogenic	Cell culture (MCF-7)	Collins, sub
Luteolin	Phytoestrogen	ER binding		Breast cancer ER (MCF-7)	Collins, sub
		E responsive cells	Estrogenic	Cell culture (MCF-7)	Collins, sub
		E responsive cells	Antiestrogenic	Cell culture (MCF-7)	Collins, sub
Miroestrol	Phytoestrogen	ER binding		Sheep uterine ER	[Shutt, 1972]
Naringenin	Phytoestrogen	ER binding		Human uterine ER in COS-7 cells	[Miksicek, 1993]
		ER binding		Breast cancer ER (MCF-7)	Collins, sub
		E responsive cells	Estrogenic	Human uterine ER in HeLa cells	[Miksicek, 1993]
		E responsive cells	Antiestrogenic	Cell culture (MCF-7)	Collins, sub
Phloretin	Phytoestrogen	ER binding		Recombinant human ER in COS-7 cells	[Miksicek, 1993]
		E responsive cells	Estrogenic	Cell culture (HeLa)	[Miksicek, 1993]
		E responsive cells	Antiestrogenic	Cell culture (MCF-7)	Collins, sub
β-Sitosterol	Phytoestrogen	ER binding		Rat hepatic ER	[Rosenblum, 1993]
		ER binding		Rat uterine ER	[Rosenblum, 1993]
		E responsive cells	Estrogenic	YES (human ER)	[Gaido, 1997]
		E responsive cells	Estrogenic	T-47D (human ER)	[MacLatchy, 1995]

Table 4-3. *Continued*

Compound	Source/use	Mode of action		Receptor system	Reference
		Interaction	Effect		
Zearalenone	Mycoestrogen	ER binding		Rat uterine ER	[Verdeal, 1980]
		ER binding		Calf uterine ER	[Kiang, 1978]
		ER binding		Breast cancer ER (MCF-7)	[Martin, 1978]
		E responsive cells	Estrogenic	Cell culture (MCF-7)	[Martin, 1978]
		E responsive cells	Antiestrogenic	Cell culture (MCF-7)	Collins, sub
Zearalanol	Mycoestrogen	ER binding		Rat uterine ER	[Verdeal, 1980]
		ER binding		Calf uterine ER	[Kiang, 1978]
Zearalenol	Mycoestrogen	ER binding		Breast cancer ER (MCF-7)	[Martin, 1978]
		ER binding		Calf uterine ER	[Kiang, 1978]
		E responsive cells	Estrogenic	Cell culture (MCF-7)	[Martin, 1978]
		E responsive cells	Antiestrogenic	Cell culture (MCF-7)	Collins, sub

[a] Compounds were only included if there was competitive binding evidence of significant steroid receptor interactions. Significant receptor binding was determined by original authors, as explained in the text. Binding evidence was cited for each tissue type that served as a receptor source, as well as each species that evidence was obtained from. When available, a single example of the category of receptor binding was also included for each known effect of a compound. Citations denoting the effect of a compound were preferentially chosen from steroid responsive cell culture evidence over steroid responsive tissue evidence. Finally, cross-taxa or wildlife examples were also included.

which some steroid antagonists promote DNA binding while failing to activate (or fully activate) transcription may hinge on the receptor-associated proteins. Thus receptor-associated proteins present another mechanism through which EDCs may act—by modifying or failing to fully stimulate normal receptor-associated protein interactions.

Receptor Translocation

Steroid receptors are stimulated to move into the nucleus following effective ligand interaction. For example, significant augmentation of nuclear ER is observed in under 2 hr following estrogen treatment. At least part of the mechanism of nuclear translocation involves rapid increases in the affinity of receptors, such as the ER for chromatin (Ignar-Trowbridge et al., 1995). ARs also translocate from a perinuclear position in the absence of androgen to a nuclear position in the presence of androgens (Simental et al., 1991; Zhou et al., 1994). Furthermore, nonsteroidal compounds such as growth factors and EDCs can stimulate nuclear translocation of steroid receptors. ERs were stimulated to move into the nucleus following treatment with chlordecone, indicating that chlordecone is effective in binding to and activating the ER (Hammond et al., 1979). Chlordecone bound the ER with 0.04%, the relative binding affinity of estradiol. ER movement into the nucleus required 10,000 times the concentration of chlordecone as estradiol, or 0.01% E_2, approximately proportional to the relative binding affinity of the two compounds (Hammond et al., 1979). Significant nuclear translocation of AR was observed in monkey kidney COS cells after exposure to M1, M2, and vinclozolin in the relative quantities that reflect their ability to compete for AR binding (Wong et al., 1995). Receptor translocation is an indication of EDC receptor interaction, however; translocation does not indicate the type of receptor interaction, because receptor translocation is part of agonistic and antagonistic interactions (see below).

Receptor Dissociation and Degradation

Ligand-receptor interactions involve an equilibrium between binding and dissociation that has an associated time frame or length of interaction. For example, in AR binding, T dissociated 3 times as fast as DHT and was also less effective in stabilizing the AR (Wilson and French, 1976; Zhou et al., 1994). A mutated form of the AR with unmodified equilibrium androgen binding was found to be androgen insensitive due to increased rates of androgen dissociation from the AR and subsequent AR degradation. Furthermore, similar levels of AR stabilization and function were observed at higher concentrations of faster-dissociating androgens and lower concentrations of slower-dissociating androgens (Kemp-painen et al., 1992; Zhou et al., 1995). This last fact is of special interest to EDC research, because EDCs often show reduced binding relative to the natural ligand, but may show prolonged occupancy of the receptor. For example, o,p'-DDT interacts with the ER in what appears to be a weaker but longer-term bond than that of E_2 (Robison et al., 1985). The prolonged action of kepone (an organochlorine pesticide) on the chicken oviduct and zearalenone (a fungal estrogen) on the calf uterus is also linked to reduced clearance or prolonged binding (Kiang et al., 1978; Palmiter and Mulvihill, 1978). Natural receptor turnover may be different by tissue type, although data on ER turnover from rat uterine cells suggests a turnover rate similar to that observed in human breast cancer cells (Nardulli and Katzenellenbogen, 1986). The relative potency of natural estrogens (e.g., the low potency of estriol compared to E_2; Gaido et al., 1997) or synthetic estrogens (e.g., the low potency of dimethylstilbestrol compared to DES; Katzenellenbogen et al., 1978) can be directly correlated to the length of receptor binding and the resulting changes in receptor levels. The phytoestrogen equol was found to be uterotrophic in the rat uterus, but the duration of uterine growth was much shorter than that induced by E_2, suggesting a shorter-term interaction with the ER than that of E_2 (Tang and Adams, 1980). The extent of receptor replenishment induced by equol was less than that of E_2 and may in fact be the major difference between the actions of equol and E_2 in this system. Tang et al. (1980) concluded that equol is weakly estrogenic in its ability to bind to ER and initiate transcription; however, equol has an overall antagonistic effect due to its failure to initiate replenishment of ER in the cytoplasm.

Heterodimers

The possibility of a compound acting as a receptor agonist in one situation and an antagonist in another is a serious complication for interpreting experimental studies (see above). One condition that transforms a receptor antagonist to a receptor agonist is hypothesized to be the state of the receptor dimer (Brown, 1994). ER agonists such as E_2 and tamoxifen stabilize a ligand-ER complex that consists of a homodimer. Pure antiestrogens, however, prevent homodimer formation (Nimrod and Benson, 1996). The existence of receptor subtypes further complicates this picture due to the demonstration of ER-α/ER-β heterodimers (Cowley et al., 1997; Pettersson et al., 1997). Wong et al. (1995) suggest a similar system for AR agonists and antagonists. When androgens are present with antiandrogens, the dimer is expected to be a heterodimer of androgen-AR and antiandrogen-AR. Without AR antagonists, the homodimer of androgen-AR and androgen-AR is fully agonistic. In the absence of androgen the homodimer of antiandrogen-AR and antiandrogen-AR is at least partially agonistic (Wong et al., 1995). The importance of homo- and het-erodimers to EDC interactions is speculative at this point because much of the evidence is indirect, such as polyacrylamide gel analysis of receptors following trypsin digestion; similar lines of evidence suggest a role for receptor-induced conformational changes.

Ligand-Induced Receptor Conformational Change

As discussed above, steroid receptor binding induces a conformational change in the receptor such that the receptor-associated proteins dissociate from the receptor, and the receptor obtains the ability to interact with other receptors and the SRE of the DNA. This conformational change has been hypothesized to be the underlying mechanisms of agonism, partial agonism, and/or antago-nism for any compound interacting with steroid receptors (Allan et al., 1992). Distinct agonist- and antagonist-induced modifications in the receptor have been demonstrated for AR, ER, and PR (Allan et al., 1992; Kallio et al., 1994; McDonnell et al., 1995). For example, the natural estrogen, E_2, the pure anti-estrogen-ICI164,384, and the partial ER agonist-4-hydroxy-tamoxifen modify the ER in three distinct ways, as revealed by products of trypsin digestion (McDonnell et al., 1995). Unlike the difference between natural and synthetic agonist-induced ER conformational changes, the natural AR agonist T and DHT induce a conformational change in the AR that is indistinguishable from the conformational change induced by the synthetic AR agonist, mibolerone (Kallio et al., 1994). However, the synthetic AR antagonists, cyproterone acetate and casodex, fail to induce the appropriate conformational change in the AR (Kallio et al., 1994) in a manner similar to antagonist action in the ER. The nature of the conformational change is unclear, and evidence of the change is largely indi-rect evidence of differences in polyacrylamide gel migration speeds and novel protein fragments resulting from trypsin digestion. However, there is evidence that the E_2-induced conformational change in the ER involves an association of the amino- and carboxyl-terminal regions (TAF1 and TAF2 transactivation regions, respectively) leading to a positive transcriptional relationship between the two regions (Kraus et al., 1995). Agonists and antagonists may both promote translocation of the receptor to the DNA and initiate DNA binding; antagonists, however, appear to alter receptor structure in such a way as to make it incompatible with transcriptional activation (McDonnell et al., 1995).

Age Specificity

Age-specific effects of EDC-receptor interactions based on specific receptor properties are unknown at this time, although the existence of developmentally restricted receptors suggests age-specific interactions. The age-specific actions of the synthetic ER agonist/antagonist tamoxifen also supports age as a critical variable for EDC receptor interactions. Tamoxifen is an ER agonist in newborn and fetal guinea pig uterus (Nguyen et al., 1986). Tamoxifen is also a potent ER inducer in the uterus and vagina of neonatal and ovariectomized adult mice; however, the responsiveness to tamoxifen differs between neonatal and adult tissue. ER mRNA expression in fetal mice was induced by tamoxifen treatment 3 times faster than in ovariectomized adult mice (Sato et al., 1996). Furthermore, the effectiveness of tamoxifen as an agonist *in vivo* during embryogenesis suggests a role for organizational effects. The organizational model suggests that EDC exposure during a specific period of embryonic development can permanently modify the organization of the reproductive, immune, and nervous systems (Guillette et al., 1995). Developmentally restricted periods of increased sensitivity combine with the fact that developing animals normally have high concentrations of binding proteins (see above) and low concentrations of steroids to make EDC exposure of an embryo significantly different from that of an adult.

Receptor Expression

Evidence for EDC-mediated changes in receptor level are sparse, although the plasticity of receptor numbers suggests that changes in receptor numbers are a possible route for EDC action. Umbreit and Gallo (1988) suggest that TCDD is able to inhibit the uterotrophic response of E_2 in rats without binding to the ER because TCDD decreases the numbers of ERs. Additional evidence supports the contention that endocrine changes are induced by TCDD without TCDD interacting directly with the steroid receptors. For example, *in utero* exposure to TCDD affects E_2 concentration and therefore ER levels leading to modified posttranscriptional or posttranslational processing (Chaffin et al., 1996). Experiments involving the concentration of ER are also pertinent to EDC action. McDonnell et al. (1991) found that at low concentrations of ER, ligand was required to promote DNA binding; however, at high levels of receptor, ER occupied the SRE in the absence of ligand. These ligand-neutral translocations failed to activate transcription (McDonnell et al. 1991).

Tissue Specificity

Excellent evidence for tissue specificity of a receptor agonist or antagonist is tamoxifen's interaction with the ER. Tamoxifen, a synthetic estrogen antagonist in breast tissue, is actually an ER agonist in the cardiovascular system, bone, and uterine tissue (Kedar et al., 1994; Love et al., 1992). ER interaction with *o,p'-*

DDT is also different by tissue type within the rat uterus (Robison et al., 1985). DNA synthesis stimulated by o,p'-DDT is only 70% that induced by E_2 in the uterine stroma and myometrium; however, E_2 and o,p'-DDT were equally effective in the luminal epithelium. Several authors point to distinct tissue-related transcription factor pools or the promoter context in which the SREs reside to explain tissue-dependent responses to EDCs (Katzenellenbogen, 1996). Again, tamoxifen is the best evidence for this type of tissue specificity. Triphenylethyl-ene-derived compounds, such as tamoxifen, interact with the TAF2 transactivation region of the human ER, which generally is required for transcription (Kraus et al., 1995). In cells where TAF2 function is independent of the ER, TAF1 function allows tamoxifen to establish partial agonist activity (Tora et al., 1989; Tzukerman et al., 1994). The tissue specificity of EDC action suggests an increased need for careful *in vivo* studies as part of any EDC research plan.

Receptor Subtypes

Recent evidence suggests that receptor subtypes may contribute to differential activity of EDCs by their tissue-specific distribution and transcriptional activation. Evidence is limited to comparisons between the recently characterized ER-β and ER-α at this time; however, there are implications for other steroid receptors. Tissue distribution of ER-α is distinct from that of ER-β (e.g., both ER-α and ER-β are found in the ovary, testis, and uterus; ER-α dominates in the epididymis, adrenal, pituitary, and kidney; ER-β dominates in the prostate, bladder, lung, and brain) (Kuiper et al., 1997). Although Kuiper's lab found little difference in the binding characteristics of ER-α and ER-β, they were able to find major differences in transactivation of the two receptors in the context of an SRE and an AP1 element. Stimulation of the two receptors with 17β–estradiol resulted in opposite reactions (Paech et al., 1997). ER-α was stimulated by 17β–estradiol, similar to existing studies; however, 17β–estradiol inhibited transcription with ER-β. Furthermore, the classic antiestrogens tamoxifen, raloxifene, and ICI 164384 were found to activate ER-β. Clearly, receptor subtypes are critical to experiments predicting EDC effects or determining the result of EDC exposure.

Species Differences

There is evidence of species differences in the binding of EDCs to steroid receptors; however, the weight of the evidence is currently unclear. The insecticide chlordecone (kepone) is a good example of a compound demonstrating strong species differences in receptor binding. Chlordecone has a 100-fold higher affinity for the ER in chicken oviducts than in rats (Hammond et al., 1979; Palmiter and Mulvihill, 1978). Species differences in EDC binding to ER have also been shown between rats and rabbits, Muller and Kim (1978) found strong differences between the ER of rabbits and rats in the ability of tetrahydronaphthol-2 to compete with E_2 for ER binding in these species. Species differences in the effects of the phytoestrogens have also been noted. The phytoestrogen β-sitosterol was active in both trout and breast cancer systems examined by Mellanen et al. (1996). However, while the trout did not respond to other phytoestrogens, the MCF-7 and T-47 breast cancer cells responded to additional phytoestrogens (Mellanen et al., 1996). Permanent infertility in the ovine system has been linked to equol, in sharp contrast to the insignificant or low uterotrophic effect observed in rats (Tang and Adams, 1980; Thompson et al., 1984). Species differences have also been noted in serum or binding protein affects on EDCs. Human serum was more effective in reducing the activity of EDCs in the YES assay than alligator serum, indicating that species differences in serum proteins may directly affect the ability of EDCs to encounter the receptor (Arnold et al., 1996). Some indirect evidence of species differences is also based on the study of phytoestrogens. No differences were detected in the activity of six phytoestrogens (biochanin-A, coumestrol, equol, genistein, formononetin, and daidzein) in an *in vitro* trout hepatocyte VTG assay (Pelissero et al., 1993). However, *in vivo* the activity of the phytoestrogens was variable in sturgeons, indicating species differences and or *in vitro/in vivo* differences in phytoestrogen potency (Pelissero et al., 1991, 1993). There are several additional lines of evidence that suggest low species differences. White et al. (1994) examined fish, avian, and mammalian cells, finding few differences in the estrogenicity of a number of alkylphenolic compounds either in effect or potency of compound across taxa. Additionally, Sumpter and Jobling (1995) tested ER binding of E_2, a phytoestrogen (genistein), a pesticide (*o,p'*-DDT) and a surfactant (nonylphenol) in two distantly related fish. There was no evidence of species specificity between trout and roach ERs in the two fish (Sumpter and Jobling, 1995). The firm demonstration of species differences in EDC receptor interactions will only be possible with more studies, and in particular with parallel studies at the *in vitro* and *in vivo* levels.

Synergy

The extrapolation of laboratory studies to wildlife populations requires the measurement of all EDCs in a given habitat, or more accurately, all EDCs present in the tissue and/or plasma of a wildlife species. Unlike most laboratory studies, a natural population is not exposed to a single contaminant, but rather wildlife species are under the simultaneous influence of a range of contaminants. Recently, the effects of this simultaneous, multiple exposure have received a great deal of attention. In an *in vitro* test, a mixture of 10 EDCs enhanced the proliferation of MCF-7 cells in the E-SCREEN in a greater than additive manner (Soto et al., 1995). Evidence of synergy in nonhuman systems comes from Sumpter and Jobling (1995) and the synergistic effects of a combination of EDCs on *in vitro* fish hepatocytes. *In vivo* evidence of synergism of EDCs has also been demonstrated in turtle sex determination. Sex-determination is a temperature-dependent phenomenon in some turtle species, with low incubation temperatures (26°C) producing males and high incubation temperatures (31°C) producing females. Female sex-reversal can be induced by administering exogenous estrogens, such as 17β-estradiol, to turtles incubated at a male-producing temperature (26°C) (Wibbels and Crews, 1995). Not only were 3-OH PCB and 4-OH PCB both effective in sex-reversing turtles incubated at male-producing temperatures, but also synergism between the two PCBs was such that a combination of the two PCBs sex-reversed animals at a total concentration that failed to produce sex-reversal with either compound alone (Bergeron et al., 1994). Recent publications involving long-term studies of PCB and dioxin mixtures indicate that mixture potency of mono-ortho PCBs is clearly nonadditive, involving synergy in the porphyrogenic response. Furthermore, the addition of PCB 153 to the mixture further exacerbated the synergism in both the porphyrogenic response and EDC-associated decreases in circulating T_4 concentrations (Birnbaum, 1997). In fact, synergistic effects of dioxin-like and nondioxin-like compounds have been reported for cytochrome P450 induction, tumor promotion, lethality, thyroid hormone metabolism, and hepatic prophyrin accumulation (reviewed in Van Birgelen and Birnbaum, 1997). Birnbaum (1997) strongly suggests that synergy displayed in the context of the Ah receptor involves additional mechanisms other than the Ah receptor; and her lab has no evidence for anything other than strict dose additivity when the response mechanisms involve activation of the Ah receptor alone. Whether synergism involves direct receptor events or other factors, the topic of synergism is perhaps the most controversial of all aspects of EDC effects on humans and wildlife. With the huge numbers of potential EDCs produced and released around the world, the stakes are high for the health of humans and wildlife, and for the companies producing these chemicals.

Conclusion

The clearest property of EDC binding to steroid receptors is the importance of context. The possibility of synergy, or any reaction among EDC mixtures, presents a major limitation in most laboratory studies—the failure to examine the binding of more than one EDC at a time. This failure also exacerbates the differences between laboratory studies and conditions faced by wildlife or humans. Teutsch et al. (1995) correctly emphasize the complexity of steroid receptor binding as a continuum from agonistic to antagonistic responses. The effect of ligand binding to a steroid receptor is not an empirical property of the ligand but rather a synthesis of the entire system, including the combination of ligands present, the species in question, the age of the individual, the season or reproductive status, the tissue in question, the cellular context, the binding proteins, the receptor, the accessory proteins, and the DNA itself. Only in this broad context can EDCs be effectively evaluated and therefore effectively regulated.

References

Allan, G.F., Leng, X., Tsai, S.Y., Weigel, N.L., Edwards, D.P., Tsai, M.J., and O'Malley, B.W., (1992). Hormone and antihormone induce distinct conformational changes which are central to steroid receptor activation. *J. Biol. Chem.* 267: 19513–19520.

————, Tsai, S.Y., Tsai, M.J., and BW, O.M. (1992). Ligand-dependent conformational changes in the progesterone receptor are necessary for events that follow DNA binding. *Proc. Natl. Acad. Sci. USA* 89: 11750–11754.

Arnold, S.F., Collins, B.M., Robinson, M.K., Guillette, L.J., Jr., and McLachlan, J.A. (1996). Differential interaction of natural and synthetic estrogens with extracellular binding proteins in a yeast estrogen screen. *Steroids* 61: 642–646.

————, Robinson, M.K., Notides, A.C., Guillette, L.J., Jr., and McLachlan, J.A. (1996). A yeast estrogen screen for examining the relative exposure of cells to natural and xenoestrogens. *Environ. Health Perspect.* 104: 544–548.

Bergeron, J.M., Crews, D., and McLachlan, J.A. (1994). PCBs as environmental estrogens: Turtle sex determination as a biomarker of environmental contamination. *Environ. Health Perspect.* 102: 780–781.

Bernal, J., and Pekonen, F. (1984). Ontogenesis of the nuclear 3,5,3'-triiodothyronine receptor in the human fetal brain. *Endocrinology* 114: 677–679.

Birnbaum, L.S. Beyond TEFs: Mixtures of dioxins and non-dioxins. In 17th International Symposium on Chlorinated Dioxins and Related Compounds. (1997). Organohalogen Compounds, Indianapolis.

Bitman, J., Cecil, H.C., Harris, S.J., and Fries, G.F. (1968). Estrogenic activity of *o,p'*-DDT in the mammalian uterus and avian oviduct. *Science* 851: 371–372.

Brown, M. (1994). Estrogen receptor molecular biology. *Breast Cancer* 8: 101–112.

Bulger, W.H., Muccitelli, R.M., and Kupfer, D. (1978). Interactions of chlorinated hydrocarbon pesticides with the 8S estrogen-binding protein in rat testes. *Steroids* 32: 165–177.

Byers, M., Kuiper, G.G., Gustafsson, J.A., and Park-Sarge, O.K. (1997). Estrogen receptor-beta mRNA expression in rat ovary: down-regulation by gonadotropins. *Mol. Endocrinol.* 11: 172–182.

Chaffin, C.L., Peterson, R.E., and Hutz, R.J. (1996). In utero and lactational exposure of female Holtzman rats to 2,3,7,8-tetrachlorodibenzo-*p*-dioxin: Modulation of the estrogen signal. *Biol. Reprod.* 55: 62–67.

Cowley, S.M., Hoare, S., Mosselman, S., and Parker, M.G. (1997). Estrogen receptors alpha and beta form heterodimers on DNA. *J. Biol. Chem.* 272: 19858–19862.

Culig, Z., Hobisch, A., Cronauer, M.V., Radmayr, C., Trapman, J., Hittmair, A., Bartsch, G., and Klocker, H. (1994). Androgen receptor activation in prostatic tumor cell lines by insulin-like growth factor-I, keratinocyte growth factor, and epidermal growth factor. *Cancer Res.* 54: 5474–5478.

Cuthill, S., Wilhelmsson, A., Mason, G.G., Gillner, M., Poellinger, L., and Gustafsson, J.A. (1988). The dioxin receptor: A comparison with the glucocorticoid receptor. *J. Steroid Biochem.* 30: 277–280.

Dellovade, T.L., Zhu, Y.S., and Pfaff, D.W. (1995). Potential interactions between estrogen receptor and thyroid receptors relevant for neuroendocrine systems. *J. Steroid Biochem. Mol. Biol.* 53: 27–31.

Eisen, C., Meyer, C., and Wehling, M. (1997). Characterization of progesterone membrane binding sites from porcine liver probed with a novel azido-progesterone radioligand. *Cell. Mol. Biol.* 43: 165–173.

Elbrecht, A., and Lazier, C.B. (1985). Selective inhibitory effects of thyroid hormones on estrogen-induced protein synthesis in chick embryo liver. *Can. J. Biochem. Cell. Biol.* 63: 1206–1211.

Evans, R.M. (1988). The steroid and thyroid hormone receptor superfamily. *Science* 240: 889–95.

Fisher, J.S., Millar, M.R., Majdic, G., Saunders, P.T., Fraser, H.M., and Sharpe, R.M. (1997). Immunolocalisation of oestrogen receptor-alpha within the testis and excurrent ducts of the rat and marmoset monkey from perinatal life to adulthood. *J. Endocrinol.* 153: 485–495.

Fry, D.M. (1995). Reproductive effects in birds exposed to pesticides and industrial chemicals. *Environ. Health Perspect.* 103 (Suppl. 7): 165–171.

Gaido, K.W., Leonard, L.S., Lovell, S., Gould, J.C., Babaï, D., Portier, C.J., and McDonnell, D.P. (1997). Evaluation of chemicals with endocrine modulating activity in a yeast-based steroid hormone receptor gene transcription assay. *Toxicol. Appl. Phamacol.* 143: 205–212.

Giambiagi, N., Pasqualini, J.R., Greene, G., and Jensen, E.V. (1984). Recognition of two forms of the estrogen receptor in the guinea-pig uterus at different stages of development by a monoclonal antibody to the human estrogen receptor. Dynamics of the translocation of these two forms to the nucleus. *J. Steroid Biochem.* 20: 397–400.

Giannoukos, G., and Callard, I.P. (1996). Radioligand and immunochemical studies of turtle oviduct progesterone and estrogen receptors: Correlations with hormone treatment and oviduct contractility. *Gen. Comp. Endocrinol.* 101: 63–75.

Glass, C.K., Holloway, J.M., Devary, O.V., and Rosenfeld, M.G. (1988). The thyroid hormone receptor binds with opposite transcriptional effects to a common sequence motif in thyroid hormone and estrogen response elements. *Cell* 54: 313–323.

Golding, T.S., and Korach, K.S. (1988). Nuclear estrogen receptor molecular heterogeneity in the mouse uterus. *Proc. Natl. Acad. Sci. USA* 85: 69–73.

Green, S. (1990). Modulation of oestrogen receptor activity by oestrogens and anti-oestrogens. *J. Steroid Biochem. Mol. Biol.* 37: 747–751.

———, and Chambon, P. (1986). A superfamily of potentially oncogenic hormone receptors. *Nature* 324: 615–617.

Guillette, L.J., Jr., Arnold, S.F., and McLachlan, J.A. (1996). Ecoestrogens and embryos: Is there a scientific basis for concern? *Animal Repro. Sci.* 42: 13–24.

———, Crain, D.A., Rooney, A.A., and Pickford, D.B. (1995). Organization versus activation: The role of endocrine-disrupting contaminants (EDCs) during embryonic development in wildlife. *Environ. Health Perspect.* 103 (Suppl. 7): 157–164.

———, Gross, T.S., Masson, G.R., Matter, J.M., Percival, H.F., and Woodward, A.R. (1994). Developmental abnormalities of the gonad and abnormal sex hormone concentrations in juvenile alligators from contaminated and control lakes in Florida. *Environ. Health Perspec.* 102: 680–688.

———, Pickford, D.B., Crain, D.A., Rooney, A.A., and Percival, H.F. (1996). Reduction in penis size and plasma testosterone concentrations in juvenile alligators living in a contaminated environment. *Gen. Comp. Endocrinol.* 1010: 32–42.

Hammond, B., Katzenellenbogen, B.S., Krauthhammer, N., and McConnell, J. (1979). Estrogenic activity of the insecticide chlordecone (Kepone) and interaction with uterine estrogen receptors. *Proc. Natl. Acad. Sci. USA* 776: 6641–6645.

Ignar-Trowbridge, D.M., Nelson, K.G., Ross, K.A., Washburn, T.F., Korach, K.S., and McLachlan, J.A. (1991). Localization of the estrogen receptor in uterine cells by affinity labeling with [3H]tamoxifen aziridine. *J. Steroid Biochem. Mol. Biol.* 39: 131–132.

———, Pimentel, M., Parker, M.G., McLachlan, J.A., and Korach, K.S. (1995). Cross talk between peptide growth factor and estrogen receptor signaling systems. *Environ. Health Perspec.* 103 (Suppl. 7): 35–39.

———, ———, ———, ———, and ——— (1996). Peptide growth factor cross-talk with the estrogen receptor requires the A/B domain and occurs independently of protein kinase C or estradiol. *Endocrinology* 137: 1735–1744.

Jobling, S., Reynolds, T., White, R., Parker, M.G., and Sumpter, J.P. (1995). A variety of environmentally persistent chemicals, including some phthalate plasticizers, are weakly estrogenic. *Environ. Health Perspec.* 103: 582–587.

Kalimi, M., and Dave, J.R. (1995) Mechanisms of action of hormones associated with reproduction. In *Reproductive Toxicology* (R.J. Witorsch, ed.), pp. 61–74. Raven, New York.

Kallio, P.J., Jänne, O.A., and Palvimo, J.J. (1994). Agonists, but not antagonists, alter the conformation of the hormone-binding domain of the androgen receptor. *Endocrinology* 134: 998–1001.

Kashon, M.L., Arbogast, J.A., and Sisk, C.L. (1996). Distribution and hormonal regulation of androgen receptor immunoreactivity in the forebrain of the male European ferret. *J. Comp. Neurol.* 376: 567–586.

Katzenellenbogen, B.S. (1996). Estrogen receptors: Bioactivities and interactions with cell signaling pathways. *Biol. Reprod.* 54: 287–293.

———, Iwamoto, H.S., Heiman, D.F., Lan, N.C., and Katzenellenbogen, J.A. (1978). Stilbestrols and stilbestrol derivatives: Estrogenic potency and temporal relationships between estrogen receptor binding and uterine growth. *Mol. Cell. Endocrinol.* 10: 103–113.

——— and Korach, K.S. (1997). A new actor in the estrogen receptor drama—enter ER-beta. *Endocrinology* 138: 861–862.

Kedar, R.P., Bourne, T.H., Powles, T.J., Collins, W.P., Ashley, S.E., Cosgrove, D.O., and Campbell, S. (1994). Effects of tamoxifen on uterus and ovaries of postmenopausal women in a randomised breast cancer prevention trial. *Lancet* 343: 1318–1321.

Kelce, W.R., Monosson, E., Gamcsik, M.P., Laws, S.C., and Gray, L.E., Jr. (1994). Environmental hormone disruptors: Evidence that vinclozolin developmental toxicity is mediated by antiandrogenic metabolites. *Toxicol. Appl. Phamacol.* 126: 276–285.

———, Lambright, C.R., Gray, L.E., Jr., and Roberts, K.P. (1997). Vinclozolin and p,p′-DDE alter androgen-dependent gene expression: *In vivo* confirmation of an androgen receptor-mediated mechanism. *Toxicol. Appl. Phamacol.* 142: 192–200.

———, Stone, C.R., Laws, S.C., Gray, L.E., Kemppainen, J.A., and Wilson, E.M. (1995). Persistent DDT metabolite p,p′-DDE is a potent androgen receptor antagonist. *Nature* 375: 581–585.

Kemppainen, J.A., Lane, M.V., Sar, M., and Wilson, E.M. (1992). Androgen receptor phosphorylation, turnover, nuclear transport, and transcriptional activation: Specificity for steroids and antihormones. *J. Biol. Chem.* 267: 968–974.

Kerkvliet, N.I., and Burleson, G.R. (1994) Immunotoxicity of TCDD and related halogenated aromatic hydrocarbons. In *Target Organ Toxicology Series: Immunotoxicology and Immunopharacology* (J.H. Dean et al., eds.), pp. 97–121. Raven, New York.

Kiang, D.T., Kennedy, B.J., Pathre, S.V., and Mirocha, C.J. ((1978) Binding characteristics of zearalenone analogs to estrogen receptors. *Cancer Res.* 38: 3611–3615.

Klotz, D.M., Beckman, B.S., Hill, S.M., McLachlan, J.A., Walters, M.R., and Arnold, S.F. (1996). Identification of environmental chemicals with estrogenic activity using a combination of *in vitro* assays. *Environ. Health Perspec.* 104: 1084–1089.

Kohno, H., Gandini, O., Curtis, S.W., and Korach, K.S. (1994). Anti-estrogen activity in the yeast transcription system: Estrogen receptor mediated agonist response. *Steroids* 59: 572–578.

Korach, K.S., Sarver, P., Chae, K., McLachlan, J.A., and McKinney, J.D. (1987). Estrogen receptor-binding activity of polychorinated hydroxybiphenyls: Conformationally restricted structural probes. *Mol. Pharmacol.* 33: 120–126.

Kraus, W.L., McInerney, E.M., and Katzenellenbogen, B.S. (1995). Ligand-dependent, transcriptionally productive association of the amino- and carboxy-terminal regions of a steroid hormone nuclear receptor. *Biochemistry* 92: 12314–12318.

Krishnan, A.V., Stathis, P., Permuth, S.F., Tokes, L., and Feldman, D. (1993). Bisphenol-A: An estrogenic substance is released from polycarbonate flasks during autoclaving. *Endocrinology* 132: 2279–2286.

Krusekopf, S., Chauchereau, A., Milgrom, E., Henderson, D., and Cato, A.C. (1991). Cooperation of progestational steroids with epidermal growth factor in activation of gene expression in mammary tumor cells. *J. Steroid Biochem. Mol. Biol.* 40: 239–245.

Kuiper, G.G., Carlsson, B., Grandien, K., Enmark, E., J, Häggblad, J., Nilsson, S., and Gustafsson, J.A. (1997). Comparison of the ligand binding specificity and transcript tissue distribution of estrogen receptors alpha and beta. *Endocrinology* 138: 863–870.

———— and Gustafsson, J.A. (1997). The novel estrogen receptor-beta subtype: Potential role in the cell- and promoter-specific actions of estrogens and anti-estrogens. *FEBS Lett.* 410: 87–90.

Kupfer, S.R., Marschke, K.B., Wilson, E.M., and French, F.S. (1993). Receptor accessory factor enhances specific DNA binding of androgen and glucocorticoid receptors. *J. Biol. Chem.* 268: 17519–17527.

Landel, C.C., Kushner, P.J., and Greene, G.L. (1995). Estrogen receptor accessory proteins: Effects on receptor-DNA interactions. *Environ. Health Perspec.* 103 (Suppl. 7): 23–28.

Landers, J.P., and Spelsberg, T.C. (1992). New concepts in steroid hormone action: Transcription factors, proto-oncogenes, and the cascade model for steroid regulation of gene expression. *Crit. Rev. Eukaryot. Gene Expr.* 2: 19–63.

Laws, S.C., Carey, S.A., Kelce, W.R., Cooper, R.L., and Gray, L.E., Jr. (1996). Vinclozolin does not alter progesterone receptor (PR) function *in vivo* despite inhibition of PR binding by its metabolites *in vitro*. *Toxicology* 112: 173–82.

Lee, Y.J., Notides, A.C., Tsay, Y.G., and Kende, A.S. (1977). Coumestrol, NBD-norhexestrol, and dansyl-norhexestrol, fluorescent probes of estrogen-binding proteins. *Biochemistry* 16: 2896–2901.

Love, R.R., Mazess, R.B., Barden, H.S., Epstein, S., Newcomb, P.A., Jordan, V.C., Carbone, P.P., and DeMets, D.L. (1992). Effects of tamoxifen on bone mineral density in postmenopausal women with breast cancer. *N. Engl. J. Med.* 326: 852–856.

MacLatchy, D.L., and Van Der Kraak, G.J. (1995). The phytoestrogen beta-sitosterol alters the reproductive endocrine status of goldfish. *Toxicol. Appl. Pharmacol.* 134: 305–312.

Markiewicz, L., Garey, J., Adlercreutz, H., and Gurpide, E. (1993). In vitro bioassays of non-steroidal phytoestrogens. *J. Steroid Biochem. Mol. Biol.* 45: 399–405.

Martin, P.M., Horwitz, K.B., Ryan, D.S., and McGuire, W.L. (1978). Phytoestrogen interaction with estrogen receptors in human breast cancer cells. *Endocrinology* 103: 1860–1867.

McDonnell, D.P., Clemm, D.L., Hermann, T., Goldman, M.E., and Pike, J.W. (1995). Analysis of estrogen receptor function *in vitro* reveals three distinct classes of antiestrogens. *Mol. Endocrinol.* 9: 659–669.

——, Nawaz, Z., and O'Malley, B.W. (1991). In situ distinction between steroid receptor binding and transactivation at a target gene. *Mol. Cell. Biol.* 11: 4350–4355.

Mellanen, P., Petänen, T., Lehtimäki, J., Mäkelä, S., Bylund, G., Holmbom, B., Mannila, E., Oikari, A., and Santti, R. (1996). Wood-derived estrogens: Studies *in vitro* with breast cancer cell lines and *in vivo* in trout. *Toxicol. Appl. Pharmacol.* 136: 381–388.

Miksicek, R.J. (1993). Commonly occurring plant flavonoids have estrogenic activity. *Mol. Pharmacol.* 44: 37–43.

—— (1994). Interaction of naturally occurring nonsteroidal estrogens with expressed recombinant human estrogen receptor. *J. Steroid Biochem. Mol. Biol.* 49: 153–160.

Mommsen, T.P., and Lazier, C.B. (1986). Stimulation of estrogen receptor accumulation by estradiol in primary cultures of salmon hepatocytes. *FEBS Lett.* 195: 269–271.

Moore, F.L., and Orchinik, M. (1991). Multiple molecular actions for steroids in the regulation of reproductive behaviors. *Neurosciences* 3: 489–496.

Mosselman, S., Polman, J., and Dijkema, R. (1996). ER beta: Identification and characterization of a novel human estrogen receptor. *FEBS Lett.* 392: 49–53.

Mueller, G.C., and Kim, U.-H. (1978). Displacement of estradiol from estrogen receptors by simple alkyl phenols. *Endocrinology* 102: 1429–1435.

Nardulli, A.M., and Katzenellenbogen, B.S. (1986). Dynamics of estrogen receptor turnover in uterine cells *in vitro* and in uteri *in vivo*. *Endocrinology* 119: 2038–2046.

Nelson, J.A. (1974). Effects of dichlorodiphenyltrichloroethane (DDT) analogs and polychlorinated biphenyl (PCB) mixtures on 17beta-(^{3}H)estradiol binding to rat uterine receptor. *Biochem. Pharmac.* 23: 447–451.

Nesaretnam, K., Corcoran, D., Dils, R.R., and Darbre, P. (1996). 3,4,3',4'-Tetrachlorobiphenyl acts as an estrogen *in vitro* and *in vivo*. *Mol. Endocrinol.* 10: 923–936.

Nguyen, B.L., Giambiagi, N., Mayrand, C., Lecerf, F., and Pasqualini, J.R. (1986). Estrogen and progesterone receptors in the fetal and newborn vagina of guinea pig: Biological, morphological, and ultrastructural responses to tamoxifen and estradiol. *Endocrinology* 119: 978–988.

Nimrod, A.C., and Benson, W.H. (1996). Environmental estrogenic effects of alkylphenol ethoxylates. *Crit. Rev. Toxicol.* 26: 335–364.

Norris, D.O. (1997). *Vertebrate Endocrinology, 3d ed.* Academic, San Diego.

Okey, A.B., and Bondy, G.P. (1978). Diethylstilbestrol binding to estrogen receptor in mammary tissue and uteri of C3H/HeJ mice. *J. Natl. Cancer Inst.* 61: 719–724.

Orchinik, M., Murray, T.F., and Moore, F.L. (1991). A corticosteroid receptor in neuronal membranes. *Science* 252: 1848–1851.

Paech, K., Webb, P., Kuiper, G.G., Nilsson, S., Gustafsson, J., Kushner, P.J., and Scanlan, T.S. (1997). Differential ligand activation of estrogen receptors ER-alpha and ER-beta at AP1 sites. *Science* 277: 1508–1510.

Pakdel, F., Le Gac, F., Le Goff, P., and Valotaire, Y. (1990). Full-length sequence and *in vitro* expression of rainbow trout estrogen receptor cDNA. *Mol. Cell Endocrinol.* 71: 195–204.

———, S, F.é., Le Gac, F., Le Menn, F., and Valotaire, Y. (1991). In vivo estrogen induction of hepatic estrogen receptor mRNA and correlation with vitellogenin mRNA in rainbow trout. *Mol. Cell Endocrinol.* 75: 205–212.

Pallardy, M., Mishal, Z., Lebrec, H., and Bohuon, C. (1992). Immune modification due to chemical interference with transmembrane signalling: Application to polycyclic aromatic hydrocarbons. *Inter. J. Immunopharmacol.* 14: 377–382.

Palmiter, R.D., and Mulvihill, E.R. (1978). Estrogenic activity of insecticide Kepone on the chicken oviduct. *Science* 201: 356–358.

Pasqualini, J.R., Sumida, C., Giambiagi, N.A., and Nguyen, B.L. (1987). The complexity of anti-estrogen responses. *J. Steroid Biochem.* 27: 883–889.

Pelissero, C., Bennetau, B., Babin, P., Le Menn, F., and Dunogues, J. (1991). The estrogenic activity of certain phytoestrogens in the Siberian sturgeon *Acipenser baeri. J. Steroid Biochem. Mol. Biol.* 38: 293–299.

———, Flouriot, G., Foucher, J.L., Bennetau, B., Dunogues, J., Le Gac, F., and Sumpter, J.P. (1993). Vitellogenin synthesis in cultured hepatocytes: An in vitro test for the estrogenic potency of chemicals. *J. Steroid Biochem. Mol. Biol.* 44: 263–272.

Pennisi, E. (1997). Differing roles found for estrogen's two receptors [news]. *Science* 277: 1439.

Pettersson, K., Grandien, K., Kuiper, G.G., and Gustafsson, J.A. (1997). Mouse estrogen receptor beta forms estrogen response element-binding heterodimers with estrogen receptor alpha. *Mol. Endocrinol.* 11: 1486–1496.

———, Svensson, K., Mattsson, R., Carlsson, B., Ohlsson, R., and Berkenstam, A. (1996). Expression of a novel member of estrogen response element-binding nuclear receptors is restricted to the early stages of chorion formation during mouse embryogenesis. *Mech. Dev.* 54: 211–23.

Pietras, R.J., and Szego, C.M. (1979). Estrogen receptors in uterine plasma membrane. *J. Steroid Biochem.* 11: 1471–1483.

Poland, A., and Knutson, J.C. (1982). 2,3,7,8-tetrachlorodibenzo-*p*-dioxin and related halogenated aromatic hydrocarbons: Examination of the mechanism of toxicity. *Ann. Rev. Pharmacol. Toxicol.* 22: 517–554.

Rabelo, E.M., and Tata, J.R. (1993). Thyroid hormone potentiates estrogen activation of vitellogenin genes and autoinduction of estrogen receptor in adult *Xenopus* hepatocytes. *Mol. Cell Endocrinol.* 96: 37–44.

Riegel, A.T., Aitken, S.C., Martin, M.B., and Schoenberg, D.R. (1987). Differential induction of hepatic estrogen receptor and vitellogenin gene transcription in Xenopus laevis. *Endocrinology* 120: 1283–1290.

Robison, A.K., Mukku, V.R., Spalding, D.M., and Stancel, G.M. (1984). The estrogenic activity of DDT: The *in vitro* induction of an estrogen-inducible protein by *o,p'*-DDT. *Toxicol. Appl. Pharmacol.* 76: 537–543.

———, Schmidt, W.A., and Stancel, G.M. (1985). Estrogenic activity of DDT: Estrogen-receptor profiles and the responses of individual uterine cell types following *o,p'*-DDT administration. *J. Toxicol. Environ. Health* 16: 493–508.

———, Sirbasku, D.A., and Stancel, G.M. (1985). DDT supports the growth of an estrogen-responsive tumor. *Toxicol. Lett.* 27: 109–113.

——— and Stancel, G.M. (1982). The estrogenic activity of DDT: Correlation of estrogenic effect with nuclear level of estrogen receptor. *Life Sci.* 31: 2479–2484.

Rosenblum, E.R., Stauber, R.E., Van Thiel, D.H., Campbell, I.M., and Gavaler,

J.S. (1993). Assessment of the estrogenic activity of phytoestrogens isolated from bourbon and beer. *Alcohol Clin. Exp. Res.* 17: 1207–1209.

Sachs, B.D., and Leipheimer, R.E. (1988). Rapid effect of testosterone on striated muscle activity in rats. *Neuroendocrinology* 48: 453–458.

Sato, T., Ohta, Y., Okamura, H., Hayashi, S., and Iguchi, T. (1996). Estrogen receptor (ER) and its messenger ribonucleic acid expression in the genital tract of female mice exposed neonatally to tamoxifen and diethylstilbestrol. *Anat. Rec.* 244: 374–385.

Shemesh, M., Lindner, H.R., and Ayalon, N. (1972). Affinity of rabbit uterine oestradiol receptor for phyto-oestrogens and its use in a competitive protein-binding radioassay for plasma coumestrol. *J. Reprod. Fertil.* 29: 1–9.

Shutt, D.A., and Cox, R.I. (1972). Steroid and phyto-estrogen binding to sheep uterine receptors *in vitro*. *J. Endocrinol.* 52: 299–310.

Simental, J.A., Sar, M., Lane, M.V., French, F.S., and Wilson, E.M. (1991). Transcriptional activation and nuclear targeting signals of the human androgen receptor. *J. Biol. Chem.* 266: 510–518.

Soto, A.M., Sonnenschein, C., Chung, K.L., Fernandez, M.F., Olea, N., and Serrano, F.O. (1995). The E-SCREEN assay as a tool to identify estrogens: An update on estrogenic environmental pollutants. *Environ. Health Perspec.* 103 (Suppl. 7): 113–122.

Spencer, T.E., and Bazer, F.W. (1995). Temporal and spatial alterations in uterine estrogen receptor and progesterone receptor gene expression during the estrous cycle and early pregnancy in the ewe. *Biol. Reprod.* 53: 1527–1543.

Sumpter, J.P., and Jobling, S. (1995). Vitellogenesis as a biomarker for estrogenic contamination of the aquatic environment. *Environ. Health Perspec.* 103 (Suppl. 7): 173–178.

Tang, B.Y., and Adams, N.R. (1980). Effect of equol on oestrogen receptors and on synthesis of DNA and protein in the immature rat uterus. *J. Endocrinol.* 85: 291–297.

Teutsch, G., Nique, F., Lemoine, G., Bouchoux, F., Cérède, E., Gofflo, D., and Philibert, D. (1995). General structure-activity correlations of antihormones. *Ann. NY Acad. Sci.* 761: 5–28.

Thompson, M.A., Lasley, B.L., Rideout, B.A., and Kasman, L.H. (1984). Characterization of the estrogenic properties of a nonsteroidal estrogen, equol, extracted from urine of pregnant macaques. *Biol. Reprod.* 31: 705–713.

Tora, L., White, J., Brou, C., Tasset, D., Webster, N., Scheer, E., and Chambon, P. (1989). The human estrogen receptor has two independent nonacidic transcriptional activation functions. *Cell* 59: 477–487.

Tran, D.Q., Kow, K.Y., McLachlan, J.A., and Arnold, S.F. (1996). The inhibition of estrogen receptor-mediated responses by chloro-S-triazine-derived compounds is dependent on estradiol concentration in yeast. *Biochem. Biophys. Res. Commun.* 227: 140–146.

Tzukerman, M.T., Esty, A., Santiso Mere, D., Danielian, P., Parker, M.G., Stein, R.B., Pike, J.W., and McDonnell, D.P. (1994). Human estrogen receptor transactivational capacity is determined by both cellular and promoter context and mediated by two functionally distinct intramolecular regions. *Mol. Endocrinol.* 8: 21–30.

Van Birgelen, A.P.J.M. and Birnbaum, L.S. (1997). Synergistic effects of mixtures of dioxin-like and non-dioxin-like compounds in rodents. In 17th International Symposium on Chlorinated Dioxins and Related Compounds. Organohalogen Compounds, Indianapolis.

Verdeal, K., Brown, R.R., Richardson, T., and Ryan, D.S. (1980). Affinity of phytoestrogens for estradiol-binding proteins and effect of coumestrol on growth of 7,12-dimethylbenz[a]anthracene-induced rat mammary tumors. *J. Natl. Cancer Inst.* 64: 285–290.

vom Saal, F.S., Montano, M.M., and Wang, M.H. (1992). Sexual differentiation in mammals. In *Chemically Induced Alterations in Sexual and Functional Development: The Wildlife/Human Connection* (T. Colborn and C.C., eds.), pp. 17–83. Scientific Publications, Princeton.

————, Nagel, S.C., Palanza, P., Boechler, M., Parmigiani, S., and Welshons, W.V. (1995). Estrogenic pesticides: Binding relative to estradiol in MCF-7 cells and effects of exposure during fetal life on subsequent territorial behaviour in male mice. *Toxicol. Lett.* 77: 343–350.

Vonier, P.M., Crain, D.A., McLachlan, J.A., Guillette, L.J., Jr., and Arnold, S.F. (1996). Interaction of environmental chemicals with the estrogen and progesterone receptors from the oviduct of the American alligator. *Environ. Health Perspec.* 104: 1318–1322.

Watson, C.S., Pappas, T.C., and Gametchu, B. (1995). The other estrogen receptor in the plasma membrane: Implications for the actions of environmental estrogens. *Environ. Health Perspec.* 103 (Suppl. 7): 40–51.

Webb, P., Lopez, G.N., Uht, R.M., and Kushner, P.J. (1995). Tamoxifen activation of the estrogen receptor/AP-1 pathway: Potential origin for the cell-specific estrogen-like effects of antiestrogens. *Mol. Endocrinol.* 9: 443–556.

Wehling, M. (1995). Looking beyond the dogma of genomic steroid action: Insights and facts of the 1990s. *J. Mol. Med.* 73: 439–447.

———— (1997). Specific, nongenomic actions of steroid hormones. *Ann. Rev. Physiol.* 59: 365–393.

————, Christ, M., and Gerzer, R. (1993). Aldosterone-specific membrane receptors and related rapid, non-genomic effects. *Trends Pharmacol. Sci.* 14: 1–4.

Welch, R.M., Levin, W., and Conney, A.H. (1969). Estrogenic action of DDT and its analogs. *Toxicol. Appl. Pharmacol.* 14: 358–367.

White, R., Jobling, S., Hoare, S.A., Sumpter, J.P., and Parker, M.G. (1994). Environmentally persistent alkylphenolic compounds are estrogenic. *Endocrinology* 135: 175–182.

Whittliff, J.L. (1975). Steroid-binding proteins in normal and neoplatic mammary cells. *Methods Cancer Res.* 11: 293–354.

Wibbels, T., and Crews, D. (1995). Steroid-induced sex determination at incubation temperatures producing mixed sex ratios in a turtle with TSD. *Gen. Comp. Endocrinol.* 100: 53–60.

Wilson, E.M., and French, F.S. (1976). Binding properties of androgen receptors: Evidence for identical receptors in rat testis, epididymis, and prostate. *J. Biol. Chem.* 251: 5620–5629.

Wong, C., Kelce, W.R., Sar, M., and Wilson, E.M. (1995). Androgen receptor antagonist versus agonist activities of the fungicide vinclozolin relative to hydroxyflutamide. *J. Biol. Chem.* 270: 19998–20003.

Zhou, Z.X., Sar, M., Simental, J.A., Lane, M.V., and Wilson, E.M. (1994). A ligand-dependent bipartite nuclear targeting signal in the human androgen receptor: Requirement for the DNA-binding domain and modulation by NH2-terminal and carboxyl-terminal sequences. *J. Biol. Chem.* 269: 13115–13123.

———, Wong, C.I., Sar, M., and Wilson, E.M. (1994). The androgen receptor: An overview. *Recent Prog. Horm. Res.* 49: 249–274.

———, Lane, M.V., Kemppainen, J.A., French, F.S., and Wilson, E.M. (1995). Specificity of ligand-dependent androgen receptor stabilization: Receptor domain interactions influence ligand dissociation and receptor stability. *Mol. Endocrinol.* 9: 208–218.

Zhou-Li, F., Skalli, M., Albaladéjo, V., Joly-Pharaboz, M.O., Nicolas, B., and André, J. (1993). Interference between estradiol and L-triiodothyronine in the control of the proliferation of a pituitary tumor cell line. *J. Steroid Biochem. Mol. Biol.* 45: 275–279.

Chapter 5

STEROID HORMONE–REGULATED PROCESSES IN INVERTEBRATES AND THEIR SUSCEPTIBILITY TO ENVIRONMENTAL ENDOCRINE DISRUPTION

Gerald A. LeBlanc

Department of Toxicology
North Carolina State University
Raleigh, NC 27695-7633

Introduction

Concern over toxicity resulting from prolonged exposure to low concentrations of toxicants was raised over 3 decades ago in Rachel Carson's *Silent Spring* (1962). Presented were graphic descriptions of reproductive failure in birds, gonadal dysfunction in mammals, and cancer in humans in support of the contention that environmental pollutants were disrupting life processes at all levels of ecological organization. Over the past 3 decades, significant effort has been expended towards the identification and characterization of the toxicity of environmental pollutants. Several pieces of environmental legislation have been enacted that require characterization of both the acute and chronic toxicity of potential environmental contaminants (Foster, 1985) to species representing various levels of organization and trophic status. More recently, concern has been raised as to whether current testing protocols are adequate for the detection of toxicity associated with alterations in endocrine function (Ankley et al., 1997).

Many environmental chemicals have been shown to elicit toxicity caused by endocrine disruption. Perhaps most notable is the ability of some chemicals to bind to the estrogen receptor and stimulate transcription of estrogen-responsive genes. Laboratory experiments have demonstrated that a variety of environmental contaminants elicit estrogen-like growth stimulation of MCF-7 cells (Soto et al., 1994; Soto et al., 1992). Incubation of trout liver hepatocytes with alkylphenolic detergents stimulated the synthesis of estrogen-regulated yolk protein, vitellogenin (Jobling and Sumpter, 1993). Administration of some polychlorinated biphenyls (PCBs) to the shell of turtle eggs increased the percentage of female hatchlings (Bergeron et al., 1994).

Field evaluations suggest that the estrogenicity of many environmental contaminants may be eliciting adverse effects on wildlife and human populations. Analyses of salmonids from the River Lea, UK (Harries et al., 1994) and the Great Lakes (Leatherland, 1992) have revealed various endocrine and reproductive dysfunctions. Vitellogenin production was observed in male trout exposed to sewage effluent in the River Lea (Harries et al., 1994), possibly due to alkylphenols in the effluent. The alligator population level in Lake Apopka, Florida, has declined steadily over the past decade, despite the maintenance of stable populations in other Florida lakes (Woodward et al., 1993). Female alligators sampled from the lake exhibited abnormal ovarian morphology (Guillette et al., 1994). Males had significantly depressed plasma testosterone levels, poorly organized testes, and diminutive phalli (Guillette et al., 1994). Lake Apopka is adjacent to a U.S. EPA Superfund site that was significantly contaminated with DDT and dicofol (Guillette et al., 1994).

Estrogenic or antiestrogenic effects of environmental contaminants have also been implicated in several human abnormalities. Breast cancer incidence has increased steadily since the 1940s (Feuer and Wun, 1992). Risk of breast cancer increases with increased cumulative estrogen exposure (Henderson et al., 1993). An epidemiological study linked breast cancer incidence with serum levels of the DDT metabolite DDE (Wolff et al., 1993). Such observations led to the hypothesis that environmental estrogens are a preventable cause of breast cancer (Davis et al., 1993). Sperm counts in human males were reported to have decreased approximately 50% over 50 years (Carlsen et al., 1992). The commensurate increase in estrogenic pollutants has been postulated as contributory to this decrease (Sharpe and Skakkebaek, 1993). Endometriosis, the extrauterine growth and proliferation of endometrial cells, is estimated to affect over 6 million women of reproductive age in the United State alone (Wheeler, 1992). The etiology of the disease is unknown, but likely involves hormonal influences. A study of rhesus monkeys chronically treated with 2,3,7,8-tetrachlorodibenzo-p-dioxin (TCDD) demonstrated a dose-dependent increase in incidence of this disease (Rier et al., 1993). TCDD can elicit a variety of endocrine-disrupting effects, including reduced estrogen responsiveness (Safe et al., 1991). TCDD and related halogenated aromatic hydrocarbons are ubiquitous environmental contaminants that may contribute to the incidence of endometriosis among humans. Residents of Seveso, Italy, who were exposed to high levels of TCDD following a chemical plant explosion in 1976 have been identified as viable candidates to assess the relationship between TCDD exposure and endometriosis in humans (Bois and Eskenazi, 1994) and may eventually elucidate the environmental etiology of this condition.

The importance of steroid hormones in reproduction and development of vertebrates is unequivocal. Accordingly, xenobiotics that elicit steroid hormone agonistic or antagonistic properties can be detrimental to these organisms. However, vertebrates represent less than 5% of the known animal species (Barnes, 1968). The role of steroid hormones in the remaining >95% percent, the invertebrates, is less well understood, as is the susceptibility of these organisms to endocrine-disrupting environmental chemicals. Invertebrates serve many important societal roles, such as direct food sources (i.e., crustaceans), pollinators of crops (bees), and pest control (predaceous insects). Invertebrates are ecologically indispensable in the trophic transfer of nutrients and carbon. Invertebrates also provide a wealth of biodiversity that serves as an inexhaustible reservoir of genetic material for current and future use in biotechnology. Finally, the abundance of many invertebrate genera in terrestrial and aquatic environments renders them useful as biological sentinels of environmental insult. Identification and characterization of the effects of endocrine-disrupting chemicals on invertebrate species may prove useful in recognizing and abating such effects before effects are elicited in vertebrate populations, including humans.

In the present chapter, the role of steroid hormones in invertebrates and the susceptibility of invertebrates to endocrine-disrupting chemicals is examined. The role of vertebrate-type sex steroids in invertebrates is given special emphasis since processes driven by these hormones have been found to be particularly susceptible to disruption by xenobiotics in vertebrates. An assessment of the role of vertebrate-type sex steroids in invertebrates will provide insights into whether invertebrates are susceptible to some of the same modes of endocrine disruption as characterized in vertebrates. Finally, the utility of invertebrates as environmental sentinels for the detection of endocrine-disrupting chemicals is considered. Results suggest that this diverse group of organisms is grossly undervalued as monitors of environmental health.

Steroid Hormones in Invertebrates

Ecdysteroids

Ecdysteroids (Fig. 5-1) are known primarily for their occurrence in arthropods (i.e., crustaceans and insects). Ecdysteroids may also regulate biological processes in other invertebrates, including annelids, nematods, and molluscs (Koolman, 1982). Like vertebrate-type sex steroid hormones, ecdysteroids control various processes involved in growth, differentiation, and reproduction. Perhaps best characterized are the roles of ecdysteroids in molting and

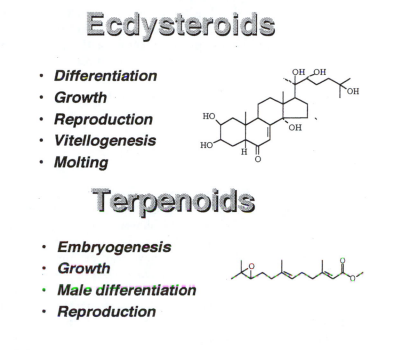

Ecdysteroids

- *Differentiation*
- *Growth*
- *Reproduction*
- *Vitellogenesis*
- *Molting*

Terpenoids

- *Embryogenesis*
- *Growth*
- *Male differentiation*
- *Reproduction*

Figure 5-1.
Ecdysteroids and terpenoids, invertebrate-type hormones that are responsible for many of the regulatory processes associated with vertebrate-type sex steroids in the vertebrates.

reproduction. The molting hormone 20-hydroxyecdysone is synthesized by the Y-gland in crustaceans and is the major determinant controlling periodic loss of the exoskeleton (molting) allowing for growth of the organism (Chang and O'Connor, 1988). 20-Hydroxyecdysone also stimulates vitellogenesis in female crustaceans (Charniaux-Cotton and Payen, 1988; Quackenbush, 1986), an estrogen-regulated process in oviparous vertebrates. Definitive reviews of ecdysteroids in invertebrates can be found elsewhere (e.g., Charniaux-Cotton and Payen, 1988; Koolman, 1982).

Vertebrate-Type Sex Steroids

Functional roles of vertebrate-type sex steroids in invertebrates are not unequivocal. However, as discussed in this chapter, evidence strongly suggests that these hormones function similarly in both vertebrates and some invertebrates. Estrogens, androgens, and progestogens have been measured in invertebrate species representing various genera. Many of the analyses for these steroids in invertebrates have utilized radioimmunoassay (RIA) using steroid-specific antibodies. While this approach is strongly indicative of the presence of immunoreactive steroid hormone, the precise identity of the steroid remains elusive in the absence of steroid identification by gas chromatography–mass spectrometry (GC-MS). Estrogens, androgens, and progestogens are commonly referred to as vertebrate-type sex steroids; however, their origin clearly predates that of the vertebrates.

Estrogens

Estrogens have been detected in various insect species including the flesh fly *(Sarcophaga bullata)* (Delinger et al., 1987), cockroach *(Nauphoeta cinerea)* (Takac et al., 1988), silkworm *(Bombyx mori)* (Ohnishi et al., 1985), and locust *(Locusta migratoria)* (Novak and Lampert, 1989). Presence of 17β-estradiol in the locust was confirmed by GC-MS. Estrogens have also been detected in crustaceans [American lobster *(Homarus americanus)* (DeLoof and DeClerk, 1986)], molluscs [snail *(Euhadra prelionphala)* (Takeda, 1980)], dogwhelk *(Nucella lapillus)* (Spooner et al., 1991), slug *(Arion aterrufus)* (Gottfried et al., 1967)] and echinoderms [starfish *(Sclerasterias mollis)* (Barker and Xu, 1993; Xu, 1991)], sea star *(Asterias vulgaris)* (Hines et al., 1992), sea cucumber *(Stichopus mobii)* (Donahue, 1940), sea urchin *(Strongylocentrotus fransiscanus)* (Botticelli et al., 1960). Estrogens most commonly have been detected in gonads, pyloric caeca, and eggs.

Enzymes responsible for the anabolic and catabolic metabolism of estrogen have also been measured in invertebrates. Aromatase, the enzyme responsible for the conversion of testosterone to estradiol, was measured in the gonads, reproductive tract, accessory glands, brain, and fat body of the adult flesh fly (Delinger et al., 1987). Gonadal homogenates from the cricket *(Orthoptera)*, sperm preparations from the sea urchin *(Arbacia punctulata)*, and tissue homogenates of the oyster *(Crassostrea virginica)* were capable of converting estradiol to estrone (Dube and Lemonde, 1970; Hathaway, 1965; Xu, 1991). Glucosylation of estradiol was demonstrated in cultured ovaries of the silkworm (Ogiso et al., 1986). Glucosylation contributes to the metabolic inactivation and elimination of estradiol. Insects thus possess the metabolic capacity for both the synthesis and elimination of estrogens.

Androgens

Testosterone and other androgens have been measured in the hemolymph of larval fleshfly (De Clerck et al.,1983, 1984; DeLoof and DeClerck, 1986) and the migratory locust (Novak and Lampert, 1989) by GC-MS. Testosterone has also been measured in a variety of other insect species by RIA (Brueggemeier et al., 1988; Delinger et al., 1987; DeLoof and DeClerk, 1986; Takac et al., 1988); the dogwhelk (Spooner et al., 1991); the testes and serum of the American lobster (Burns et al., 1984); and the ovaries of the crab *(Nephrops norvegicus)* (Fairs et al., 1989).

Both the biosynthesis and metabolic inactivation of testosterone has been shown in crustaceans. Synthesis of testosterone from androstenedione was demonstrated using tissue homogenates from the crab *(Cancer pagurus)* (Swevers et al., 1991). Daphnids *(Daphnia magna)* metabolize testosterone to products that comigrate with androstenedione and dihydrotestosterone during thin-layer chromatography (Baldwin and LeBlanc, 1994a, 1994b). These metabolites and other apolar metabolites of testosterone selectively accumulate within the daphnids (Baldwin and LeBlanc, 1994b). Daphnids extensively hydroxylate testosterone at multiple locations on the steroid ring, and these metabolites are largely eliminated from the organisms (Baldwin and LeBlanc,

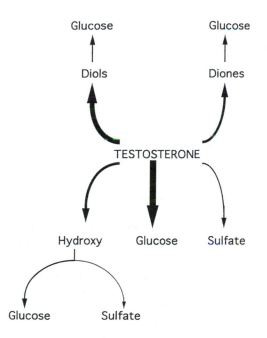

Figure 5-2.
Testosterone biotransformation strategies utilized by the crustacean *Daphnia magna*. Intensity of the arrows indicates relative abundance of the products formed. Hydroxylated, glucosylated, and sulfated derivatives are preferentially eliminated, while diols and diones are preferentially retained by the organisms. Summarized from Baldwin and LeBlanc, 1994b; Baldwin et al., 1995; Parks and LeBlanc, 1996.

1994b). Daphnids also extensively glucose-conjugate, and to a lesser extent, sulfate-conjugate testosterone and its hydroxyl metabolites (Baldwin and LeBlanc, 1994b), and these metabolites are eliminated from this species. Thus, crustaceans are equipped for the dehydrogenation, hydroxylation, and conjugation of testosterone with the capability to produce over 30 metabolites of testosterone (Fig. 5-2). Similarly, starfish are equipped with the enzyme system necessary for the biosynthesis of testosterone from acetate (summarized in Brueggemeier et al., 1988), a multistep pathway in vertebrates (Stryer, 1981).

Progestogens

Progesterone has been measured in the hemolyph of the flesh fly (De Clerck et al., 1983, 1984; Stryer, 1981), the Colorado potato beetle (Diederik et al., 1984), and the cockroach (Takac et al., 1988). The capacity for progestogen biosynthesis from the precursors cholesterol or pregnenolone has been demonstrated in the crab *(Cancer pagurus)* (Swevers et al., 1991), the sea star *(Asterias rubens)* (Voogt et al., 1991), and ovaries of the spiny lobster *(Panlirus japonicus)* (Kanazawa and Teshima, 1971; Yano, 1985). Hydroxylation of progesterone in the 20α and 17α positions has been measured in crab (Swevers et al., 1991) and sea star (Voogt et al., 1991), respectively. Hydroxylation of progesterone in the 17α position is necessary for the synthesis of glucocorticoid hormones from progesterone and, as discussed below, stimulates vitellogenin production in some crustaceans. Fatty acyl esterification of progesterone was measured using a hepatopancreas preparation from the decapod crustacean *Penaeus monodon* (Young et al., 1992). Fatty acyl esters generally serve as storage forms of steroidal compounds (Zhang and Kubo, 1992).

Steroid Hormone Receptors in Invertebrates

The presence of steroid hormones or the capability for steroid hormone metabolism in invertebrates does not establish a functional regulatory role for these molecules comparable to that observed in vertebrates. Steroid hormone action in vertebrates is mediated by hormone-specific receptors that transduce the regulatory signal instituted by the hormone to the steroid hormone-responsive genes. In the absence of a cytoplasmic or nuclear receptor, classic steroid hormone regulation of gene expression does not occur. Evidence exists suggesting the presence of an estrogen receptor in the pyloric caeca of the female starfish (DeWaal et al., 1982). A high affinity (Kd = 0.23 × 10^{-10} M) 17β-estradiol-binding protein was identified in this tissue. The pyloric caeca functions in digestion and the storage of nutrients and provides nourishment for developing oocytes (Oudejans and Van der Sluis, 1979). An estrogen receptor in this organ suggests estrogen involvement in these processes.

Function of Vertebrate-Type
Sex Steroid Hormones in Invertebrates

Evidence strongly implicates vertebrate-type sex steroids as having a functional role in the development of gonads, secondary sex characteristics, eggs, and embryos in some invertebrate phyla.

Androgens

Administration of testosterone has been shown to: (a) stimulate the conversion of ovaries to testes in the female crab (Sarojini, 1963), (b) initiate hypertrophy and hyperplasia of the androgenic gland of the paneid crustacean *Parapenaeopsis hardwickii* (Nagabhushanam and Kulkarni, 1981), (c) increase testis size in the paneid (Nagabhushanam and Kulkarni, 1981), and (d) mobilize glycogen from the midgut to the testis of the paneid (Nagabhushanam and Kulkarni, 1981). The androgenic gland of malacostracan crustaceans is responsible for the secretion of hormone that stimulates development of the testes and male sex characteristics (Charniaux-Cotton and Payen, 1988). Testicular development following administration of testosterone to paneid shrimp may be secondary to the stimulatory effects of the testosterone on the androgenic gland. Sarojini (1963) administered testosterone to female crabs in an effort to determine whether testosterone might be the hormone secreted by the androgenic gland that was responsible for the development of secondary sex characteristics and testicular maturation. Testosterone-injected females did not develop male secondary sex characteristics; however, testosterone did cause the ovaries of the crabs to develop into testes with spermatagonia and spermatocytes. Thus, testosterone could, at least in part, mimic the action of the secretions of the androgenic gland. Androgenic secretions of the androgenic gland have been variously characterized as protein (Juchault et al., 1984), steroid (Sarojini, 1964), and terpenoid (Berreur-Bonnenfant et al., 1973; Ferezou et al., 1978). The terpenoid farnesylacetone has been considered primarily responsible for androgenic activity of the androgenic gland; however, one cannot exclude the possibility that other secretions contribute to this activity. For example, farnesylacetone inhibits ovarian development in females and stimulates protein synthesis in the testis of males, but does not stimulate spermatogenesis (Berreur-Bonnenfant and Lawrence, 1984; Ferezou et al., 1977). Yet, extracts from the androgenic gland have been shown to stimulate spermatogenesis (Sarojini, 1963).

Physiological functions of steroidal androgens have also been indicated in molluscs. Androgen administration to gravid female gastropods *Deroceras reticu latus* and *Limax flavus* increased the rate of embryonic development (Takeda, 1979). Androgen administration to castrated snails *(Euhadra prelion-*

phala) stimulated the development of secondary male sex characteristics (Takeda, 1980). Testosterone administration to adult female dogwelks *(Nacella lapillus)* stimulated the development and growth of male genitalia (Spooner et al., 1991).

The freshwater crustacean *Daphnia magna* can reproduce either partheno-genetically or sexually. Experiments were conducted to determine whether treatment of parthenogenetic females and their offspring with testosterone would masculinize the organisms or otherwise interfere with parthenogenic reproduction (LeBlanc and McLachlan, unpublished). Androgen treatment had no discernible effect on secondary sex characteristics; however, fecundity of these organisms was severely compromised. Testosterone, at low micromolar aqueous concentrations, significantly increased both the number of eggs aborted and the number of dead offspring released by the organisms. Androgen specificity of this effect was suggested when the androgen androstenedione (LeBlanc and McLachlan, unpublished) was shown to elicit effects on fecundity identical to that of testosterone, while the estrogen diethylstilbestrol did not elicit such effects (Baldwin et al. 1995).

Estrogens

Though limited, existing data suggest that estrogens regulate female reproductive processes in some invertebrates in a manner similar to that observed in vertebrates. Estrogen levels fluctuate in echinoderms during the reproductive cycle (Xu and Barker, 1990a, 1990b, 1990c) in a manner suggesting involvement in oogenesis. Indeed, administration of 17β-estradiol to female starfish *(Sclerasterias mollis)* caused an increase in the size of oocytes and ovarian protein levels (Barker and Xu, 1993). Similar effects were observed when esterone was administered to the starfish *(Asterina pectinifera)* (Barker and Xu, 1993). These observations in conjunction with the demonstration of estrogen receptor strongly implicates estrogens as functional hormones in echinoderms.

Administration of estradiol or estrone, but not androgens or progestogens, increased the number of eggs produced by female gastropods *Deroceras reticulatus* and *Limax flavus* (Takeda, 1979). Egg production was reduced by administration of metopirone, which is an inhibitor of steroid synthesis in vertebrates. This inhibition implies that endogenous steroids are involved in egg production in gastropods.

Diethylstilbestrol (DES), a synthetic nonsteroidal estrogen, was administered to pregnant mothers for nearly 3 decades as a prophylactic against spontaneous abortion. While treated mothers tolerated the drug well, offspring exhibited a variety of reproductive dysfunctions, including cancers of the reproductive tract (Herbst et al., 1971, 1972), anatomical changes in the reproductive tract (Kaufman et al., 1977; Stillman, 1982), menstrual irregularities (Bibbo et al., 1977), and infertility (Haney and Hammond, 1983). Administration of DES to rodents induced similar abnormalities of the reproductive tract as observed with humans (McLachlan et al., 1980; Newbold and McLachlan, 1982). DES was administered to the crustacean *Daphnia magna* in an effort to discern any effect of excess estrogen on parthenogenic reproduction in this species (Baldwin et al., 1995). While survival and reproduction of first generation daphnids exposed to DES was not affected, fecundity of offspring of the organisms was significantly reduced. Since the second-generation daphnids were also continuously exposed to DES, it is not known whether the effects of DES on second-generation daphnid reproduction occurred during perinatal development of these organisms, as observed in vertebrates, or during postnatal growth and development.

Progestogens

Progesterone functions in conjunction with estradiol in vertebrates to regulate various aspects of female reproduction (Hadley, 1996). Similarly, progesterone levels fluctuate in echinoderms in concert with estrone levels in a manner that suggests their involvement in reproductive, as well as nonreproductive, processes (Xu, 1991; Xu and Barker, 1990a). While estrogens increase ovarian protein content when administered to starfish, progesterone reduces ovarian protein content (Barker and Xu, 1993).

Progesterone was also shown to decrease ovarian protein content in the marine penaeid crustacean *Parapenaeopsis hardwickii* while increasing oocyte development. Progesterone similarly enhanced egg maturation and spawning in the prawn *Penaeus japonicus* (Yano, 1985). These observed effects of progesterone on egg maturation led to speculation that progesterone serves as a vitellogenin-stimulating hormone in crustaceans (Yano, 1985). Vitellogenin is a precursor of egg yolk protein and is necessary for oocyte maturation. Subsequent investigations demonstrated that the progesterone metabolite 17α-hydroxyprogesterone was more potent than progesterone and stimulated a 10-fold increase in serum vitellogenin levels in *P. japonicus* (Yano, 1987).

Terpenoids

The terpenoid hormones, though nonsteroidal in structure, warrant inclusion in a discussion of steroid hormones in invertebrates. The terpenoid hormones consist of a farnesyl molecule with modifications that confer specific hormonal activity in some invertebrates (Fig. 5-1). Farnesyl compounds are synthesized via the melvalonic acid pathway and are steroid hormone precursors. Arthropods utilize these molecules as hormonal messengers that regulate several of the processes associated with steroid hormones in vertebrates. Examples of terpenoid hormones in invertebrates include juvenile hormone and farnesylacetone

Juvenile hormone (JH-III; 10,11 epoxy methyl farnesenate) regulates a variety of processes, including aspects of embryogenesis, growth, and reproduction in insects (Bowers, 1990). Juvenile hormone is produced by the corpora allata in insects (Wigglesworth, 1970) and appears to be specific to this class of invertebrates. Farnesylacetone is produced by the androgenic gland of malacostracan crustaceans (Ferezou et al., 1978). Farnesylacetone inhibits ovarian protein synthesis and stimulates oocyte degeneration (Charniaux-Cotton and Payen, 1988). Farnesylacetone may function to prevent the development of oocytes in pseudohermaphroditic male crustaceans (Charniaux-Cotton and Payen, 1988).

Phylogenic Analysis of Steroid Hormone Utilization by Invertebrates

The basic steroid is an ancient molecule that is synthesized by both prokaryotic and eukaryotic organisms (Sandor and Mehdi, 1979). However, the utilization of steroid hormones as regulators of reproduction and development appears to have diverged in invertebrate evolution after the Cnidaria (e.g., jelly fish, coral, sea anemone) (Fig. 5-3). The central importance of steroid hormones in regulating reproduction and development in chordates is well established. Similarly, abundant evidence, as presented herein, demonstrates an important role of vertebrate-type sex steroids in reproduction of echinoderms (starfish, sea urchins, etc.). Clearly, the use of vertebrate-type sex steroids as regulatory hormones was fully exploited by the Deuterostome lineage of organisms (Fig. 5-3). Evidence also supports a role for vertebrate-type sex steroids in the regulation of reproduction in the Protostome lineage; however, the magnitude of importance appears less in comparison to the Deuterostomes. Among the Protostomes, ecdysteroids and terpenoids evolved as important regulators of growth, development, and reproduction (Chang and O'Connor, 1988; Charniaux-Cotton and Payen, 1988; Quackenbush, 1986). Thus, while Protostome reproduction and development can clearly be affected by vertebrate-type sex steroids, the regulatory role of vertebrate-type sex steroids appears rivaled or surpassed by other steroidal and nonsteroidal hormones.

Some Protostomes have evolved unique strategies for the usage of vertebrate-type sex steroids. Androgens, estrogens, and progestogens all have been shown to regulate various aspects of growth and reproduction in certain para-

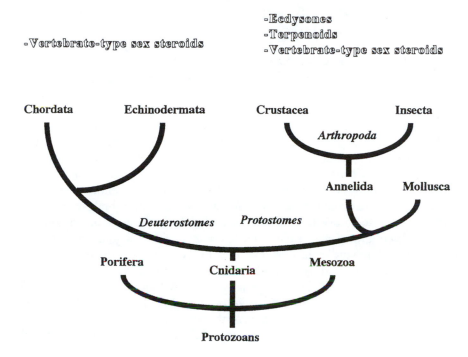

Figure 5-3.
Phylogeny of hormone usage among chordates and invertebrates. Vertebrate-type sex steroid evolved to be primarily responsible for regulating processes such as development, growth, and reproduction among the Deuterostomes. Similar regulatory processes among the Protostomes appear to be shared by edysteroids, terpenoids, and vertebrate-type sex steroids.

sites of vertebrates (Dobson, 1966; El Mofty and Smyth, 1964; Rothschild and Ford, 1966). The steroid hormones, however, are not derived from the parasite itself but rather from its host. Such strategies allow the parasite to coordinate it's reproduction with that of it's host, resulting in the generation of parasite progeny that can occupy the progeny of the host. Aquatic beetles of the families *Dytiscidae* and *Gyrinidae* secrete a defensive agent that is repellent and toxic to predators (Miller and Mumma, 1976). Included in the defense secretion are androgens (testosterone, 1,2-dehydrotestosterone) and estrogens (estrone, 17β-estradiol) (Schildknecht, 1970; Schildknecht et al., 1967). Testosterone was shown to anesthetize fish when present in the aqueous mediate, albeit at a high concentration (10 mg/l). Perhaps exposure to the concentrated steroids of the defense secretion stupefies the predator for a sufficient time so as to allow for the escape of the beetle.

Effects of Endocrine-Disrupting Xenobiotics on Invertebrates

Clearly, steroid hormone usage among invertebrates is as complex as that of the vertebrates, with components (i.e., ecdysteroids, juvenoids) unique to invertebrates. Invertebrates are thus likely to be similarly susceptible to the adverse effects of endocrine-disrupting chemicals in the environment. Molecular targets of effect among invertebrates would be the same as those targets in vertebrates, such as receptor agonists/antagonists and modulators of hormone homeostasis. However, specific targets (i.e., ecdysone and juvenoid receptors) exist in invertebrates that may confer susceptibility to some endocrine-disrupting chemicals to which vertebrates are not susceptible. Alternatively, some endocrine-disrupting chemicals that elicit effects via interaction with vertebrate-type steroid hormone receptors in vertebrates may also bind to ecdysone or juvenoid receptors of invertebrates and elicit effects through these regulatory pathways. Receptor-binding studies have shown that many endocrine-disrupting chemicals are capable of binding multiple steroid hormone receptors (Kelce et al., 1995; Laws et al., 1995). Future studies may show binding proclivity of these compound to invertebrate-specific receptors as well. Reported effects of known endocrine-disrupting toxicants on invertebrates supports the corollary that invertebrates and vertebrates both are targets of such environmental chemicals.

Tributyltin

Tributyltin is a biocide used as an antifoulant agent in marine paints. In addition to its high acute toxicity, from which its utility as a biocide is derived, more occult effects associated with chronic exposure to low levels of this compound have resulted in the enactment of severe restrictions in the use of this chemical. In particular, tributyltin has been shown to be causally associated with the development of imposex in neogastropods (Smith, 1981). Imposex is defined as the imposition of morphological reproductive characteristics of one sex onto the other. Among the neogastropods such as the dogwelk (*Nucella lapillus*) (Gibbs et al., 1991) and the mud snail (*Nassarius obsoletus*) (Smith, 1981), tributyltin-induced imposex is characterized by development of a penis and vas deferens by females (Gibbs et al., 1991). Imposex can interfere with the ability of the female to release eggs from the ovary, and in severe instances oogenesis is suppressed, seminiferous tubules develop within the ovary, and speratogenesis can occur (Gibbs et al., 1988, 1990).

Treatment of female gastropods (*Nucella lapillus*) with tributyltin caused significant penis development and increased testosterone titers while having no effect on progesterone or 17β-estradiol levels (Spooner et al., 1991). A causal association between elevated endogenous testosterone levels and penis development was suggested by administering testosterone to female *N. lapillus*

(Spooner et al., 1991). Testosterone did stimulate penis development, although the effect was slight and variable. These observations, taken together with the previously described stimulation of male sex characteristics in molluscs by testosterone (Takeda, 1980), suggest that tributyltin stimulates imposex by elevating endogenous androgen levels (or decreasing the estrogen to androgen ratio) in female gastropods. Additional studies are required to conclusively establish whether elevated testosterone levels are responsible for imposex in these organisms.

Studies of the effects of tributyltin on testosterone metabolism have indicated that tributyltin decreases the metabolic inactivation of testosterone while enhancing the conversion of testosterone to other androgenic steroid hormones. Exposure of periwinkles (*Littorina littorea*) to concentration of tributyltin along with administration of [^{14}C]testosterone resulted in decreased elimination of testosterone polar metabolites and increased tissue retention of testosterone and androgenic metabolites of testosterone (Ronis and Mason, 1996). Similarly, exposure of daphnids (*D. magna*) to tributyltin significantly increased the rate of conversion of [^{14}C]testosterone to [^{14}C]androstenedione (Oberdorster et al., 1995). These results suggest that tributytin causes imposex by altering steroid metabolism and increasing endogenous androgen levels.

Tributyltin-induced imposex highlights the need for constant vigilance for unique responses of individual groups of organisms to the toxicity of endocrine-disrupting chemicals. Toxicity tests performed with the standard array of aquatic organisms recommended for use in environmental toxicity assessments unequivocally demonstrated *no observed effect concentrations* (NOEC) for the more sensitive laboratory species to tributyltin at concentrations ranging from low parts per billion (μg/l) to high parts per trillion (ng/l) (World Health Organization, 1990). However, imposex can develop in some gastropods at parts per quadrillion (pg/l) concentrations (Gibbs et al., 1987). Thus, environmental concentrations of tributyltin deemed acceptable based on laboratory studies can lead to reproductive impairment of some gastropod species. While it is impractical to assess the toxicity of environmental chemicals to all classes of organisms, the need to identify species that have unique sensitivities to endocrine-disrupting chemicals is imperative. The incorporation of such species into the standard testing regimens and the exploitation of such species as sentinels of environmental contamination would ensure maximum protection against such insults.

4-Nonylphenol

Nonylphenols are degradation products of alkylphenol ethoxylates, which are nonionic surfactants used in a variety of applications including cleaning products, textiles, agricultural chemicals, and paper products (Talmage, 1994). Nonylphenols have been shown to be estrogenic in cultured human cells (Soto et al., 1991) and in fish (Jobling and Sumpter, 1993). Exposure of rainbow trout to 4-nonylphenol caused a significant reduction in testis growth (Jobling et al., 1996). In daphnids, 4-nonylphenol was shown to be a reproductive toxicant, to inhibit testosterone glucosylation, and to increase the production of apolar testosterone metabolites (Baldwin et al., 1996). While causality was not established, metabolic and reproductive effects in daphnids occurred at the same exposure concentrations of the surfactant.

4-Nonylphenol caused metabolic androgenization as demonstrated by the decreased metabolic clearance of testosterone and enhanced production of androgenic oxido-reduced derivatives of testosterone. We have recently observed that the imidazole fungicide propiconazole similarly altered the metabolism of exogenously administered testosterone to daphnids favoring the accumulation of androgenic derivatives of testosterone and the suppression of the metabolic clearance of the androgens (LeBlanc et al., 1996). At this time it is not clear whether these compounds are differentially affecting the metabolic enzymes (i.e., induction of reductases/dehydrogenases and inhibition of hydroxylases/conjugases) or whether metabolic elimination is inhibited by these compounds, resulting in the shunting of the testosterone to the reductase/dehydrogenase pathways. Regardless, these observations suggest that many environmental chemicals have the potential to elicit similar effects on endocrine homeostasis in invertebrates. A comparison of lowest observed effect concentrations of a variety of xenobiotics with respect to decreased fecundity and altered steroid hormone metabolic capabilities in daphnids revealed a significant relationship (Fig. 5-4) with metabolic effects occurring at concentrations below those that significantly affected fecundity. These results suggest that: (a) substantial alterations in steroid hormone metabolic processes in invertebrates may cause decreased fecundity, and, (b) toxicant effects on steroid metabolism may serve as a predictive indicator of reproductive toxicity. Additional studies are warranted to identify the putative regulator of steroid metabolism in invertebrates that is susceptible to perturbation by environmental chemicals and potentially exploit it as a biomarker of endocrine disruption.

Triazine Herbicides

Triazine herbicides such as atrazine and simazine are widely used herbicides that exhibit low acute toxicity to mammals. The triazines have been shown to elicit a variety of endocrine-disrupting effects in mammals, including: (a) modulation of enzymes involved in the biotransformation of steroid hormones (Ugazio et al., 1991), (b) interference with steroid hormone-receptor complex formation (Simic et al., 1991), (c) stimulation of thyroxine release by thyroid cells (Cooper et al., 1995), (d) interference with androgen metabolism both in treated adults (Babic-Gojmerac et al., 1989; Kniewald et al., 1979) and in

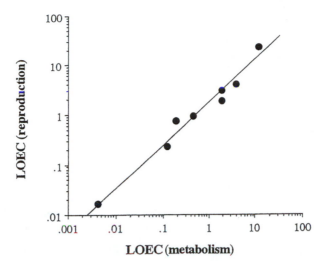

Figure 5-4.
Relationship between *lowest observed effect levels* (LOEC) of xenobiotics that significantly altered testosterone metabolism and fecundity of exposed *Daphnia magna*. A
significant relationship exists between the two parameters with metabolic LOECs consistently being lower than reproductive LOECs. Severe effects on steroid metabolism
(i.e., >50% alteration) are required before toxicant effects on reproduction are evident
(data from LeBlanc et al., 1995).

adults exposed peri/neonatally (Kniewald et al., 1987), (e) increases in the
incidence of mammary tumors in male rats (Pinter et al., 1990), and (f) lowering of serum levels of the ovarian hormones estradiol and progesterone
(Cooper et al., 1995).

While information of the direct endocrine effects of triazine herbicides on
invertebrates are lacking, effects of these compounds on the physiology of
invertebrates are strongly suggestive of endocrine disruption. Exposure of
daphnids *(Daphnia pulex)* to concentrations of atrazine as low as 1.0 mg/l significantly reduced fecundity in a concentration-dependent manner (Fig. 5-5)
(Schober and Lampert, 1977). In contrast, a 20-fold higher concentration was
required to significantly reduce survival of the organisms. Similar results were
obtained among *D. magna*, with reduced fecundity occurring at concentrations as low as 0.25 mg/l (Macek et al., 1976). Reproductive impairment and
developmental impairment of second-generation scud *(Gammarus faciatus)*
from exposure to atrazine also has been observed (Macek et al., 1976). These
effects on invertebrates are not limited to atrazine. Similar experiments performed with simazine and *D. pulex* have shown that, like atrazine, this
compound caused reproductive impairment in daphnids and also reduced
molt frequency (Fitzmayer et al., 1982).

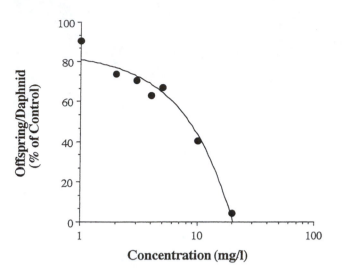

Figure 5-5.
Effects of the triazine herbicide atrazine on fecundity of the water flea *(Daphnia pulex)*. Data are from Schober and Lampert, 1977.

Cadmium

Cadmium has been reported to elicit a variety of endocrine and endocrine-associated effects in invertebrates as well as vertebrates. In daphnids *(D. magna)*, cadmium exposure reduced ecdysteroid titers resulting in impaired molting and growth (Bodar et al., 1990). At these same exposure concentrations, cadmium also affected reproduction by increasing the number of progeny per brood and reducing the size of individual progeny (Bodar et al., 1988). Other reproductive effects of cadmium in invertebrates include impaired gonadal follicle development in mussels (Kluytmans et al., 1988) and defects in embryonic development of the sea star *(Asterias rubens)* (Den Besten, 1991). Cadmium also has been reported to significantly lower progesterone and testosterone levels in the pyloric caeca of the sea star (Den Besten et al., 1991).

Pentachlorophenol

The biocide pentachlorophenol has been shown to reduce circulating thyroid hormone levels in mammals (Jekat et al., 1994) and alter hepatic phase I and II metabolic enzymes that are associated with steroid hormone metabolic clearance (Kimbrough and Linder, 1978; NTP, 1989). Experiments in daphnids have shown that pentachlorophenol alters the metabolic clearance of steroid hormones at concentrations that also reduce fecundity (Parks and LeBlanc, 1996). Studies with the sea urchin have shown that pentachlorophenol alters embryonic development and differentiation (Ozretic and Krajnovic-Ozretic, 1985).

Organochlorine Pesticides

Organochlorine pesticides are known to elicit a variety of endocrine-disrupting effects in vertebrates including estrogenicity (*o,p'*-DDT, methoxychlor, chlordecone (Johnson et al., 1992); endosulfan, toxaphene, dieldrin (Soto et al., 1994) and anti-androgenicity (*p,p'*-DDE [Kelce et al., 1995]). Organochlorine pesticides have been shown to elicit reproductive effects in vertebrates, including infertility, decreased frequency of implanted ova, ovarian and testicular abnormalities, altered estrus cyclicity, and egg shell thinning (Hall, 1987; Public Health Service, 1993a, 1993b).

Many of these compounds have also been shown to elicit reproductive toxicities, which are suggestive of endocrine-disrupting effects, in invertebrates. For example, continuous exposure of grass shrimp *(Paleomonetes pugio)* to concentrations of endrin as low as 0.03 μg/l delayed the onset of spawning and reduced the number of viable embryos produced per female (Tyler-Schroeder, 1979). These reproductive effects were elicited at concentrations approximately two orders of magnitude below lethal concentrations of endrin. Chronic exposure to toxaphene reduced fecundity of *Daphnia magna* at concentrations significantly below those that reduced survival (Sanders, 1980). Similarly, DDT was shown to severely reduce fecundity of the pond snail *(Lymnaea stagnalis L.)* at concentrations that were nonlethal to the organisms (Woin and Bronmark, 1992). Dieldrin was found to elicit a variety of reproductive effects on the copepod *Eurytemora affinis* at sublethal exposure levels (Daniels and Allan, 1981). Similar to endrin's effects on grass shrimp and testosterone's effects on daphnids, discussed previously, dieldrin both delayed and reduced the number of viable offspring produced by copepods. Using an integrative approach to assess the combined sublethal effects of dieldrin on the rate of population growth of this species, the authors concluded that this pesticide would adversely affect populations at concentrations approximately one order of magnitude below those concentration that are acutely lethal.

Polychlorinated Biphenyls (PCBs)

Reported endocrine-disrupting effects of some PCB isomers in vertebrates include altered sex ratios in turtles (Bergeron et al., 1994), reduced male rat fertility following early postnatal exposure (Sager et al., 1987), and altered expression of steroid hormone-metabolizing enzymes (Dieringer et al., 1979).

Exposure of sea stars (*A. rubens*) to PCBs via contaminated food organisms caused a significant decrease in progesterone and testosterone levels in the pyloric caeca and increased testosterone levels in the gonads (Den Besten et al., 1991). These changes in steroid hormone homeostasis may be causally associated with altered embryo development observed in sea stars similarly fed PCB-contaminated mussels (Den Besten et al., 1989). Oocytes obtained from PCB-treated females and fertilized *in vitro* were found to undergo aberrant embryonic development characterized by multipolar cleavage, archenteron malformations, and inferior differentiation.

Conclusions

Clearly, invertebrates are susceptible to chemical toxicity resulting from perturbations in endocrine function. Comparative laboratory studies have shown that invertebrates are often comparably, or more, sensitive to the effects of environmental exposure to endocrine-disrupting chemicals than are vertebrates (Maki, 1979). In at least one instance (tributyltin), invertebrates have been recognized as a sensitive environmental sentinel of environmental contamination (Gibbs et al., 1987). Review of the reported effects of known endocrine-disrupting chemicals on invertebrates reveals that alterations in steroid metabolism, sexual differentiation, embryonic development, fecundity, and molting are all sensitive indicators of endocrine-disrupting toxicity to invertebrates. Through ubiquity and sensitivity, invertebrates thus have great potential to serve as monitors of environmental contamination. However, our limited understanding of normal endocrine processes in these diverse organisms renders analyses of changes due to chemical exposure difficult to assess under noncontrolled (i.e., environmental) conditions.

Acknowledgment

Research conducted by the author was supported by the Air Force Office of Scientific Research, Air Force Systems Command grant F49620-94-1-0266.

References

Ankley, G., Johnson, R., Detenbeck, N., Bradbury, S., Toth, G., and Folmar, L. (1997). Development of a research strategy for assessing the ecological risk of endocrine disruptors. *Rev. Toxicol.* 1: 71–107.

Babic-Gojmerac, T., Kniewald, Z., and Kniewald, J. (1989). Testosterone metabolism in neuroendocrine organs in male rats under atrazine and deethylatrazine influence. *J. Steroid Biochem.* 33: 141–146.

Baldwin, W.S., Graham, S. E., Shea, D., and LeBlanc, G.A. (1996). Metabolic androgenization of female *Daphnia magna* by the xenoestrogen 4-nonylphenol. Submitted at 17th Annual Meeting of the Society of Toxicology and Chemistry, Washington, DC.

—— and LeBlanc G.A. (1994a). Identification of multiple steroid hydroxylases in *Daphnia magna* and their modulation by xenobiotics. *Environ. Toxicol. Chem.* 13: 1013–1021.

—— and —— (1994b). In vivo biotransformation of testosterone by phase I and II detoxication enzymes and their modulation by 20-hydroxyecdysone in *Daphnia magna*. *Aquatic Toxicol.* 29: 103–117.

——, Milam D.L., and LeBlanc G.A. (1995). Physiological and biochemical perturbation in *Daphnia magna* following exposure to the model environmental estrogen diethylstilbestrol. *Environ. Toxicol. Chem.* 14: 945–952.

Barker M.F., and Xu, R.A. (1993). Effects of estrogens on gametogenesis and steroid levels in the ovaries and pyloric caeca of *Sclerasterias mollis* (Echinodermata: Asteroidea). *Invert. Reprod. Develop.* 24: 53–58.

Barnes R.D. (1968). *Invertebrate Zoology*, pp. 1–4. Saunders, Philadelphia.

Bergeron, J.M., Crews, D., and McLachlan, J.A. (1994). PCBs as environmental estrogens: Turtle sex determination as a biomarker of environmental contamination. *Environ. Health Perspect.* 102: 780–781.

Berreur-Bonnenfant, J., and Lawrence, F. (1984). Comparative effect of farnesylacetone on macromolecular synthesis in gonads of crustaceans. *Gen. Comp. Endocrin.* 54: 462–468.

——, Meusy, J. J., Ferezou, J.P., Devys, M., Quesneau-Thierry, A., and Barbier, M. (1973). Recherches sur la secretion de la glande androgene des Crustaces malacostraces: Purification d'une substance a activite androgene. *CR Acad. Sci. Paris* 277: 971–974.

Bibbo, M., Gill, W.B., Azizi, F., Blough, R., Fang, V.S., Rosenfield, R.L., Schumacher, G.F.B., Sleeper, K., Sonek, M.G., and Wied, G.L. (1977). Follow-up study on male and female offspring of DES-exposed mothers. *Obstet. Gynecol.* 49: 1–8.

Bodar, C.W.M., Van Leeuwen, C.J., Voogt, P.A., and Zandee, D.I. (1988). Effect of cadmium on the reproduction strategy of *Daphnia magna*. *Aquatic Toxicol.* 12: 301–310.

——, Voogt, P.A., and Zandee, D.I. (1990). Ecdysteroids in *Daphnia magna*: Their role in moulting and reproduction and their levels upon exposure to cadmium. *Aquatic Toxicol.* 17: 339–350.

Bois, F.Y., and Eskenazi, B. (1994). Possible risk of endometriosis for Seveso, Italy, residents: An assessment of exposure to dioxin. *Environ. Health Perspect.* 102: 476–477.

Botticelli, C.R., Hisaw, F.J., and Wotiz, H.H. (1960). Estrogens and progesterone in the sea urchin *(Strongylocentrotus fransiscanus)* and pecten *(Pecten
hericius)*. *Proc. Soc. Exp. Biol.* 106: 887–889.

Bowers, W.S. (1990). Prospects for the use of insect growth regulators in agri-
culture. In *Advances in Invertebrate Reproduction 5* (M. Hoshi and O.
Yamashita, eds.), pp. 365–382. Elsevier Science, New York.

Brueggemeier, R.W., Yocum, G.D., and Denlinger, D.L. (1988). Estranes,
androstanes, and pregnanes in insects and other invertebrates. In *Physiological Insect Ecology* (F. Sehnal, A. Zabza, and D.L. Denlinger, eds.), pp.
885–898. Wroclaw Technical University Press, Wroclaw.

Burns, B.G., Sangalang, G.B., Freeman, H.C., and McMenemy, M. (1984). Isolation of testosterone from the serum and testes of the American lobster
(Homarus americanus). *Gen. Comp. Endocrin.* 54: 429–435.

Carlsen E., Giwercman, A., Keiding, N., and Skakkebaek, N.E. (1992). Evidence for decreasing quality of semen during the past 50 years. *Br. Med. J.*
304: 609–613.

Carson, R. (1962). *Silent Spring.* Houghton Mifflin, Boston.

Chang, E.S., and O'Connor, D. (1988). Crustacea: molting. *Endocrinology of
Selected Invertebrate Types* (H. Laufer and G.H. Downer, eds.), pp. 259–278.
Liss, New York.

Charniaux-Cotton, H. (1955). Le determinisme hormonal des caracteres
sexuels d'*Orchestia gammarella* (Crustace Amphipode). *CR Acad. Sci. Paris*
240: 1487–1489.

——— and Payen, G. (1988). Crustacean reproduction. *Endocrinology of
Selected Invertebrate Types* (H. Laufer and G.H. Downer, eds.), pp. 279–303.
Liss, New York.

Cooper R.L., Parrish, M.B., McElroy, W.K., Rehnberg, G.L., Hein, J.F.,
Goldman, J.M., Stoker, T.E., and Tyrey, L. (1995). Effect of atrazine on the
hormonal control of the ovary. *Toxicologist* 15: 294.

Daniels, R.E. and Allan, J.D. (1981). Life table evaluation of chronic exposure
to a pesticide. *Can. J. Fish Aquatic Sci.* 38:485–494.

Davis, D.L., Bradlow, H.L., Wolff, M., Woodruff ,T., Hoel, D.G., and Anton-
Culver, H. (1993). Medical Hypothesis: Xenoestrogens as preventable
causes of breast cancer. *Environ. Health Perspect.* 101: 372–377.

De Clerck, D., Diederik, H., and De Loof, A. (1984). Identification by capillary
gas chromatography-mass spectrometry of eleven non-ecdysteroid steroids
in the haemolymph of *Sarcophaga bullata*. *Insect Biochem.* 14: 199–208.

———, Eechaute, W., Leusen, I., Diederik, H., and De Loof, A. (1983). Identification of testosterone and progesterone in haemolymph of larvae of the
fleshfly, *Sarcophaga bullata*. *Gen. Comp. Endocrin.* 52: 368–378.

Delinger, D.L., Brueggemeier, R.W., Mechoulam, R., Katlic, N., Yocum, L.B., and Yocum, G.D. (1987). Estrogens and androgens in insects. In *Molecular Entomology* (J. Law, ed.), pp. 189–199. Liss, New York.

DeLoof, A., and DeClerk, D. (1986). Vertebrate-type steroids in arthropods: Identification, concentrations, and possible functions. In *Advances in Invertebrate Reproduction Vol. 4* (M. Porchet, J.C. Andries, and A. Dhainaut, eds.), pp. 117–123. Elsevier Science, Amsterdam.

Den Besten, P.J., Elenbass, J.M.L., Maas, J.R., Dieleman, S.J., Herwig, H.J., and Voogt, P.A. (1991). Effects of cadmium and polychlorinated biphenyls (Clophen A50) on steroid metabolism and cytochrome P-450 monooxygenase system in the sea star *Asterias rubens L. Aquatic Toxicol.* 20: 95–110.

———, Herwig, H.J., Zandee, D.I., and Voogt, P.A. (1989). Effects of cadmium and PCBs on reproduction of the sea star *Asterias rubens*: Aberations in the early development. *Ecotoxicol. Environ. Saf.* 18: 173–180.

DeWaal, M., Poortman, J., and Voogt, P.A. (1982). Steroid receptors in invertebrates: A specific 17β-estradiol binding protein in a sea star. *Marine Biol. Lett.* 3: 317–323.

Diederik, H., De Clerck, D., and De Loof, A. (1984). Nonecdysteroids in hemolymph of the Colorado potato beetle larvae *(Leptinotarsa decemlineata). Gen. Comp. Endocrin.* 53: 449.

Dieringer, C.S., Lamartiniere, C.A., Schiller, C.M., and Lucier, G.W. (1979). Altered ontogeny of hepatic steroid-metabolizing enzymes by pure polychorinated biphenyl congeners. *Biochem. Pharmacol.* 28: 2511–2514.

Dobson, C. (1966). The effects of pregnancy and treatment with progesterone on the host-parasite relationship of *Amplicaecum robertsi. Parasitology* 56: 417–424.

Donahue, D.J. (1940). Occurrence of estrogens in the ovaries of certain marine invertebrates. *Endocrinology* 27: 149–152.

Dube, J., and Lemonde, A. (1970). Transformations des steroids par la femelle adulte d'un insecte orthoptere, *Schistocerca gregaria* Forskal. *Gen. Comp. Endocrin.* 15: 158–164.

El Mofty, M.M., and Smyth, J.D. (1964). Endocrine control of encystation in *Opalina ranarum* parasitic in *Rana temporaria. Exp. Parasit.* 15: 185–199.

Fairs, N.J., Evershed, R.P., Quinlan, P.T., and Goad, L.J. (1989). Detection of unconjugated and conjugated steroids in the ovary, eggs, and haemolymph of the decapod crustacean *Nephrops norvegicus. Gen. Comp. Endocrin.* 4: 199–208.

Ferezou, J.P., Barbier, M., and Berreur-Bonnenfant, J. (1978). Biosynthese de la farnesylacetone-(E-E) par les glandes androgenes du crabe *Carcinus maenas. Helv. Chim. Acta* 61: 669–674.

———, Berreur-Bonnenfant, J., Tekitek, A., Rojas, M., Barbier, M., Suchy, M., Wipf, H.K., and Meusy, J.J. (1977). Biologically active lipids from the androgenic gland of the crab *Carcinus maenas*. In *Marine Natural Products Chemistry* (D.J. Faulkner and W.H. Fenica, eds.), pp. 361–366. Plenum, New York.

Feuer, E.J. and Wun, L.-M. (1992). How much of the recent rise in breast cancer can be explained by increases in mammography utilization? *Am. J. Epidemiol.* 136: 1423–1436.

Fitzmayer, K.M., Geiger, J.G., and Van Den Avyle, M.J. (1982). Effects of chronic exposure to simazine on the cladoceran, *Daphnia pulex*. *Arch. Environ. Contam. Toxicol.* 11: 603–609.

Foster, R.B. (1985). Environmental legislation. In *Fundamentals of Aquatic Toxicology* (G.M. Rand and S.R. Petrocelli, eds.), pp. 587–599. Hemisphere, New York.

Franklin, M.R. (1972). Inhibition of hepatic oxidative xenobiotic metabolism by piperonyl butoxide. *Biochem. Pharmacol.* 21: 3287–3299.

Gibbs, P.E., Bryan, G.W., and Pascoe, P.L. (1991). TBT-induced imposex in the dogwelk, *Nucella lapillus*: Geographical uniformity of the response and effects. *Marine Environ. Res.* 32: 79–87.

——, ——, ——, and Burt, G.R. (1987). The use of the dog-whelk, *Nucella lapillus*, as an indicator of tributyltin (TBT) contamination. *J. Mar. Biol. Assoc. U.K.* 67: 507–523.

——, ——, ——, and —— (1990). Reproductive abnormalities in female *Ocenebra erinacea* (Gastropoda) resulting from tributyltin-induced imposex. *J. Mar. Biol. Assoc. U.K.* 70: 639–656.

——, Pascoe, P.L., and Burt, G.R. (1988). Sex change in the female dog-whelk, *Nucella lapillus*, induced by tributyltin from antifouling paints. *J. Mar. Biol. Assoc. U.K.* 68: 715–731.

Gottfried, H., Dorfman, R.I., and Wall, P.E. (1967). Steroids of invertebrates: Production of oestrogens by an accessory reproductive tissue of the slug *Arion aterrufus* (Linn.). *Nature* 215: 409.

Guillette, L.J., Gross, T.S., Masson, G.R., Matter, J.M., Percival, F., and Woodward, A.R. (1994). Developmental abnormalities of the gonad and abnormal sex hormone concentrations in juvenile alligators from contaminated and control lakes in Florida. *Environ. Health Perspect.* 102: 681–688.

Hadley, M.E. (1996). *Endocrinology*, pp. 412–437. Prentice Hall, Upper Saddle River.

Hall, R.J. (1987). Impact of pesticides on bird populations. *Silent Spring Revisited* (G.J. Marco, R.M. Hollingworth, and W. Durham, eds.), pp. 85–112. American Chemical Society, Washington, DC.

Haney, A.F., and Hammond, C.B. (1983). Infertility in woment exposed to diethylstilbestrol *in utero*. *J. Reprod. Med.* 28: 851–856.

Harries, J.E., Sheahan, D.A., Rycroft, R., Matthiessen, P., Sumpter, J.P., Jobling, S., and Routledge, E.J. (1994). Evidence of oestrogenic activity in U.K.

inland waters. Estrogens in the Environment III: Global Health Implications, National Institute of Environmental Health Sciences, Research Triangle Park.

Hathaway, R.R. (1965). Conversion of estradiol-17β by sperm preparation of sea urchin and oysters. *Gen. Comp. Endocrin.* 5: 504–508.

Henderson, B.E., Ross, R.K., and Pike, M.C. (1993). Hormonal chemoprevention of cancer in women. *Science* 259: 633–638.

Herbst, A.L., Kurman, R.J., Scully, R.E., and Postkanzer, D.C. (1972). Clear cell adenocarcinoma of the genital tract in young females: Registry report. *N. Engl. J. Med.* 287: 1259–1264.

———, Ulfelder, H., and Poskanzer, D.C. (1971). Adenocarcinoma of the vagina: Association of materna stilbestrol therapy with tumor appearance in young women. *N. Engl. J. Med.* 287: 878–81.

Hines, G.A., Watts, S.A., Sower, S.A., and Walker, C.W. (1992). Sex steroid levels in the testes, ovaries, and pyloric caeca during gametogenesis in the sea star *Asterias vulgaris. Gen. Comp. Endocrin.* 87: 451–460.

Jekat, F.W., Meisel, M.L., and Winterhoff, H. (1994). Effects of pentachlorophenol (PCP) on the pituitary and thyroidal hormone regulation in the rat. *Toxicol. Lett.* 71: 9–25.

Jobling, S., Sheahan, D., Osborne, J.A., Matthiessen, P., and Sumpter, J.P. (1996). Inhibition of testicular growth in rainbow trout *(Oncorhynchus mykiss)* exposed to estrogenic alkylphenolic chemicals. *Environ. Toxicol. Chem.* 15: 194–202.

——— and Sumpter, J.P. (1993). Detergent components in sewage effluent are weakly oestrogenic to fish: An in vitro study using rainbow trout *(Oncorhynchuys mykiss)* hepatocytes. *Aquatic Toxicol.* 27: 361–466.

Johnson, D.C., Sen, M., and Dey, S.K. (1992). Differential effects of dichlorodiphenyltrichloroethane analogs, chlordecone, and 2:3:7:8-tetrachlorodibenzo-*p*-dioxin on establishment of pregnancy in hypophysectomized rat. *Proc. Soc. Exp. Biol. Med.* 199: 42–48.

Juchault, P., Legrand, J.J., and Maissiat, J. (1984). Present state of knowledge on the chemical nature of the androgenic hormone in higher crustaceans. In *Biosynthesis Metabolism and Mode of Action of Invertebrate Hormones* (J. Hoffman and M. Porchet, eds.), pp. 155–160. Springer-Verlag, Berlin.

Kanazawa, A., and Teshima, S. (1971). In vivo conversion of cholesterol to steroid hormones in the spiny lobster, *Panulirus japonicus. Bull. Japan. Soc. Sci. Fish.* 37: 891–898.

Kaufman, R.H., Binder, G.L., Gray, P.M., and Adam, E. (1977). Upper genital tract changes associated with exposure *in utero* to diethylstilbestrol. *Am. J. Obster. Gynecol.* 128: 51–59.

Kelce, W.R., Stone, C.R., Laws, S.C., Gray, L.E., Kemppainen, J.A., and Wilson,

E.M. (1995). Persistent DDT metabolite p,p'-DDE is a potent androgen receptor antagonist. *Nature* 375: 581–585.

Kimbrough, R.D., and Linder, R.E. (1978). The effect of technical and purified pentachlorophenol on the rat liver. *Toxicol. Appl. Pharmacol.* 46: 151–162.

Kluytmans, J.H., Brands, F., and Jandee, D.I. (1988). Interactions of cadmium with the reproductive cycle of *Mytilus edulis* L. *Mar. Environ. Res.* 24: 189–192.

Kniewald, J., Mildner, P., and Kniewald, Z. (1979). Effects of s-triazine herbicides on hormone-receptor complex formation, 5α-reductase and 3α-hydroxysteroid dehydrogenase activity at the anterior pituitary level. *J. Steroid Biochem.* 11: 833–838.

———, Peruzovic, M., Gojmerac, T., Milkovic, K., and Kniewald, Z. (1987). Indirect influence of s-triazines on rat gonadotropic mechanism at early postnatal period. *J. Steroid Biochem.* 27: 1095–1100.

Koolman, J. (1982). Ecdysone metabolism. *Insect Biochem.* 12: 225–250.

Laws, S.C., Carey, S.A., and Kelce, W.R. (1995). Differential effects of environmental toxicants on steroid receptor binding. *Toxicologist* 15: 294.

Leatherland, J.F. (1992). Endocrine and reproductive function in Great Lakes salmon. In *Chemically-Induced Alterations in Sexual and Functional Development: The Wildlife/Human Connection* (T. Colborn and C. Clement, eds.), pp. 21, 129–145. Princeton Scientific Publishing, Princeton.

LeBlanc, G.A., Lopez, L., and McLachlan, J.B. (1996). Comparative study of the endocrine-disrupting and reproductive toxicity of the fungicides propiconazole and clotrimazole. Submitted at 17th Annual Meeting of the Society of Environmental Toxicology and Chemistry, Washington, DC.

———, Parks, L.G., Baldwin, W.S., McLachlan, J.B., and Oberdorster, E. (1995). Altered steroid hormone metabolic profiles as a biomarker of reproductive toxicity of endocrine-disrupting chemicals. *Proc. Second SETAC World Congr.* PH-207.

Macek, K.J., Buxton, K.S., Sauter, S.S., Gnilka, S., and Dean, J.W. (1976). Chronic toxicity of atrazine to selected aquatic invertebrates and fishes. U.S. Environmental Protection Agency. EPA-600/3-76-047.

Maki, A.W. (1979). Correlation between *Daphnia magna* and fathead minnow *(Pimephales promelas)* chronic toxicity values for several classes of test substances. *J. Fish. Res. Board Can.* 36: 411–421.

McLachlan, J.A., Newbold, R.R., and Bullock, B.C. (1980). Long-term effects on the female mouse genital tract associated with prenatal exposure to diethylstilbestrol. *Cancer Res.* 40: 3988–3999.

Miller, J.R., and Mumma, R.A. (1976). Physiological activity of water beetle defensive agents. I. Toxicity and anesthetic activity of steroids and norsesquiterpenes administered in solution in the minnow *Pimephales promelas*

Raf. *J. Chem. Ecol.* 2: 115–130.

Nagabhushanam, R., and Kulkarni, G.K. (1981). Effect of exogenous testosterone on the androgenic gland and testis of a marine penaeid prawn, *Parapenaeopsis hardwickii* (Miers) (Crustacea, Decapoda, Penaeidae). *Aquacult.* 23: 19–27.

Newbold, R.R., and McLachlan, J.A. (1982). Vaginal adenosis and adenocarcinoma in mice exposed prenatally or neonatally to diethylstilbestrol. *Cancer Res.* 42: 2003–2011.

Novak, F.J.S., and Lampert, J.G.D. (1989). Pregnenolone, testosterone, and estradiol in the migratory locust *Locusta migratoria*: A gas chromatographical-mass spectrometrical study. *Gen. Comp. Endocrin.* 76: 73–82.

NTP (1989). NTP technical report on the toxicology and carcinogenesis studies of pentachlorophenol (CAS no 87-86-5) in B6C3F1 mice (feed studies). National Toxicology Program, Research Triangle Park.

Oberdorster, E., LeBlanc, G.A., and Rittschof, D. (1995). Acute and chronic effects of tributyltin on *Daphnia magna*: Alterations of testosterone metabolism. Fifth COMTOX Symposium on Toxicology and Clinical Chemistry of Metals, Vancouver, British Columbia.

Ogiso, M., Fujimoto, Y., Ikekawa, N., and Ohnishi, E. (1986). Glucosidation of estradiol-17β in the cultured ovaries of the silkworm, *Bombyx mori*. *Gen. Comp. Endocrin.* 61: 393–401.

Ohnishi, E., Ogiso, M., Wakabayashi, K., Fujimoto, Y., and Ikekawa, N. (1985). Identification of estradiol in the ovaries of the silkworm, *Bombyx mori*. *Gen. Comp. Endocrin.* 60: 35–40.

Oudejans, R.C.H.M., and Van der Sluis, I. (1979). Storage and depletion of lipid components in the pyloric caeca and ovaries of the sea star *Asterias rubens* during its annual reproductive cycle. *Mar. Biol.* 53: 239–247.

Ozretic, B., and Krajnovic-Ozretic, M. (1985). Morphological and biochemical evidence of the toxic effect of pentachlorophenol on the developing embryos of the sea urchin. *Aquatic Toxicol.* 7: 255–263.

Parks, L.G., and LeBlanc, G.A. (1996). Inhibition of steroid hormone biotransformation/elimination as a biomarker of pentachlorophenol chronic toxicity. *Aquatic Toxicol.* 34: 291–303.

Pinter, A., Torok, G., Borzsonyi, M., Surjan, A., Calk, M., Kelecsenvi, Z., and Kocsis, Z. (1990). Long-term carcinogenicity bioassay of the herbicide atrazine in F344 rats. *Neoplasia* 37: 533–544.

Public Health Service (1993a). Toxicology Profile for 4.4'-DDT, d,d'-DDE, 4,4'-DDD. U.S. Department of Health and Human Services, Agency for Toxic Substances and Disease Registry, Washington, DC.

——— (1993b). Toxicology profile for methoxychlor. U.S. Department of Health and Human Services, Agency for Toxic Substances and Disease Registry, Washington, DC.

Quackenbush, L.S. (1986). Crustacean endocrinology: A review. *Can. J. Fish Aquatic Sci.* 43: 2271–2282.

Rier, S.E., Martin, D.C., Bowman, R.E., Dmowski, W.P., and Becker, L. (1993). Endometriosis in rhesus monkeys *(Macaca mulatta)* following chronic exposure to 2,3,7,8-tetrachlorodibenzo-*p*-dioxin. *Fund. Appl. Toxicol.* 21: 433–441.

Ronis, M.J.J., and Mason, A.Z. (1996). The metabolism of testosterone by the periwinkle *(Littorina littorea) in vitro* and *in vivo*: Effects of tributyltin. Submitted, *Mar. Environ. Res.*, 42: 161–166.

Rothschild, M., and Ford, B. (1966). Hormones of the vertebrate host controlling ovarian regression and copulation of the rabbit flea. *Nature* 211: 261–266.

Safe, S., Astroff, B., Harris, M., Zacharewski, T., and Dickerson, R. (1991). 2,3,7,8-tetrachlorodibenzo-*p*-dioxin (TCDD) and related compounds as antiestrogens: Characterization and mechanism of action. *Pharmacol. Toxicol.* 69: 400–409.

Sager, D.B., Shih-Scaroeder, W., and Girand, D. (1987). Effect of early postnatal exposure to polychlorinated biphenyls (PCBs) on fertility in male rats. *Bull. Environ. Contam. Toxicol.* 38: 946–953.

Sanders, H.O. (1980). Sublethal effects of toxaphene on daphnids, scuds, and midges. U.S. Environmental Protection Agency, Washington, DC.

Sandor, T., and Mehdi, A.Z. (1979). Steroids and evolution. In *Hormones and Evolution* (E.J.W. Barrington, ed.), pp. 1–72. Academic, New York.

Sarojini, S. (1963). Comparison of the effects of androgenic hormone and testosterone propionate on the female ocypod crab. *Curr. Science* 9: 411–412.

——— (1964). A note on the chemical nature of the crustacean androgenic hormone. *Curr. Science* 33: 55–56.

Schildknecht, H. (1970). Defensive substances of water beetles (Dytiscidae). *Angew. Chem. Internat. Edit.* 9: 1–9.

———, Birringer, H., and Maschwitz, U. (1967). Testosterone as protective agent of the water beetle Ilybius. *Angew. Chem. Internat. Edit.* 6: 558–559.

Schober, U., and Lampert, W. (1977). Effects of sublethal concentrations of the herbicide atrazin on growth and reproduction of *Daphnia pulex. Bull. Envion. Contam. Toxicol.* 17: 269–277.

Sharpe, R.M., and Skakkebaek, N.E. (1993). Are oestrogens involved in falling sperm count and disorders of the male reproductive tract? *Lancet* 341: 1392–1395.

Simic, B., Kniewald, Z., Davies, J.E., and Kniewald, J. (1991). Reversibility of the inhibitory effect of atrazine and lindane on cytosol 5α–dihydrotestosterone receptor complex formation in rat prostate. *Bull. Environ. Contam. Toxicol.* 46: 92–99.

Smith B.S. (1971). Male characteristics on female mud snails caused by antifouling bottom paints. *J. Appl. Toxicol.* 1: 22–25.

Soto, A.M., Chung, K.L., and Sonnenschein, C. (1994). The pesticides endosulfan, toxaphene, and dieldrin have estrogenic effects on human estrogen-sensitive cells. *Environ. Health Perspect.* 102: 380–383.

———, Justicia H., Wray, J.W., and Sonnenschein, C. (1991). p-Nonyl-phenol: An estrogenic xenobiotic released from "modified" polystyrene. *Environ. Health Perspect.* 92: 167–173.

———, Lin, T.-M., Justicia, H., Silvia, R.M., and Sonnenschein, C. (1992). An "in culture" bioassay to assess the estrogenicity of xenobiotics (E-screen). In *Chemically-Induced Alterations in Sexual and Functional Development: The Wildlife/Human Connection, Vol. XXI.* (T. Colborn and C. Clement, eds.), pp. 295–309. Princeton Scientific Publishing, Princeton.

Spooner, N., Gibbs, P.E., Bryan, G.W., and Goad, L.J. (1991). The effects of tributyltin upon steroid titres in the female dogwhelk, *Nucella lapillus*, and the development of imposex. *Marine Environ. Res.* 32: 37–49.

Stillman, R.J. (1982). *In utero* exposure to diethylstilbestrol: Adverse effects on the reproductive tract and reproductive performance in male and female offspring. *Am. J. Obstet. Gynecol.* 142: 905–921.

Stryer, L. (1981). *Biochemistry,* pp. 457–484. Freeman, San Francisco.

Swevers, L., Lambert, J.G.D., and Loof, A.D. (1991). Metabolism of vertebrate-type steroids by tissues of three crustacean species. *Comp. Biochem. Physiol.* 99B: 35–41.

Takac, P., Pavol,V., Kozanek, M., Huckova,A., and Slovak, M. (1988). Estradiol, progesterone, testosterone, and dihydrotestosterone concentrations in some tissues of cockroach *Nauphoeta cinerea:* In *Physiological Insect Ecology* (F. Sehnal, A. Zabza , and D.L. Denlinger, eds.), pp. 899–905. Wroclaw Technical University Press, Wroclaw.

Takeda, N. (1979). Induction of egg-laying by steroid hormones in slugs. *Comp. Biochem. Physiol.* 62A: 273–278.

——— (1980). Hormonal control of head-wart development in the snail *Euchadra peliomphala. J. Embryol. Exp. Morph.* 60: 57–69.

Talmage, S.S. (1994). *Environmental and Human Safety of Major Surfactants: Alcohol Ethoxylates and Alkylphenol Ethoxylates.* Lewis, Boca Raton.

Tyler-Schroeder, D.B. (1979). Use of grass shrimp, *Paleomonetes pugio*, in a life-cycle toxicity test: In *Aquatic Toxicology and Hazard Assessment.* American Society for Testing and Materials, Philadelphia.

Ugazio, G., Bosio, A., Nebbia, C., and Soffietti, M.G. (1991). Age- and sex-related effects on hepatic drug metabolism in rats chronically exposed to dietary atrazine. *Res. Comm. Chem. Pathol. Pharmacol.* 73: 231–243.

Voogt, P.A., Den Besten, P.J., and Jansen, M. (1991). Steroid metabolism in relation to the reproductive cycle in *Asterias rubens* L. *Comp. Biochem. Physiol.* 99B: 77–82.

———, van Ieperenm, S., van Rooyen, M.B.J.C.B., Wynne, H.J., and Jansen M. (1991). Effect of photoperiod on steroid metabolism in the sea star *Asterias rubens* L. *Comp. Biochem. Physiol.* 100B: 37–43.

Wheeler, J.M. (1992). Epidemiology and prevalence of endometriosis. *Infertil. Reprod. Med. Clin.* 3: 545–549.

Wigglesworth, V.B. (1970). *Insect Hormones.* Freeman, San Francisco.

Woin, P., and Bronmark, C. (1992). Effect of DDT and MCPA (4-chloro-2-methylphenoxyacetic acid) on reproduction of the pond snail, *Lymnaea stagnalis* L. *Bull. Environ. Contam. Toxicol.* 48: 7–13.

Wolff, M.S., Paolo, G., Toniolo, P., Lee, E.W., Rivera, M., and Dubin, N. (1993). Blood levels of organochlorine residues and risk of breast cancer. *J. Natl. Cancer Inst.* 85: 648–652.

Woodward, A.R., Percival, H.F., Jennings, M.L., and Moore, C.T. (1993). Low clutch viability of American alligators on Lake Apopka. *Fla. Sci.* 56: 52–63.

World Health Organization (1990). Tributyltin compounds. *Environ. Health Criteria 116.* Geneva.

Xu, R.A. (1991). Annual changes in the steroid levels in the testis and the pyloric caeca of *Sclerasterias mollis* (Hutton) (Echinodermata: Asteroidea) during the reproductive cycle. *Invert. Reprod. Develop.* 20: 147–152.

———, and Barker, M.F. (1990a). Annual changes in the steroid levels in the ovaries and the pyloric caeca of *Sclerasterias mollis* during the reproductive cycle. *Comp. Biochem. Physiol.* 95A: 127–133.

——— and ——— (1990b). Effect of diets on steroid levels and reproduction in the starfish, *Sclerasterias mollis. Comp. Biochem. Physiol.* 96A: 33–39.

——— and ——— (1990c). Photoperiodic regulation of oogenesis in the starfish *Sclerasterias mollis* (Hutton 1872).(Echinodermata: Asteroides). *J. Exp. Mar. Biol. Ecol.* 141: 159–168.

Yano, I. (1985). Induced ovarian maturation and spawing in greasyback shrimp, *Metapenaeus ensis,* by progesterone. *Aquaculture* 47: 223–229.

——— (1987). Effect of 17α-hydroxy-progesterone on vitellogenin secretion in Kuruma prawn, *Penaeus japonicus. Aquaculture* 61: 49–57.

Young, N.J., Quinlan, P.T., and Goad, L.J. (1992). Progesterone metabolism *in vitro* in the decapod crustacean, *Penaeus monodon. Gen. Comp. Endocrin.* 87: 300–311.

Zhang, M., and Kubo, I. (1992). Characterization of ecdysteroid-22-O-acyltransferase from tobacco budworm, *Heliothis virescens. Insect Biochem. Molec. Biol.* 22: 599–603.

Chapter 6

CONTAMINANT-ALTERED THYROID FUNCTION IN WILDLIFE

John F. Leatherland

Department of Biomedical Sciences
Ontario Veterinary College
University of Guelph
Guelph, Ontario N1G 2W1
Canada

Introduction

The thyroid hormones play key regulatory roles in several very important aspects of the life history of vertebrate animals, including organogenesis of the embryo, reproductive physiology, metabolic regulation, and, in poikilotherms, thermeogenesis. Thyroid hormones are very simple molecules, being essentially iodinated thyronine compounds, with iodide being attached to the 3- and 5-positions of the phenolic rings. The three main forms of iodinated tyrosines in the blood of vertebrates are the tetraiodothyronine (3,3′,5:5′-thyroxine, T_4), the triiodothyronines (3,3′,5-triiodothyronine, T_3), and 3,3′,5′-triiodothyronine (reverse T_3, rT_3). Of these, T_3 is the biologically active form, and it exerts its hormonal effect by binding to specific T_3 receptors (TRs) that are attached to chromatin in the nucleus of the target cells, and by so doing regulating the expression of specific genes. Some of the iodinated thyronines, including some of the forms of diiodothyronine, may also act at extranuclear sites (e.g., the mitochondria or cell membranes) (Lanni et al., 1994).

In addition to its affinity for nuclear receptor proteins, T_3 as well as T_4 (which acts as a precursor for T_3) binds to specific transport proteins in the blood and cerebrospinal fluid (CSF). The binding of the thyronine compounds to the transport and receptor proteins has similar characteristics, in that for both proteins the iodinated thyronine occupies a pocket of specific size, three-dimension shape, and electrical charge in the protein (Oppenheimer et al., 1995; Wagner et al., 1995).

Under normal (euthyroid) conditions, there is a balance established between the synthesis, release, and clearance of thyroid hormones, so that the very large pools of protein-bound reserve thyroid hormones are retained in the blood as well as in some organ systems, such as the brain. A dynamic equilibrium state is established that provides considerable stability to the overall state of thyroid hormone availability, a so-called thyroid hormone homeostasis. Any shift in this equilibrium state results in changes in the dynamics that exist between the production, delivery, use, and excretion of thyroid hormones, thus impacting on the overall thyroid hormone economy and often resulting in compensational changes of varying degrees. The impact of endocrine-disrupting toxicants on the overall thyroid hormone economy, as it relates to wildlife populations, will form the subject of this chapter.

Some persistent environmental anthropogenic toxicants, such as some congeners of polychlorinated biphenyls (PCBs), polybrominated biphenyls (PBBs), and dioxins bear a striking resemblance to the thyroid hormones (Porterfield, 1994)(Fig. 6-1). Therefore, in theory, they could occupy the same binding pocket on the thyroid hormone transport and/or T_3 receptor (TR) proteins. The attachment of xenobiotic toxicants to the thyronine-binding sites to transport proteins would reduce the thyroid hormone–carrying capacity of the blood, or CSF. Similarly, attachment of xenobiotics to T_3 binding sites on the TR would interfere with normal interactions between T_3 and its receptor, with one of two possible outcomes. If the ligand binding prevents the binding of T_3 but does not activate the TR, it will exert an antagonistic effect, whereas if the ligand binding activates the TR, it will have an agonistic action resulting in inappropriate gene expression. In either case, the thyroid hormone homeostasis (economy) would be adversely impacted, resulting in either compensatory responses and/or impaired thyroid hormone function. It must also be emphasized that T_3-TR binding is only one of the conditions necessary for the the expression of the genes that are regulated by T_3. Gene expression itself involves structural changes in coactivator and corepressor elements (e.g., Weiss et al., 1996; Wondisford, 1996), and any factor that interferes with these events could also compromise the regulatory function of thyroid hormones. Moreover, there is increasing evidence of posttranscriptional actions of thyroid hormones (Rall, 1995), and thus the action of toxicants at this level of hormone action could also impact on thyroid hormone function.

In addition to potential interactions of toxicants with transport proteins and at the TR level, some toxicants are known to affect the synthesis and metabolism of thyroid hormones, and thus may disturb thyroid hormone homeostasis and economy by altering the production or clearance of iodinated thyronines. Taken together, such environmental thyroid hormone–disrupting agents, even at very low levels, have the potential to adversely affect the complex developmental processes of vertebrates, and may also have significant actions on metabolism and reproduction.

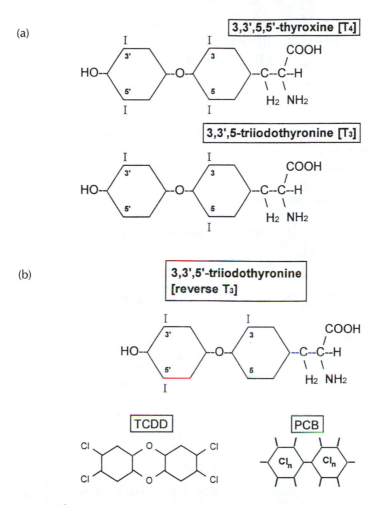

Figures 6-1a, b.
Schematic diagrams showing (a) the structure of the two principal iodinated thyronine compounds released from the thyroid gland, thyroxine (T_4), and triiodothyronine (T_3), and (b) of reverse T_3 (rT_3). For comparison, the general structure of PCBs and of 2,3,7,8-tetrachlorodibenzo-*p*-dioxin (TCDD) are presented [not to the same scale] (Fig. 6-1b).

In this chapter, I review the several processes that are involved in thyroid hormonogenesis, the delivery of the hormone to target cells, and the action of thyroid hormones at the target sites, with particular emphasis on those sites that have been demonstrated to be susceptible to purturbation by toxicants. I also review the studies that are available in wildlife species (for the purposes of this chapter, defined as studies on species other than human beings or laboratory rodents).

Thyroid Hormone Homeostasis and Economy: Potential Sites of Impact of Environmental Thyroid Hormone–Disrupting Toxicants

As mentioned above, T_3 acts on target cells by regulating the expression of specific genes, and some iodothyronines may also have nonnuclear sites of action. The regulatory action of the thyroid hormones at the molecular level is dependent on their uninterrupted synthesis by the thyroid gland and their efficient delivery to the target cells. There are many individual steps involved in this process, each of which if impeded will impact thyroid hormone home-ostasis and economy. It is beyond the scope of this chapter to present an exhaustive comparative review of thyroid hormone homeostasis and economy in the major classes of vertebrates. However, in order to comprehend the range of possible sites of impact of environmental contaminants, it is essential to

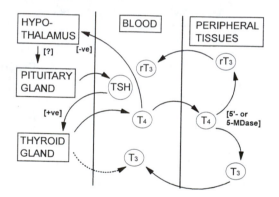

Figure 6-2.
Schematic diagram illustrating some of the major components of the hypothalamus––pituitary gland––thyroid gland axis involved in thyroid hormone homeostasis (economy). In all vertebrates investigated to date, the hypothalamus is found to secrete hypophysiotropic hormones that regulate the secretion of TSH from the pituitary gland. The nature of this regulation is highly species variable, with examples of both inhibitory and stimulatory regulation of TSH secretion being known. TSH is probably the major factor in stimulating the synthesis and release of thyroid hormones. Both thyroxine (T_4) and triiodothyronine (T_3) are released under the influence of TSH, but T_3 is released in much lower amounts than T_4. T_4 plays an essential role in the nega-tive feedback regulation of blood T_4 levels, with the feeback operating at the level of the hypothalamus, and possibly also the pituitary gland. T_4 enters nonthyroidal tissues, probably by diffusion, although the mechanism is not well understood. Within specific cells, T_4 is converted either to T_3 or reverse T_3 (rT_3) under the influence of 5'- or 5-monodeiodinase, respectively. The triiodothyronines may then leave the cell to contribute to the pool of iodinated thyronines in the blood, or, in the case of T_3, move to the nucleus to activate chromatin-bound thyroid hormone receptors.

understand the several key elements involved in the regulation of the synthesis, secretion, transport, metabolism, and action of thyroid hormones.

Control of Blood Thyroid Hormone Concentrations

The two principal thyroid hormones present in the blood are T_4 and T_3, together with very small amounts of mono- and diiodotyrosines (MIT and DIT); the latter may also have some biological activity (Lanni et al., 1994). In most vertebrates studied to date, T_4 is present at higher concentrations in the blood than T_3 and is the major factor in the negative feedback control of total blood thyroid hormone levels, operating via the hypothalamus-pituitary gland axis (Fig. 6-2). However, T_4 has much lower biological activity than T_3 because of its lower affinity for the TR in the target cell nucleus; in fact, most of the apparent biological activity of T_4 (including its role in the negative feedback regulation of thyroid-stimulating hormone [TSH] secretion) can be explained on the basis of intracellular conversion of T_4 to T_3 prior to binding to the TR.

With the possible exception of lampreys and hagfishes, the pituitary hormone TSH is the principal regulator of thyroid tissue activity in all classes of vertebrates, and in turn, TSH secretion is regulated by hypothalamic hormones. The nature of the hypothalamic regulation of TSH secretion shows considerable species variation and may be stimulatory (as in mammals) or inhibitory (as in teleost fish). As will be shown below, disruption of this negative feedback regulation of TSH secretion by toxicants results in thyroid enlargement (goiter) associated with either reduced or sometimes increased thyroid hormone output.

T_3 and T_4 Synthesis and Release

The thyroid hormones are synthesized by attaching iodide to the 3- and/or 5-positions on the phenolic rings of tyrosine to produce MIT and DIT, followed by the coupling of these iodinated tyrosine compounds to produce the iodinated thyronine compounds, T3 and T4 (Fig. 6-3). The follicular organization of the thyroid gland, with the thyroid cells (thyrocytes) forming a tight epithelium surrounding a fluid-filled lumen, is essential for this iodination process to occur efficiently, and this is probably why the follicular form of the thyroid has been highly conserved by all classes of vertebrates. The follicle lumen contains a colloidal, large molecular weight, iodinated glycoprotein called thyroglobulin. The iodinated tyrosine and thyronine compounds are incorporated into the thyroglobulin molecule at very specific sites.

The iodine, an essential element for thyroid hormone synthesis, is garnered, as iodide, from the diet by terrestrial vertebrates, and from the diet and ambient water by amphibians and fishes. Iodide is taken up from extracellular fluid by

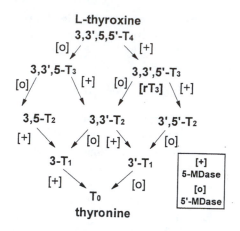

Figure 6-3.
Cascade of pathways for the metabolism of the iodinated thyronine compounds, showing the sites of action of 5'- and 5-monodeiodinase (5'-MDase and 5-MDase, respectively).

the cell membrane on the base of the thyrocytes. This region of the membrane is rich in sodium-iodide cotransporters that use the energy released by the movement of sodium down its electrochemical gradient. An iodide gradient is maintained within the thyrocyte by the activity of sodium–potassium dependent ATPases located in the apical regions of the thyrocyte cell membrane (Golstein et al., 1992, 1995; Carrasco, 1993; Dai et al., 1996).

The apical membrane of the thyrocytes contains thyrocyte-specific peroxidase (the so-called thyroid peroxidase); the enzyme-active site of these membrane-associated peroxidase molecules is directed toward the lumen. Thus, as iodide moves from the thyrocyte into the lumenal compartment, it is oxidized, under the influence of the thyroid peroxidases, and in the presence of H_2O_2: forms either iodine or a free radical of iodine. A second active site on the thyroid peroxidase catalyses the oxidation of the phenyl ring of specific tyrosine elements in the thyroglobulin, and the iodine or its free radical combines with the phenyl ring, thus forming mono- and diiodotyrosine (MIT and DIT) (Xiao et al., 1996). Following on behind the formation of the iodinated tyrosine is the peroxidase-catalyzed coupling together with the iodotyrosine units, MIT and DIT, to form the major thyroid hormones, T_4 and T_3 (Golstein et al., 1992; Carrasco, 1993). At this stage, the T_3 and T_4 are still components of the thyroglobulin molecule, and it is only after enzyme-catalyzed hydrolysis occurs that the hormones are released from the thyroglobulin and can then be released from the thyroid gland.

Under the influence of TSH, the thyrocytes pinocytose droplets of thyroglobulin and the droplets are incorporated into the cytoplasm in the form of membrane-bound vesicles. Lysosomes, containing proteases (which are also synthesized by the thyrocytes), fuse with the vesicles, and the thyroglobulin is hydrolyzed, resulting in the chemical release of the iodinated tyrosine (MIT and DIT) and thyronine compounds (T_4 and T_3). The MIT and DIT are largely deiodinated intracellularly (Fig. 6-3) and the components used in the formation of new iodinated tyrosine compounds. T_4 and T_3, on the other hand, are released from the basal regions of the thyrocytes, probably by simple diffusion, although there may also be transmembrane porter molecules, as yet poorly understood.

Thyroid Hormone Transport in the Blood and Movement Between Body Compartments

Most of the T_4 and T_3 in the blood is bound reversably to transport proteins (Lissitzky, 1990), which generally have a higher affinity for T_4 than for T_3. The bound and free hormone components are in equilibrium. Thus, as free hormone is removed (e.g., by entering the target cells, conjugation, or other metabolism), fractions of the bound hormone are released, thus maintaining the free hormone component. There are considerable species difference in the nature of the transport proteins (Refetoff et al., 1970; Galton, 1986; Larrson et al., 1985; Robbins and Edelhoch, 1986; Eales and Shostak, 1987; Licht et al., 1991; Licht and Pavgi, 1992), and with the exception of the thryoid hormone transport proteins in a few mammalian species, most have received only cursory examination. In humans, thyronine-binding globulin (TBG) and transthyretin (TTR)—formerly called thyroxine-binding prealbumin (TBPA)—are present, and in laboratory rodents, TTR and albumin appear to play the transporter roles.

The attachment of the transport of thyroid hormones to blood proteins has several advantages: (1) It allows more of these lipid-soluble hormones to be carried in the blood than would be otherwise possible; (2) it decreases the fractional turnover of the hormones; (3) it buffers against any sudden change in the availability of free serum (plasma) T_3 because of the equilibrium that is established between the free and bound pool; (4) it provides a reserve of hormones in excess of immediate needs that will protect the animal from any sudden change in hormone release; and (5) it has been shown to facilitate the uniform distribution of hormone to all cells within an organ system (e.g., the liver) (Hayashi and Refetoff, 1995). A sixth function has also been proposed, namely, as a protection against the loss of iodine. By binding the small iodothyronine compounds to the much larger carrier proteins, the potential for loss of the smaller molecules is vastly reduced. Indeed, it has been speculated that the lower affinity binding of T_4 in fish species (Refetoff et al., 1970) is related to the fact that fish (even freshwater species) have a more abundant source of iodine in their natural environment (Hayashi and Refetoff, 1995). These transport proteins play an important role in the overall thyroid hormone economy of vertebrates, and changes in the serum levels of the transport proteins, or competition for the binding sites by toxicants (see below), can shift the ratio of the bound and "free" serum hormones and thus have a major influence on the kinetics of thyroid hormone clearance.

The lipid soluble thyroid hormones probably enter and leave cells by diffusion (cell membranes are generally more permeant to T_4 than T_3), although kinetic studies have found evidence of a saturable transporter system in some tissues (Lissitsky, 1990). The hormones also move readily between body compartments, and T_4 (but not T_3) has been shown to traverse the placenta and blood brain barrier in experimental mammals. The movement of the thyroid hormones between different body compartments is another important part of the thyroid hormone homeostasis model, and if the hormone movement dynamics are altered, directly or indirectly by toxicants, there is the potential for a major shift in overall thyroid hormone partitioning and kinetics.

Metabolism of Thyroid Hormones

Although both T_4 and T_3 are released from the thyrocytes, most of the T_3 present in the serum of the vertebrate species studied to date is probably derived from the conversion of T_4 into T_3 by the removal of one iodide from the outer tyrosine ring (5' position), catalyzed by the enzyme, 5'-monodeiodinase (5'-MDase); 5'-MDase is present in several organ systems, including the liver, kidney, and brain (Lissitsky, 1990). Whereas some tissues produce sufficient T_3 for their own immediate needs (e.g., brain), other tissues produce T_3 at levels in excess of their own immediate needs, and the hormone leaves the cells and contributes to the serum T_3 pool (e.g., liver) (Lissitsky, 1990).

As well as 5'-Mdase, the enzyme 5-monodeiodinase (5-MDA) is also found in several tissues. It catalyzes the removal of iodide from the inner tyrosine ring (5 position) to produce reverse T_3 (rT_3) (Fig. 6-1b). Although rT_3 has no known physiological role, it has been argued that the conversion of T_4 to biologically inert rT_3 facilitates the degradation of T_4 via a pathway that does not require the production of biologically potent T_3.

In addition to the monodeiodination of T_4, the 5- and 5'-MDAs sequentially remove iodide from the inner and outer tyrosine rings, respectively, to further deiodinate the two triiodothyronine compounds to form thyronine (Fig. 6-3) (Lissitsky, 1990). This deiodination is an important means of metabolism of the iodinated thyronine compounds, prior to excretion. Other routes of metabolism and excretion included oxidative deamination, decarboxylation, and excretion in the feces and urine as conjugated (sulphated or glucuronidated) thyronine compounds.

Clearly, because peripheral production of T_3 is central to overall thyroid hormone homeostasis, any interactions of toxicants with the enzymes that regulate either the production of T_3 directly, or the redirection of T_4 metabolism via rT_3 production, will have the potential to disrupt overall thyroid hormone stability.

Cellular Sites of Action of T_3

The TRs are tightly associated with the chromatin of the target cells. Unlike the related steroid hormone receptors, which are predominantly homodimers, the TRs have to form heterodimers with the retinoic X receptors (RXRs) in order to allow a response to the presence of T_3 (Zhang and Pfahl, 1993; Kopp et al., 1996) (Fig. 6-4). The interaction of the TR component of this heterodimer with T_3 either activates (e.g., GH gene) or inactivates (e.g., TSH α and β subunit genes) gene expression, depending on the particular gene. The activation of the gene involves the interaction of the TR with the thyroid hormone response elements (TREs) in the promoter regions of the target genes. The inhibition of gene activation is less well understood but probably involves the enhanced action by repressor proteins in combination with the suppression of activator proteins (Williams and Brent, 1995; Kopp et al., 1996). Thus, the thyroid hormones act as gene switches, bringing about changes in the expression of specific genes. In addition, binding of T_3 to mitochondria has also been postulated as evidence of similar receptors in that organelle. However, the evidence for specific receptors is controversial, as is the available evidence for direct effects of T_3 on mitochondrial function, because of the very high levels of T_3 required to demonstrate these actions (Lissitsky, 1990). Nevertheless, the evidence is increasing for nonnuclear sites of action of the thyroid hormones.

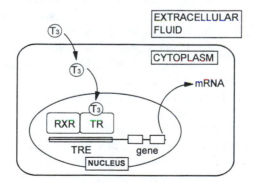

Figure 6-4.
Simplified diagram of the processes accompanying the action of triiodothyronine (T_3) at the thyroid hormone receptor (TR) level. The TR is one component of a heterodimer (the retinol X receptor [RXR] being the other) receptor. T_3 attachment to the TR initiates a series of events that permits the receptor protein to bind to the thyroid response element (TRE) on the chromatin, and the subsequent activation of the specific gene to promote mRNA transcription.

If toxicants compete with the binding sites on TR receptors, or other possible extracellular sites of action of T_3, or change the responsiveness of co-repressors or co-activators of the genes of the target cells, clearly they will impact normal regulatory physiology. Given the plethora of regulatory roles that have been attributed to the thyroid hormones, the impact of toxicants at the target cell level would have potentially devastating effects, including possible teratogenic effects associated with toxicant exposure during key periods of embryogenesis. However, it is also possible that toxicant exposures do not completely block the interactions of thyroid hormones with their target cells, and that the responses to the toxicant may be nonlife threatening. For example, for those tissues that have a particular need for T_3 during development, such as the brain, impaired T_3 binding to the target cells, although not resulting in brain deformity as measured by morphological changes, might result in impaired brain function that can be identified by behavioral studies (see Chap. 2).

Principal Functions of the Thyroid Hormones Related to the Potential for Actions of Environmental Thyroid Hormone–Disrupting Toxicants

The thyroid hormones are key regulatory factors in almost all organ systems investigated to date; a detailed description of these functions is beyond the scope of this chapter. However, in general, their regulatory actions can be separated into those that are involved in three major domains, namely, development (differentiation and growth), reproduction, and the maintenance of metabolic processes (Lissitzky, 1990; Oppenheimer et al., 1995). Of particular significance are the regulatory actions of the thyroid hormones during early ontogeny in all vertebrate classes. The central importance of thyroid hormones in vertebrate development is perhaps best exemplified by two well-studied processes, one normal (metamorphosis) and one pathological (cretinism). In the class Amphibia, the metamorphosis of the larval form (tadpoles) to adults involves an extensive set of orchestrated morphological and physiological changes that involve every one of the animal's organ systems. That the thyroid hormones are the main regulatory factors of tadpole metamorphosis is shown by the fact that precocious metamorphosis can be induced by administering exogenous thyroid hormone to amphibian larvae, whereas thyroidectomy of the larvae prevents the onset of metamorphosis. The term cretinism is used to describe the syndrome of neurological pathologies that result from hypothyroidism during the fetal development period of humans. In its most extreme form, cretinism is expressed as severe mental retardation (Porterfield, 1994). Although cretinism, per se, describes the irreversible neurological problems, cretins may also be afflicted by dwarfism and suffer musculoskeletal and reproductive abnormalities (Ingbar and Woeber, 1981). These "developmental" regulatory actions of the thyroid hormones differ from their regulatory actions on reproductive and metabolic processes in one important way: Thyroid hormone deficiencies during fetal ontogeny and infant growth result in developmental abnormalities that are irreversible, whereas most of the signs and symptoms of hypothyroidism in adults can be reversed by hormone therapy. Consequently, any impact of environmental agents, even within a relatively small window of exposure, could potentially have life-limiting effects on the organism. Similar exposure of the organism after developmental processes have ended would potentially have less severe effects, and they may well be reversible.

Although the embryos of all vertebrates are potentially vulnerable to environmental contaminants, those of egg-laying species may be more prone to exposure than mammals. The eggs of fish, amphibians, reptiles, and birds are rich in lipids, and thus are potential sites of accumulation of lipophilic toxicants, such as PCBs and other PHAHs. As the developing embryo absorbs the yolk, the toxicants burden is also absorbed. As a result, an embryo may suffer chronic exposure to toxicants for the entire developmental period. The effects of such exposure has been well demonstrated in piscivorous birds in some regions of the Great Lakes. These populations suffer extremely high embryo deformity and mortality, probably as a result of the PHAH contaminants that are accumulated by the breeding females from their contaminated fish diets (see Colborn et al., 1993). Moreover, the high mortalities of some fish stocks in the Great Lakes region have similarly been attributed to toxicant-induced early mortality (see Colborn et al., 1993).

An additional potential hazard for the embryos of egg-laying vertebrates, and one that may act synergistically with the toxicant exposure, is the presence of significant levels of hormones (steroids and thyroid hormones), of maternal origin, in the yolk of eggs. It is unlikely that such hormones are involved directly in the regulation of normal development. In fact, normal embryonic development appears to depend on the ability of the embryo to metabolize and excrete these "exogenous" hormones (see Leatherland, 1994). Consequently, toxicants that interfere with the metabolism of thyroid or steroid hormones could potentially have direct and/or indirect actions on the normal development of the embryos of egg-laying species.

Actions of Toxicants on Thyroid Hormone Homeostasis and Economy

Most of our knowledge of the actions of chemicals that affect the secretion of thyroid hormones, transport of those hormones to target tissues, or the action of thyroid hormones at target sites has been derived from studies of mammals, usually rodents. The details of these interactions are summarized here for purposes of comparison with similar studies of wildlife species.

Inhibition of Iodide Transport, Iodination of Thyroglobulin, and Coupling of Iodotyrosines

Several chemicals are known to inhibit the synthesis of thyroid hormones. This results in a lowering of plasma T_4 concentration, resulting in an increased secretion of TSH, which in turn acts on the thyroid, causing hypertrophy and hyperplasia and ultimate enlargement of the thyroid, commonly called a goiter. Chemicals that induce goiters are termed goitrogens, and the known goitrogens act either to inhibit iodine uptake or to inhibit the iodination of protein and/or the coupling of the iodotyrosine compounds, MIT and DIT.

The known iodide uptake inhibitors are generally monovalent ions, the best studied of which are thiocyanates and perchlorates. Cyanogenic glucosides and other cyanide derivatives, found in some plant families, also inhibit iodide uptake, although the action of these compounds follows the release of thiocyanates during digestion of the molecule. The second major group of goitrogens inhibit the iodination of proteins and/or the coupling of tyrosine residues in the thyroglobulin. Three major families of chemicals are known to act in this manner: (a) compounds that contain the thionamide group [e.g., 6-n-propylthiouracil, thiourea, 1-methyl 2-mercaptoimidazole, p-aminobenzoic acid, sulfonamides]; (b) aminoheterocyclic compounds [e.g., p-amino salicylic acid, 1-butyl-3(p-tolylsulfonyl) urea]; and (c) substituted phenols [e.g., resorcinol, salycimide]. Almost all agents that inhibit organic binding of iodide also inhibit the coupling of iodinated tyrosine compounds in the thyroglobulin, suggesting that the site of action in the two processes is similar. L-vinyl 2-thiooxazolidone, better known as the goitrin, is chemically similar to thionamides. Some of these chemicals are thought to act via competitive inhibition of thyroid peroxidases (Taurog, 1970). Other factors, such as cyanides and azides, can inhibit thyroid hormonogensis by indirectly inhibiting the action of thyroid peroxidases, perhaps by preventing the production of hydrogen peroxide that is required for the iodination process.

In addition to these known goitrogens, there is indirect evidence for the presence of other, so far unidentified, goitrogens that are produced by bacteria and that have been suggested as agents in the formation of goitres in both human beings and wildlife species (Gaitan et al., 1980; Leatherland, 1994).

Altered Binding of Thyroid Hormones to Blood and CSF Proteins, and to the T_3 Receptor

Blood Proteins

Several drugs are known to inhibit the binding of thyroid hormones to the carrier proteins, and they are used clinically to diagnose and treat thyroid hormone diseases (Ingbar and Woeber, 1981). The binding of thyroid hormones to TBG (a transporter of both T_4 and T_3) in human beings is inhibited by pheny-toin, terachlorothyronine, salicylate, anilinonaphthalene-sulfonic acid, and dichlorodiphenyldichloroethane (o,p'-DDD). Similarly, T_4 binding to TTR (a transporter solely of T_4) is inhibited by barbital, salicylate, 2,4-dinitrophenol, and penicillin.

In a similar manner, numerous experimental studies, and an increasing number of theoretical structural models, support the thesis that some environmental contaminants can bind to the thyroid hormone transporter plasma proteins. Competition between these toxicants and thyroid hormones for binding sites can potentially impact on the thyroid hormone reserves of the blood, and therefore of the delivery of thyroid hormones to target tissues. This mode of action has been either demonstrated or postulated for several congeners of the PCB, PBB, and dioxin families, as well as perfluoro-n-decanoic acid and photomirex (Bastomsky, 1974, 1977; Bastomsky and Murthy, 1976; Bastomsky et al., 1976; Villeneuve et al., 1979; Bahn et al., 1980; Collins and Capen, 1980; McKinney et al., 1985a, b, 1987; Spear and Moon, 1985; Brouwer and Van den Berg, 1986; Brouwer et al., 1995; Rickenbacher et al., 1986; Gorski and Rozman, 1987; Henry and Gasiewicz, 1987; McKinney and Pedersen, 1987; Gorski et al., 1988; Gutshall et al., 1988; Van den Berg et al., 1988; Brouwer, 1989; McKinney, 1989; Spear et al., 1990, 1994; Morse et al., 1992, 1993; van Birgelen et al., 1992; Ness et al., 1993; McKinney and Waller, 1994; Goldey et al., 1995; Sewall et al., 1995; Corey et al., 1996; Kohn et al., 1996; Seegal, 1996; Leatherland, 1997). The experimental evidence in support of the hypothesis consistently shows a lowering of plasma or serum total T_4 concentration, associated with either no effect, or minimal effects, on plasma T_3 concentration, an enlargement of the thyroid gland, and an increased glucuronidation and clearance of T_4. Such observations are consistent with a reduced binding of T_4 to transport proteins; a reduction in binding leads to an increased availability of the free (unbound) T_4 fraction for metabolism and clearance and thus a marked decline in plasma or serum total T_4 concentrations. There may be little effect on the free T_4 concentration, although some studies have shown a lowering of free T_4 in rats dosed with specific congeners of PCB and dioxin (e.g., van Birgelen et al., 1992).

The reduction in total plasma T_4 concentrations results in a release of the negative feedback brake that regulates pituitary TSH secretion, and thus an increase in the size of the thyroid gland by both hyperplasia and hypertrophy. Plasma T_3 levels would not be affected to the same extent as T_4 since there is a much lower level of T_3 in the blood, and most of the T_3 that is present originates from the monodeiodination of T_4 by peripheral (nonthyroidal) tissues. It is interesting to note that mixtures of polyhalogenated aromatic hydrocarbon (PHAH) compounds appear to be more potent in inducing these thyroid responses than single congeners, or even commercial mixes of a single family of toxicants (e.g., Aroclor mixes of PCBs). For example, rats fed diets containing "naturally contaminated" Great Lakes salmon, and seals fed fish diets that were also "naturally contaminated" with PHAH mixes, demonstrated thyroid responses that were considerably more pronounced than could be induced by exposure to single toxicants (reviewed by Leatherland, 1997).

Few comparable studies have been made in wildlife (i.e., nonhuman, nonrodent) species. Among mammalian wildlife species, Brouwer et al. (1989) demonstrated, in the common seal, that the carrying capacity for both vitamin A and T_4 was influenced by the dietary level of "naturally occurring" PCBs, and is thus consistent with the hypothesis that some PCB congeners act by reducing the blood transport capacity for thyroid hormones. Conversely, when captive mink were fed Great Lakes fish diets that were contaminated with PCB and methyl mercury, although the growth rates of kits nursed by the PCB-fed females were significantly impaired (Wren et al., 1987b) (thus indicative of a response to the PCB exposure), there was no evidence of changes in either plasma total thyroid hormone concentrations or thyroid histology of the females (Wren et al., 1987a). Moreover, a study of the pathobiology of a population of beluga whales (Delphinapterus leucas), which inhabits the Gulf of St. Lawrence in Eastern Canada, yielded little direct evidence of a major disruption of thyroid function. This population exhibited several, very severe indications of health problems, possibly linked to their diet of PHAH-contaminated fish, but thyroid lesions were present at relatively low levels (Béland et al., 1993; deGuise et al., 1995).

There is no simple explanation as to why there are species differences in the actions of PCBs and other PHAHs on thyroid hormone homeostasis. One possible explanation is that these organochlorine toxicants have a greater affinity for some transport proteins than for others. The transport proteins differ among species, and this may explain, at least in part, some diversity of response.

In studies of wild and captive birds, a reduction in plasma total T_4 concentration, sometimes associated with thyroid enlargement, has been reported in gulls, pigeons, and guillemots fed diets contaminated with PCB, p,p'-DDE, or dieldrin (Jefferies and French, 1969, 1972; Jefferies and Parslow, 1972, 1976). In addition, comparative histopathology of the thyroid of gulls feeding in different regions of the Great Lakes basin, and thus exposed to different levels of dietary organochlorine contaminants (Moccia et al., 1986; Fox, 1993), similarly provides indirect support for the hypothesis that these toxicants reduced the animal's blood T_4 carrying capacity.

Only two studies have been made of the effect of organochlorine contaminants on thyroid function in fishes (Leatherland and Sonstegard, 1978, 1979, 1980). These studies found a lowering of plasma T_4 and T_3 concentrations in coho salmon (Oncorhynchus kisutch) and rainbow trout (Oncorhynchus mykiss) fed PCB- and Mirex-contaminated diets. However, the response was only evident in the fish that were fed a level of contaminant that caused a reduction in feeding rate and poor growth. Moreover, there was no evidence of toxicant-induced thyroid enlargement in either species. A possible explanation is the fact that in salmonid fishes, in comparison with mammals and birds, far less of the plasma or serum total thyroid hormone is bound to plasma proteins, which are thought to be prealbumin fractions (Eales and Shostak, 1987; Leatherland, 1994). Consequently, toxicant actions on thyroid hormone binding could be less evident in these species than it is in some mammals.

CSF proteins

In addition to the thyroid hormone transporters in the blood, there is evidence for TTR playing an important role in enabling the transfer of T_4 across the blood brain barrier into the CSF of some mammals (Chanoine and Braverman, 1992). Recent findings by Palha et al. (1997) would contradict this hypothesis, since the T_4 transfer rates of the brain of TTR-null mice were similar to those of wild-type mice. However, the mutant mice did have low brain T_4 levels because of the absence of T_4-TTR complexing, and it may be that the TTR is less important for transfer of T_4 across the choroid plexus and more important in ensuring an appropriate level of the T_4 to act as the substrate for T_3 production within the brain. Since TTR represents up to 25% of the proteins of the CSF, it would provide an abundant supply of T_4 for this purpose. If this is the case, then any condition, or toxicant, that inhibits the binding of T_4 to the CSF would likely restrict the availability of the hormone to the brain, resulting in a potentially reduced *in situ* production of T_3 from the T_4. Since this *in situ* T_3 production is essential for normal brain development (Porterfield, 1994), and probably also for normal brain function, this potential site of action of environmental toxicants has grave significance. What is the evidence for such an action? To date, there is no direct evidence, although experimental work with rodents has shown a markedly enhanced 5'-MDase activity in the brain of pups born of females that had been treated with PCBs (Morse et al., 1993). This response is very similar to that of hypothyroid rats and/or pups born of severely hypothyroid rats (Calvo et al., 1990; Silva and Larsen, 1982; Porterfield and Hendrich, 1993; Serrano-Lozano et al., 1993; Porterfield, 1994) and suggests that the PCBs are creating a similar developmental environment for the developing brain. The increased 5'-MDase in response to either PCB-exposure or severe hypothryoidism is likely to be a compensatory response to the decreased delivery of the T_4 precursor to the brain, and thus provides indirect evidence of an impaired CSF transport capability.

T_3 receptor

The nature of the interaction of thyroid hormones with transport proteins is very similar to the nature of the interaction of T_3 with the TR in the nucleus of target cells (Oppenheimer et al., 1995; Wagner et al., 1995), with the hormone occupying a pocket in the protein. Occupation of the pocket by a ligand results in a conformational change in the TR, which in turn stimulates a co-activator and suppresses a co-repressor, resulting in the activation of the target gene. Consequently, those toxicants that are able to bind to transporter proteins in the blood or CSF are also likely to interact in some manner with the TR, acting either as T_4 agonists or antagonists.

Some of the PHAHs are potential candidates for interaction with transport and/or TR, based on their three-dimensional structure and electrical charge. Theoretically, the responses of tissues to TCDD are consistent with the binding of the toxicant to the TR family of receptors (Lucier et al., 1993). Moreover, there is some evidence for such actions of dioxins in experimental situations. TCDD appears to have a mild T_3 agonistic action (Pazdernik and Rozman, 1985) and increases the expression of c-erb-A genes (Bombick et al., 1988).

An additional piece of evidence that links the binding of PHAHs with TRs is the observation that the administration of thyroid hormones in supra-physiological levels to PHAH-treated animals can modulate the action of the toxicant on the experimental animals. For example, in mice, TCDD myelotoxicity was modulated by administration of T_4 in a manner that suggested competition for binding sites by TCDD and T_4 (Hong et al., 1987). Similarly, in a bird, the spotted munia *(Lonchura punctulata)*, the administration of T_4 ameliorated the effects of DDT as regards growth, blood hemoglobin content, and plasma protein and cholesterol concentrations (Kar et al., 1995).

Altered Capacity to Synthesize T_3 from T_4

There are three known types (I, II, and III) of deiodinases that catalyze the monodeiodination of T_4. They are distinguished in part by their tissue specificity, affinity for the T_4 substrate, and product (T_3 or rT_3), but also in part as to whether the enzyme activity is inhibited by goitrogens, such as propylthiouracil and iopanoic acid. Thus, clearly, some types of deiodinases are known to be strongly inhibited in the presence of toxicants, and such inhibition could, theoretically, have adverse effects on thyroid hormone homeostasis and economy.

Raasmaja et al. (1996) reported a decreased type I 5'-MDase activity in the liver and an increased type II 5'-MDase activity in the brown adipose tissue of TCDD-exposed rats, suggesting a direct or indirect action of the toxicant on T_3 production potential., Contrary evidence was presented by Spear et al., (1990), who reported no change in hepatic type I 5'-MDase in PBB-treated rats, despite the fact that the treated animals exhibited distinct signs of disturbed thyroid hormone homeostasis. However, in that study, the enzyme measurements were made on crude homogenates of previously frozen tissue. It is very likely that enzyme activity had been destroyed by the freezing and thawing process; work in our laboratory demonstrates a 50% to 75% loss of hepatic 5'-MDase activity in tissues that are treated in this manner (unpublished data).

Three studies have also reported the impact of heavy metal toxicants on monodeiodination by tissues. Yoshida et al. (1987) showed that Cd, either added to the incubation medium of the *in vitro* enzyme assays, or administered to rats prior to the tissues being taken for assay, reduced 5′-MDase activity. The enzymes require a thiol source for the conversion; the thiol probably acts as an an electron donor enabling the release of iodide from the tyrosine ring. For the *in vitro* assays, dithiothreitol is provided as a source of thiol, and Yoshida et al. (1987) found an interesting interaction between Cd concentration and dithio-threitol requirements for the enzyme assay. Thus, the effects of Cd may be exerted via this electron transfer step in metabolism.

Two other studies showed that Pb lowered serum T_3 concentration and 5′-MDase activity in cockerels and catfish *(Clarias batrachas)*, and also lowered plasma T_4 concentrations in the catfish (Chaurasia et al., 1995, 1996). The site and mechanism of action of the heavy metal is not known, although it should be noted that the levels of exposure to lead were high for both studies (1.5 mg per bird—1-day-old broiler chicks, body weight not given—10–15 ppm in the ambient medium, respectively).

Thyroid Tumors in Wildlife Specie as Indicators of Environmental Antithyroid Agents

The prevalence of benign thyroid lesions in populations provides a gross indicator of the summation of distruptions in thyroid hormone economy, and it has the advantage that it is not necessary to know the etiological cause of the condition. Epidemiological studies (i.e., those pertaining to human beings) have provided evidence of the presence of environmental agents, other than iodide deficiency, that induce an enlargement of the thyroid gland. Epizootiological studies (i.e., those pertaining to nonhuman animal populations) of thyroid lesions in wildlife species are rare. Human studies excepted, most reports of thyroid lesions in other animals are in fishes (Leatherland et al., 1998) and a single study of herring gulls in the Great Lakes (Moccia et al., 1986; Fox, 1994). Only one systematic epizootiological study has been made in fishes, namely, the 100% prevalence of thyroid hyperplasia and hypertrophy in Great Lakes coho, chinook *(Oncorhynchus tshawytscha)*, and pink salmon *(Oncorhynchus gorbuscha)* (see Leatherland, 1993, 1994 for literature reviews).

The thyroid lesions in Great Lakes salmon were initially attributed to iodide deficiency, although later studies found no evidence of such a deficiency (see Leatherland, 1993, 1994, for references). Moreover, interlake comparison of the prevalence and size of the lesions suggest a correlation between the degree of eutrophication of the lakes and the extent of the thyroid problem, suggesting that factors in the lakes, rather than iodide deficiency, were likely causitive agents. The nature of these agents remains to be determined, but based on PCB and Mirex feeding trials in coho salmon and rainbow trout, PHAHs do not induce thyroid enlargement in oncorhynchid species (Leatherland and Sonstegard, 1978, 1979, 1980). Moreover, feeding Great Lakes salmon diets (containing "natural" mixes of PHAHs) to coho salmon did not induce any thyroid enlargement (Leatherland and Sonstegard, 1982).

The interlake comparisons that point to a link between eutrophication and thyroid lesion might suggest a problem related to the microflora (bacteria and/or algae) of the lakes. This would be consistent with findings of goitrogenic bacterial products (Gaitan et al., 1978, 1980) linked to thyroid lesions in human populations. Moreover, preliminary work in our laboratory (E. Gaitan, J. F. Leatherland, and R. A. Sonstegard) has shown that Lake Erie water contains a factor(s) that inhibits the sythesis of iodinated compounds by porcine thyroid slices incubated *in vitro*.

In contrast, the prevalence and size of the thyroid lesions reported in Great Lakes herring gulls do appear to be correlated with the level of PHAH contamination of their diet, and thus these lesions may be of the same form as those found in PHAH-dosed rodents (see above, also Leatherland, 1997).

Although the nature of environmental goitrogens has yet to be determined, there is one laboratory study in fish *(Rivulus marmoratus)* that has demonstrated a potent goitrogenic and possibly carcinomatous effect of N-methyl-N'-nitrosoguadine (25 ppm for 4 months) (Park et al., 1993); however, no other chemical agent having this type of effect in wildlife species has been found.

Summary and Conclusions

This chapter explores the theoretical and experimental evidence for the action of toxicants on thyroid hormone homeostasis and economy in wildlife species, compared with such evidence in human beings and laboratory rodents. The regulation of the delivery of appropriate amounts of thyroid hormones to tissues under euthyroid conditions involves processes of synthesis, transport, metabolism, and excretion of the hormones. The overall thyroid hormone budget depends on the maintenance of a very stable thyroid hormone reserve in the blood, with an equilibrium developed between free and carrier-protein–bound hormone. Minor disruption of this interactive and usually stable system will give rise to compensatory responses, and the possible correction of the disruption; more intensive disruption would effect changes that may cause considerable inbalances to be established, resulting in hypo- or hyperthyroid conditions. There is abundant evidence to show that several key parts of the overall thyroid hormone homeostasis are susceptible to the action of a wide range of chemicals, some of them pharmaceutical agents, but some of which are now part of the "natural" background contaminant spectrum of many ecosystems. The most susceptible stages include the acquisition of iodide and the incorporation of iodide into the iodinated tyrosine and thyronine compounds, the transport of thyroid hormones in the blood, the distribution of thyroid hormones among body compartments, the conversion of the biologically inactive thyroid hormone (T_4) to the biologically active form (T_3), and the binding of T_3 to the TR in target cells.

Some of the known organic toxicants (e.g., PHAHs) bear a striking resemblence to thyroid hormones and can theoretically mimic the natural hormones, thus acting as agonists or antagonists. In addition, some heavy metals appear to interfere with the electron donor events associated with iodination of amino acids. Much of the available evidence to link environmental toxicants with thyroid dysfunction has been derived from experimental studies using rodent models, with some additional work in human beings. Relatively little work has been undertaken to examine the effects of toxicants on thyroid function in wildlife species, in large measure because of the difficulties of carrying out field studies of this type. However, the studies that are published are consistent with the evidence that some organic toxicants hinder the transport of thyroid hormones in the blood. Indirect evidence from herring gull and Pacific salmon populations in the Great Lakes indicates the presence of environmental factors that adversely influence thyroid hormone homeostasis. In the herring gulls, the agents are likely to be PHAHs, but the factor(s) associated with the thyroid dysfunction in the Great Lakes Pacific salmon remains to be identified. In addition, the high embryo mortalities of some populations of birds and fish in contaminated areas might also be interpreted to support an impaired thyroid hormone economy thesis, although direct evidence to support this is not available, and the estrogenic properties of some of the contaminants could theoretically affect development via nonthyroid hormone pathways.

References

Bahn, A.K., Mills, J.L., Snyder, P.J., Gann, P.H., Houten, L., Bialik, O. Hollmann, L., and Utiger, R.D. (1980). Hypothyroidism in workers exposed to polybrominated biphenyls. *New England J. Med.* 302: 31–33.

Bastomsky, C.H. (1974). Effects of a polychlorinated biphenyl mixture (Aroclor 1254) and DDT on biliary thyroxine excretion in rats. *Endocrinology* 95: 1150–1155.

——— (1977). Enhanced thyroxine metabolism and high uptake goiters in rats after a single dose of 2,3,7,8-tetrachlorodibenzo-p-dioxin. *Endocrinology* 101: 292–296.

———, and Murthy, P.V.N. (1976). Enhanced *in vitro* hepatic glucuronidation of thyroxine in rats following cutaneous application or ingestion of polychlorinated biphenyls. *Can. J. Physiol. Pharmacol.* 54: 23–25.

———, ———, and Banovak, K. (1976). Alterations in thyroid metabolism produced by cutaneous application of microscope immersion oil: Effects due to polychlorinated biphenyls. *Endocrinology* 98: 1309–1314.

Béland, P., DeGuise, S., Girard, C., Legacé, A., Martineau, D., Michaud, R., Muir, D.C.G., Norstrom, R.J., Pelletier, E., Ray, S., and Shugart, L.R. (1993). Toxic compounds and health and reproductive effects in St. Lawrence beluga whales. *J. Great Lakes Res.* 19: 766–775.

Bombick, D.W., Jankum, J., Tullis, K., and Matsumura, F. (1988). 2,3,7,8-Tetra-chlorodibenzo-*p*-dioxin causes increases in expression of c-erb-A and levels of protein-tyrosine kinases in selected tissues of responsive mouse strains. *Proc. Natl. Acad. Sci. USA* 85: 4128–4132.

Brouwer, A. (1989).Inhibition of thyroid hormone transport in plasma of rats by polychlorinated biphenyls. *Arch. Toxicol.* (Suppl. 13): 440–445.

———, Ahlborg, U.G., Van den Berg, M., Birnbaum, L.S., Boersma, E.R., Bosveld, B., Denison, M.S., Gray, L.E., Hagmar, L., Holene, E., Huisman, M., Jacobson, S.W., Jacobson, J.L., Koopman-Esseboom, C, Koppe, J.G., Kulig, B.M., Morse, D.C., Muckle, G., Peterson, R.E., Sauer, P.T.T., Seegal, R.F., Smits-Van Prooije, A.E., Touwen, B.C.L., Weisglas-Kuperus, N., and Winneke, G. (1995). Functional aspects of developmental toxicity of poly-halogenated aromatic hydrocarbons in experimental animals and human infants. *Eur. J. Pharmacol. Envir. Toxicol. Pharmacol.* 293: 1–40.

———, and Van den Berg, K.J. (1986). Binding of a metabolite of 3,4,3',4'-tetrachlorobiphenyl to transthyretin reduces serum vitamin A transport by inhibiting the formation of the protein complex carrying both retinol and thyroxine. *Toxicol. Appl. Pharmacol.* 85: 301–312.

———, ———, Blaner, W.S., and Goodman, DeW.S. (1986). Transthyretin (prealbumen) binding of PCBs, a model for the mechanism of interference with vitamin A and thyroid hormone metabolism. *Chemosphere* 15: 1699–1706.

———, Reijnders, P.J.H., and Koeman, J.H. (1989). Polychlorinated biphenyl (PCB)-contaminated fish induces vitamin A and thyroid hormone defi-ciency in the common seal *(Phoca vitulina)*. *Aquat. Toxicol.* 15: 99–106.

Calvo, R., Obregón, M.J., Ruiz de Oña, C., Escobar del Rey, F., and Morreale de Escobar, G. (1990). Congenital hypothyroidism, as studied in rats. Crucial role of maternal thyroxine but not of 3,5,3'-triiodothyronine in the protec-tion of the fetal brain. *J. Clin. Invest.* 86: 889–899.

Carrasco, N. (1993). Iodide transport in the thyroid gland. *Biochim. Biophys. Acta* 1154: 65–82.

Chanoine, J.P., and Braverman, L.E. (1992). The role of transthyretin in the transport of thyroid hormone to cerebrospinal fluid and brain. In *Acta Medica Austriaca, 4th Thyroid Symposium-Brain and Thyroid 19* (L.E. Braverman, T.J. Visser, O. Eber, and W. Langsteger, eds.), pp. 25–28. Jahrgang Sonderheft, Grza, Austria.

Chaurasia, S.S., Gupta, P., Kar, A., and Maiti, P.K. (1995). Lead impairs hepatic type I-5′-monodeiodinase activity and thyroid function in cockerels. *Current Sci.* 69: 698–700.

——, ——, ——, and —— (1995). Lead induced thyroid dysfunction and lipid peroxidation in the fish *Clarias batrachus* with special reference to hepatic type I-5′-monodeiodinase activity. *Bull. Environ. Contam. Toxicol.* 56: 649–654.

Colborn, T., vom Saal, F.S., and Soto, A.M. 1993. Developmental effects of endocrine-disrupting chemicals in wildlife and humans. *Environ. Health Perspec.* 101: 378–384.

Collins, W.T., and Capen, C.C. (1980). Ultrastructural and functional alterations of the rat thyroid gland produced by polychlorinated biphenyls compared with iodide excess and deficiency, and thyrotropin and thyroxine administration. *Virchows Arch.* B 33: 213–231.

Corey, D.A., Deku, L.M.J., Bingman, V.P., and Meserve, L.A. (1996). Effects of exposure to polychlorinated biphenyl (PCB) from conception on growth, and development of endocrine, neurochemical, and cognitive measures in 60 day old rats. *Growth Devel. Ageing* 60: 131–143.

Dai, G., Levy, O., and Carrasco, N. (1996). Cloning and characterization of the thyroid iodide transporter. *Nature, Lond.* 379: 458–460.

deGuise, S., Martineau, D., Béland, P., and Fournier, M. (1995). Possible mechanisms of action of environmental contaminants on St. Lawrence beluga whales *(Delphinapterus leucas). Environ. Health Perspec.* 103 (Suppl. 4): 73–77.

Eales, J.G., and Shostak, S. (1987). Total and free thyroid hormones in plasma of tropical marine teleost fish. *Fish Physiol. Biochem.* 3: 127–131.

Fox, G.A. (1993). What have biomarkers told us about the effects of contaminants on the health of fish-eating birds in the Great Lakes? The theory and literature review. *J. Great Lakes* Res. 19: 722–736.

Gaitan, E., Medina, P., DeRouen, T.A., and Zia M.S. (1980). Goiter prevalence and bacterial contamination of water supplies. *J. Clin. Endocrinol. Metab.* 51: 957–961.

——, Merino, H., Rodriguez, G., Medina, P., Meyer, J.D., DeRouen, T.A., and MacLennan, R. (1978). Epidemiology of endemic goitre in western Columbia. *Bull. World Health Organ.* 56: 403–416.

Galton, V.A. (1980). Binding of thyroid hormones in serum and liver cytosol of *Rana catesbeiana* tadpoles. *Endocrinology* 107: 61–69.

Goldey, E.S., Kehn, L.S., Lau, C., Rehnberg, G.L., and Crofton, K.M. (1995). Developmental exposure to polychlorinated biphenyls (Aroclor 1254) reduces circulating thyroid hormone concentrations and causes hearing deficits in rats. *Toxicol. Appl. Pharmacol.* 135: 77–88.

Golstein, P.E., Abramow, M., Dumont, J.E., and Beauwens, R. (1992). The iodide channel of the thyroid: A plasma membrane vesicle study. *Am. J. Physiol.* 263 (*Cell Physiol.* 32): C590–C597.

——, Sener, A., and Beauwens, R. (1995). The iodide channel of the thyroid. II. Selective iodide conductance inserted into liposomes. *Am. J. Physiol.* 268 (*Cell Physiol.* 37), C111–C118.

Gorski, J.R., Muzi G., Weber, L.W.D., Pereira, D.W., Arceo, R.J., Iatropoulos, M.J., and Rozman, K. (1988). Some endocrine and morphological aspects of the acute toxicity of 2,3,7,8-tetrachlorodibenzo-*p*-dioxin (TCDD). *Toxicol. Pathol.* 16: 313–320.

——, and Roxman, K. (1987). Dose-response and time course of hypothy-roxinemia and hypoinsulinemia and characterization of insulin hypersentivity in 2,3,7,8-tetrachlorodibenzo-*p*-dioxin (TCDD)-treated rats. *Toxicology* 44: 297–307.

Gutshall, D.M., Pilcher, G.D., and Langley, A.E. (1988). Effect of thyroxine supplementation on the response to perfluoro-*n*-decanoic acid (PFDA) in rats. *J. Toxicol. Environ. Health* 24: 491–498.

Hayashi, Y., and Refetoff, S. (1995). Genetic abnormalities of thyroid hormone transport serum proteins. In *Molecular Endocrinology: Basic Concepts and Clinical Correlations.* (B.D. Weintraub, ed.), pp. 371–387. Raven, New York.

Henry, E.C., and Gasiewicz, T.A. (1987). Changes in thyroid hormones and thyroxine glucuronidation in hamsters compared with rats following treatment with 2,3,7,8-tetrachlorodibenzo-*p*-dioxin. *Appl. Pharmacol.* 89: 165–174.

Hong, L.H., McKinney, J.D., and Luster, M.I. (1987). Modulation of 2,3,7,8-tetrachlorodibenzo-*p*-dioxin (TCDD)-mediated myelotoxicity by thyroid hormones. *Biochem. Pharmacol.* 36: 1361–1365.

Ingbar, S.H., and Woeber, K.A. (1981). The thyroid gland. In *Textbook of Endocrinology.* (R.H. Williams, ed.), pp. 117–247. Saunders, Philadelphia.

Jefferies, D.J., and French, M.C. (1969). Avian thyroid effects of p,p′-DDT on size and activity. *Science* 166: 1278–1280.

—— and —— 1972. Changes induced in the pigeon thyroid by p,p′-DDE and dieldrin. *J. Wildlife Manage.* 36: 24–30.

—— and Parslow, J.L.F. 1972. Effects of one polychlorinated biphenyl on size and activity of the gull thyroid. *Bull. Environm. Contam. Toxicol.* 8: 306–310.

—— and —— (1976). Thyroid changes in PCB-dosed guillemots and their indication of one of the mechanisms of action of these materials. *Environ. Pollut.* 10: 293–311.

Kar, A., Nayudu, P., and Maiti, P.K. (1995). Involvement of thyroxine in the regulation of DDT toxicity in spotted munia *(Lonchura punctulata). Ind. J. Ann. Sci.* 65: 998–1000.

Kohn, M.C., Sewall, C.H., Lucier, G.W., and Portier, C.J. (1996). A mechanistic model of effects of dioxin on thyroid hormones in rat. *Toxicol. Appl. Pharmacol.* 136: 29–48.

Kopp, P., Kitajima, K., and Jameson, J.L. (1996). Syndrome of resistance to thyroid hormone: Insights into thyroid hormone action. *Proc. Soc. Exp. Biol. Med.* 211: 49–61.

Lanni, A., Moreno, M., Horst, C., Lombardi, A., and Goglia, F. (1994). Specific binding sites for 3:3'-diiodo-L-thyronine (3,3'-T2) in rat liver mitochondria. *FEBS Lett.* 351: 237–240.

Larsson, M., Pettersson, T., and Carlström, A. (1985). Thyroid hormone binding in serum of 15 vertebrate species: Isolation of thyroxine-binding globulin and prealbumin analogs. *Gen. Comp. Endocrinol.* 58: 360–375.

Leatherland, J.F. (1993). Field observations on reproductive developmental dysfunction in introduced and native salmonids from the Great Lakes. *J. Great Lakes Res.* 19: 737–751.

——— (1994). Reflections on the thyroidology of fishes: From molecules to humankind. *Guelph Ichthyol.* Rev. 2: 1–67.

——— (1998). Changes in thyroid hormone economy following consumption of environmentally contaminated Great Lakes fish. *Toxicol. Indust. Health* 14: 41–57.

———, Down, N.E., Falkmer, S., and Sonstegard, R.A. (1998). Endocrine glands and reproductive system. In *Pathobiology of Spontaneous and Induced Neoplasms in Fishes: Comparative Characterization, Nomenclature and Literature.* (C. Dawe and J. Harshbarger, eds.), Academic, New York (in press).

——— and Sonstegard, R.A. (1978). Lowering of serum thyroxine and tri-iodothyronine levels in yearling coho salmon, *Onchorhynchus kisutch*, by dietary Mirex and PCB's. *J. Fish. Res. Bd. Can.* 35: 1285–1289.

——— and ——— (1979). Effect of dietary Mirex and PCB (Aroclor 1254) on thyroid activity and lipid reserves in rainbow trout, *Salmo gairdneri* Richardson. *J. Fish Dis.* 2: 43–48.

——— and ——— (1980). Effect of dietary Mirex and PCB's in combination with food deprivation and testosterone administration on thyroid activity and bioaccumulation of organochlorines in rainbow trout *Salmo gairdneri* Richardson. *J. Fish Dis.* 3: 115–124.

——— and ——— (1982). Bioaccumulation of organochlorines by yearling coho salmon (Oncorhynchus kisutch Walbaum) fed diets containing Great Lakes coho salmon, and the pathophysiology of the recipients. Comp. Biochem. Physiol. 72C: 91–99.

Licht, P., Denver, R.J., and Herrera, B.E. (1991). Comparative survey of blood thyroxine binding proteins in turtles. J. Exp. Zool. 259: 43–52.

——— and Pavgi, S. (1992). Identification and purification of a high-affinity thyroxine-binding protein that is distinct from albumin and prealbumin in the blood of a turtle, *Trachemys Scripta. Gen. Comp. Endocrinol.* 85: 179–192.

Lissitzky, S. (1990). Thyroid hormones. In *Hormones from Molecules to Disease.* (E.-E. Baulieu and P.A. Kelly, eds.), pp. 340–374. Chapman & Hall, New York.

Lucier, G.W., Portier, C.J., and Gallo, M.A. (1993). Receptor mechanisms and dose-response models for the effects of dioxins. *Environ. Health Perspec.* 101: 36–44.

McKinney, J.D. (1989). Multifunctional receptor model for dioxin and related compound toxic action: Possible thyroid hormone-responsive effector-linked site. *Environ. Health Perspec.* 82: 323–336.

———, Chae, K., Oatley, S.J., and Blacke, C.C.F. (1985a). Molecular interactions of toxic chlorinated dibenzo-*p*-dioxins and dibenzofurans with thyroxine binding prealbumins. *J. Med. Chem.* 28: 375–381.

———, Fawkes, J., Jordan, S. Chae, K. Oatley, S., Coleman, R.E., and Briner, W. (1985b). 2,3,7,8-Tetrachlorodibenzo-*p*-dioxin (TCDD) as a potent and persisten thyroxine agonist: A mechanistic model for toxicity based on molecular reactivity. *Environ. Health Perspec.* 61: 41–53.

———, Fannin, R., Jordan, S., Chae, K., Richenbacher, U., and Pedersen, L. (1987). Polychlorinated biphenyls and related compound interactions with specific binding sites for thyroxine in rat liver nuclear extracts. *J. Med. Chem.* 30: 79-86.

—— and Pedersen, L.G. (1987). Do residue levels of polychlorinated biphenyls (PCBs) in human blood produce mild hypothyroidism? *J. Theor. Biol.* 129: 231–241.

—— and Waller, C.L. (1994). Polychlorinated biphenyls as hormonally active structural analogues. *Environ. Health Perspec.* 102: 290–297.

Moccia, R.D., Fox, G.A., and Britton, A. (1986). A quantitative assessment of thyroid histopathology of herring gulls (*Larus argentatus*) from the Great Lakes and a hypothesis on the causal role of environmental contaminants. *J. Wildl. Dis.* 22: 60–70.

Morse, D.C., Groen, D., Veerman, M., Van Amerongen, C.J., Köeter, H.B. W.M., Smits-Van Prooije, A.E., and Brouwer, A. (1993). Interference of polychlorinated biphenyls in hepatic and brain thyroid hormone metabolism in fetal and neonatal rats. *Toxicol. Appl. Pharmacol.* 122: 27–33.

———, Koeter, H.B.W.M., Smits van Prooijen, A.E., and Brouwer, A. (1992). Interference of polychlorinated biphenyls in thyroid hormone metabolism: Possible neurotoxic consequences in fetal and neonatal rats. *Chemosphere* 25: 165–168.

Ness, D.K., Schantz, S.L., Moshtaghian, J., and Hansen, L. (1993). Effects of perinatal exposure to specific PCB congeners on thyroid hormone congeners on thyroid hormone concentrations and thyroid histology in the rat. *Toxicol. Lett.* 68: 311–323.

Oppenheimer, J.H., Schwartz, H.L., and Strait, K.A. (1995). An integrated view of thyroid hormone action *in vivo*. In *Molecular Endocrinology: Basic Concepts and Clinical Correlations.* (B.D. Weintraub, ed.), pp. 249–267. Raven, New York.

Palha, J.A., Hays, M.T., Morreale de Escobar, G., Episkopou, V., Gottesman, M.E., and Saraiva, M.J.M. (1997). Transthyretin is not essential for thyroxine to reach the brain and other tissues in transthyretin-null mice. *Am. J. Physiol.* 272 (*Endocrinol. Metab.* 35): E485–E493.

Park, E.H., Chang, H.H., Lee, K.C., Kweon, H.S. Heo, OS., and Ha, K.W. (1993). High frequency of thyroid tumor induction by N-methyl-N¢-nitro-N-nitrosoguanidine in the hermaphroditic fish Rivulus marmoratus. *Jap. J. Cancer Res.* 84: 608–615.

Pazdernik, T.L., and Rozman, K.K. (1985). Effect of thyroidectomy and thyroxine on 2,3,7,8-tetrachlorodibenzo-*p*-dioxin-induced immunotoxicity. *Life Sci.* 36: 695–703.

Porterfield, S.P. (1994). Vulnerability of the developing brain to thyroid abnormalities: environmental insults to the thyroid system. *Environ. Health Perspec.* 102 (Suppl. 2): 125–130.

———— and Hendrich, C.E. (1993). The role of thyroid hormones in prenatal and neonatal neurological development current perspectives. *Endocr. Rev.* 14: 94–106.

Raasmaja, A., Viluksela, M., and Rozman, K.K. (1996). Decreased liver type I 5′-deiodinase and increased brown adipose tissue type II 5′-deiodinase activity in 2,3,7,8-tetrachlorodibenzo-*p*-dioxin (TCDD)-treated Long-Evans rats. *Toxicology* 114: 199–205.

Rall, E. (1995). Posttranscriptional effects of thyroid hormones. In *Molecular Endocrinology: Basic Concepts and Clinical Correlations.* (B.D. Weintraub, ed.), pp. 241–247. Raven, New York.

Refetoff, S, Robin, N.I., and Fang, V.S. (1970). Parameters of thyroid function in serum of 16 selected vertebrate species: A study of PBI, serum T_4, free T_4, and the pattern of T_4 and T_3 binding to serum proteins. *Endocrinology* 86: 793–805.

Rickenbacher, U., McKinney, J.D., Oatley, S.J., and Blake, C.C.F. (1986). Structurally specific binding of halogenated biphenyls to thyroxine transport protein. *J. Med. Chem.* 29: 641–647.

Robbins, J., and Edelhoch, H. (1986). Thyroid hormone transport proteins: Their nature, biosynthesis, and metabolism. In *Werner's The Thyroid, 5th ed.* (S.H. Ingbar and L.E. Braverman, eds.), pp. 116–127. Lippincott, Philadelphia.

Seegal, R.F. (1996). Epidemiological and laboratory evidence of PCB-induced neurotoxicity. *Crit. Rev. Toxicol.* 26: 709–737.

Serrano-Lozano, A., Montiel, M., Morell, M., and Morata, P. (1993). 5′ Deiodinase activity in brain regions of adult rats: Modifications in different situations of experimental hypothyroidism. *Brain Res. Bull.* 30: 611–616.

Sewall, C.H., Flagler, N., Vanden Heuvel, J.P., and Lucier, G.W. (1995). Alterations in thyroid function in female Sprague-Dawley rats following chronic treatment with 2,3,7,8-tetrachlorodibenzo-*p*-dioxin. *Toxicol. Appl. Pharmacol.* 132: 237–244.

Silva, J.E., and Larsen, P.R. (1982). Comparison of iodothyronine 5′-deiodinase and other thyroid-hormone-dependent enzyme activities in the cerebral cortex of hypothyroid neonatal rats: Evidence for adaptation to hypothyroidism. *J. Clin. Invest.* 70: 1110–1123.

Spear, P.A., Higueret, P., and Garcin, H. (1990). Increased thyroxine turnover after 3,3′,4,4′,5,5′-hexabromobiphenyl injection and lack of effect on peripheral triiodothyronine production. *Can. J. Physiol. Pharmacol.* 68: 1079–1084.

——, ——, and —— (1994). Effects of fasting and 3,3′,4,4′,5,5′-hexabromobiphenyl on plasma transport of thyroxine and retinol: Fasting reverses elevation of retinol. *J. Toxicol. Environ. Health* 42: 173-183.

—— and Moon, T.W. (1985). Low dietary iodine and thyroid anomolies in ring doves, *Streptopelia risoria*, exposed to 3,4,3′,4′-tetrachlorobiphenyl. *Arch. Environ. Contam. Toxicol.* 14: 547–553.

Taurog, A. (1970). Thyroid peroxidase and thyroxine biosynthesis. *Rec. Progr. Hormone Res.* 26: 189–241.

van Birgelen, A.P.J.M., van der Kolk, J., Poiger, H., van den Berg, M., and Brouwer, A. (1992). Interactive effects of 2,2′,4,4′,5,5′-hexachlorobiphenyl and 2,3,7,8-tetra-chlorodibenzo-*p*-dioxin on thyroid hormone, vitamin A, and vitamin K metabolism in the rat. *Chemosphere* 25: 1239–1244.

Van den Berg, K.J., Zurcher, C., and Brouwer, A. (1988). Effects of 3,4,3′,4′-tetrachlorobiphenyl on the thyroid function and histology in marmoset monkey. *Toxicol. Lett.* 41: 77–86.

Villeneuve, D.C., Valli, V.E., Chu, I., Secours, V., Ritter, L., and Becking, G.C. (1979). Ninety-day toxicity of photomirex in the male rat. *Toxicology* 12: 235–250.

Wagner, R.L., Apriletti, J.W., McGrath, M.E., West, B.L., Baxter, J.D., and Fletterick, R.J. (1995). A structural role for hormone in the thyroid hormone receptor. *Nature, Lond.* 378: 690–697.

Weiss, R.E., Hayashi, Y., Nagaya, T., Petty, K.J., Murata, Y., Tunca, H., Seo, H., and Refetoff, S. (1996). Dominant inheritance of resistance to thyroid hormone not linked to defects in the thyroid hormone receptor alpha or beta genes may be due to a defective cofactor. *J. Clin. Endocrinol. Metab.* 81: 4196–4203.

Wondisford, F.E. (1996). Thyroid hormone beyond the receptor. *J. Clin. Endocrinol. Metab.* 81: 4194–4195.

Williams, G.R., and Brent, G.A. (1995). Thyroid hormone response elements. In *Molecular Endocrinology: Concepts and Clinical Correlations.* (B.D. Weintraub, ed.), pp. 217–239. Raven, New York.

Wren, C.D., Hunter, D.B., Leatherland, J.F., and Stokes, P.M. (1987a). The effects of polychlorinated biphenyls and methylmercury, singly and in combination, on mink. I: Uptake and toxic responses. *Arch. Environ. Contam. Toxicol.* 16: 441–447.

——, ——, ——, and —— (1987b). The effects of polychlorinated biphenyls and methylmercury, singly or in combination on mink. II: Reproduction and kit development. *Arch. Environ. Contam. Toxicol.* 16: 449–454.

Xiao, S., Dorris, M.L., Rawitch, A.B., and Taurog A. (1996). Selectivity in tyrosyl iodination sites in human thyroglobulin. *Arch. Biochem. Biophys.* 334: 284–294.

Yoshida, K., Sugihira, N., Suzuki, M., Sakurada, T., Saito, S., Yoshinaga, K., and Saito, H. (1987). Effect of cadmium on T_4 outer ring monodeiodination by rat liver. *Environ. Res.* 42: 400–405.

Zhang, X.K., and Pfahl, M. (1993). Regulation of retinoid and thyroid hormone action through homodimeric and heterodimeric receptors. *Trends Endocrinol. Metab.* 4: 156–162.

Chapter 7

BIOMARKERS OF IMMUNOTOXICITY: AN EVOLUTIONARY PERSPECTIVE

Michel Fournier,[1,4] Pauline Brousseau,[2] Helen Tryphonas,[3] and Daniel Cyr[4]

[1] Département des Sciences
   Biologiques
   Université du Québec à Montréal
   Montréal, Québec

[2] Department of Biological
   Sciences
   Concordia University
   Montreal, Quebec

[3] Toxicolology Research Division
   Bureau of Chemical Safety
   Food Directorate, Health
   Protection Branch, Health Canada
   Ottawa, Ontario

[4] Department of Fisheries
   and Oceans
   Institut Maurice-Lamontagne
   Mont-Joli, Quebec

Introduction

The immune system, developed during evolution through interactions between hosts and infectious agents, uses its effector mechanisms against all potentially pathogenic agents: viruses and other microorganisms, parasites and neoplasic cells (Roitt et al., 1993). Without this system, which defines and determines the "self" and the "nonself," animal life would likely not exist or, at least, not exist as we know it. The considerable importance of the immune system justifies the work devoted to immunotoxic phenomena and their consideration in evaluating the risk of chemical substances.

The immune system is directly integrated into the organism and interacts intimately with the functioning of many organs and organ systems. It is extremely vulnerable throughout to injury by xenobiotics. Although major changes in the immune system are rapidly expressed in significant morbidity, and even mortality of the organisms involved, they are often preceded by subtle changes in some of its components that could be used as early indicators of toxicity or as biomarkers (Dean et al., 1982). This is all the more interesting because these effects generally occur at levels that are lower than those causing acute toxicity (Koller and Exon, 1985).

This chapter presents a brief summary of the structure and function of the immune system in selected species, a description of the main immunotoxic effects of environmental contaminants, and a review of potentially useful immunologic biomarkers in different groups of organisms including invertebrates (earthworms and molluscs) and vertebrates (fish and marine mammals).

Structure and Function of the Immune System

The more evolved an animal, the more complex is its immune system (Table 7-1). Schematically, two forms of immunity can be distinguished, namely, nonspecific immunity and specific immunity.

Nonspecific Immunity

Nonspecific immunity is based on the involvement of specialized cells that recognize foreign bodies without any specificity. It basically consists of two types of responses: phagocytosis or the ingestion and destruction of foreign bodies, and the inflammatory response or the cascade of reactions that occur as a response to a localized injury or infection, which, among others, involve cells that are capable of phagocytosis. Phagocytosis and the inflammatory response can also be promoted by certain elements produced by other cells (opsonins or lymphokines produced by lymphocytes in mammals). Depending on the species, the cells responsible for nonspecific immunity can be circulating cells (neutrophil granulocytes in mammalian blood), cells present in organs or tissues where they can be bound (hepatic or splenic macrophages in mammals, for example), or mobile (pulmonary macrophages in mammals), with different types of cells most commonly present simultaneously. The cells involved in the inflammatory response also have the characteristic of being infiltrating, that is, of penetrating into a tissue or an organ from which they are normally absent.

In acoelomate invertebrates that do not have a circulatory system, phagocytosis is accomplished by mobile cells, the amebocytes. In coelomate invertebrates, two cell types are involved, hemocytes in the blood (or its analog) and coelomocytes in the coelom (Grassé et al., 1970).

Certain cells involved in nonspecific immunity, the natural killer cells (NK) and natural cytotoxic cells (NCC), are able to recognize the changes that occur on the surface of cells infected with certain viruses and tumor cells, and to spontaneously induce lysis of these cells. In vertebrates, these are located in the blood and spleen. It appears that a primitive form of this NK activity exists in some invertebrates (Franceschi et al., 1991; Porchet-Hennere et al., 1992), but very little work has been devoted to it.

Specific Immunity

The specific immune response is organized around two major components of the immune system conveniently grouped into cellular immunity (*cell-mediated immunity*, CMI) and humoral immunity (*humoral-mediated immunity*, HMI). In vertebrates, cellular immunity is essentially based on specific cells, or T lymphocytes. It is involved in the regulation of immune system function, in delayed hypersensitivity, in graft versus host reactions, and in resistance to infection. Humoral immunity is linked to the production of antibodies by specialized cells, or B lymphocytes. Five classes of antibodies, or immunoglobulins, have been identified in vertebrates (Table 7-1). Their role is basically to protect the organism from infectious diseases through different mechanisms: antigen neutralization, opsonization, cytotoxic lymphocyte binding, binding to the complement system, etc. After maturation in the primary lymphoid organs (thymus for T lymphocytes, bursa of Fabricius or equivalent organ for

Table 7-1. Characteristics of immune systems in invertebrates and the different groups of vertebrates (simplified from Weeks et al., 1992)

Group	Categories of immunoglobulins	Lymphoid tissues	Importance of the role of macrophages
Invertebrates	—	—	High
Fish	IgM	thymus, spleen, kidney, liver	High
Amphibians	analog of IgG, IgM	thymus, spleen, kidney, lymph nodes,[a] bone marrow[a]	Moderate
Reptiles	IgG, IgM	thymus, spleen, kidney, lymph node analogs, bone marrow, liver	Moderate
Birds	analog of IgG, IgM, IgA	thymus, bursa of Fabricius,spleen, lymph nodes, bone marrow	Low
Mammals	IgG, IgM, IgA, IgE, IgD	thymus, bone marrow, spleen, lymph nodes, liver (in the fetus)	Moderate

[a] In anuran amphibians.

B lymphocytes), the B and T lymphocytes migrate by means of the circulation to the secondary lymphoid organs such as the lymph nodes or the spleen, where they are likely to come into contact with an antigen carried by the lymph or the blood cells (Roitt et al., 1993).

With invertebrates, various experiments, particularly with allografting and/or xenografting, have shown the existence of signs of cellular immunity in spongiae (Humphreys and Reinherz, 1994), cnidaria (Bigger, 1980; Lubbock, 1980), annelids (Cooper, 1968, 1969), echinoderms (Karp and Hildemann, 1976), and insects (Thomas and Ratcliffe, 1982; Karp, 1990). As for humoral immunity, no immunoglobulin has ever been detected in the invertebrates. On the other hand, there are proteins that have an analogous role, the agglutinins (Roitt et al., 1993), whose existence has been demonstrated in all groups of invertebrates and whose specific role is not always clearly understood, even though they are involved in the recognition of the nonself by hemocytes in mollusks (Mullainadhan and Renwrantz, 1986) and are increasingly synthesized when foreign elements are introduced into certain annelids (Stein et al., 1982; Wojdani et al., 1982; Laulan et al., 1985).

Despite the apparent disparity of the mechanisms involved in the immune response, some areas of agreement nonetheless emerge. In particular, all animals have specialized cells in one form or another to fight against foreign elements (immunocompetent cells or immunocytes). As a result, research on immunotoxic biomarkers has basically focused on the study of these cells and their response to toxic agents.

Influence of Hormones on the Immune System

The immune system of vertebrates is highly sensitive to endocrine influences. As such, it is likely that endocrine disrupters can exert considerable effects on immune function by indirectly or directly altering any aspect of endocrine-immune interactions.

Glucocorticoids

Glucocorticoids are widely used as both immunosuppressive agents and as anti-inflammatory agents. The immunosuppressive effects of glucocorticoids occur at many different levels such as macrophage activation, inhibition of cytokine production by T lymphocytes, retrafficking of leukocytes, and induction of apoptosis (programmed cell death) in T and B cell progenitors and in mature T cells. Apoptosis is an important physiological process in the immune system which eliminates unwanted subpopulations of T cells. In thymocytes, the process of apoptosis is stimulated by dexamethasone, an analogue of cortisol.

Dexamethasone induction of apoptosis results initially in a stimulation of clusterin (Bertuzzi et al., 1991). While the role of clusterin in apoptosis has been the subject of debate, it is expressed early in apoptotic cells and may be a mechanism of protection by apoptotic to escape recognition by the immune system (Tomei, 1991). Recent studies have reported that interleukin (Il)-2, -4, and -9 can all inhibit dexamethasone-induced apoptosis in cultured T cells. Interestingly, only Il-4 and Il-9 could continue to stimulate cell division in the presence of dexamethasone, suggesting the presence of different cellular mechanisms that operate for either cell growth and for apoptosis (Louahed et al., 1996).

In rainbow trout, acute stress has been associated with a decrease in phagocytic activity by the macrophages of the pronephros (Narnaware et al., 1994). Both cortisol and norepinephrine injections prevented stress-induced decreases in phagocytosis (Narnaware and Baker, 1996). Since acute stress results in a decline in numbers of lymphocytes, an effect opposed by cortisol and norepinephrine, it has been proposed that stress may cause a retrafficking of immune cells towards other organs and that cortisol and epinephrine act to prevent this type of redistribution of immune cells (Naranaware and Baker, 1996).

Sex Steroids

It has long been established that there are important differences in immune function between males and females (Grossman, 1985; Olsen and Kovacs, 1996). Clazori (1889) first observed that the thymus gland of rabbits castrated prior to puberty were larger than in intact controls. In humans, autoimmune diseases such as rheumatoid arthritis, lupus erythematosus, Hashimoto's thyroiditis, and biliary cirrhosis are all more prevalent in women than in men and are indicative of an increased sensitivity due to female sex hormones or the result of protective effects of androgens (Grossman, 1985; Olsen and Kovacs, 1996). Studies have also shown that females contain much higher titres of immunoglobulin than men—all of which suggest that sex steroids play an important role in immune function.

Androgens

A role for androgen in immune function has been suggested by observations that hypogonadal males, due to Klinefelter's syndrome, have a higher tendency to develop autoimmune diseases such as lupus erythematous (Bizzaro et al., 1987; Olsen and Kovacs, 1995). The activity of lupus erythematous in these hypogonadal males can be decreased if they are administered androgens (Olsen and Kovacs, 1995).

The mammalian thymus gland contains receptors for androgens, estrogens, and progestins. Ochidectomy in several mammalian species causes an increase in the size of the thymus glands (Henderson, 1904; Fitzpatrick et al., 1985; Olsen et al., 1991). In mice, the increase in thymus size after orchidectomy results in part from an increased proliferation of immature thymocytes (Olsen et al., 1994). This effect occurs rapidly following orchidectomy, but eventually the proliferative index returns to normal while the weight of the thymus remains elevated. The mechanism of action of androgens on the thymocytes appears to be mediated via androgen receptors whose concentration are similar to those found in androgen target tissues of the male reproductive tract (Viselli et al., 1995). Interestingly, however, the addition of testosterone or its active metabolite dihydrotestosterone in the presence of serum from intact males do not stimulate rat thymocyte cell division *in vitro* unless the cells are cultured in the presence of serum from orchidectomized rats.

In birds, the central organ for lymphocyte formation is the bursa of Fabricius. In Japanese quail, orchidectomy delays the pubertal onset of bursal involution, whereas testosterone implants stimulate the process (Mase and Oishi, 1991). In mammalian bone marrow, orchidectomy increases pre–B cell populations, whereas treatment of androgen decreases this population, suggesting a role for androgens in maturation of B cells (Wilson et al., 1995).

In mammals, peripheral B cells are also influenced by androgens. Orchidectomy results in an increased spleen weight that contains an increased B cell population (Viselli et al., 1995). Interestingly, peripheral B cells do not appear to contain androgen receptors, thereby suggesting an indirect effect of androgen on the peripheral B cells.

Estrogens

Females are more likely to develop autoimmune diseases than their male counterparts (Grossman, 1985; Olsen and Kovacs, 1996). Animal models such as the NZB/NZW F1 mouse hybrid, which develops an autoimmune disease similar to human lupus erythematous (Roubinian et al., 1977), and the NOD mouse, which develops Type I diabetes, also exhibit sexual dimorphism. The effects of estradiol on the progression of these autoimmune diseases is not clear. It has been suggested that supra-physiological doses of estradiol may cause toxicity, while physiological doses may not accelerate the progression of the autoimmune diseases in the NZB/NZW and NOD mice, but that the sexual dichotomy associated with these diseases is in fact the absence of testosterone (Verheul et al., 1995). Walker et al. (1994) have reported that the administration of the antiandrogen flutamide to NZB/NZW mice accelerates the time of death, supporting the hypothesis that it is the lack of androgen that is responsible for the gender-specific sensitivities.

The actions of estrogens on the immune system are complex and differ as a result of the concentration used and state of maturity of the animal. As with androgens, estradiol stimulates the involution of the thymus by decreasing the number of immature thymocytes while increasing the number of mature thymocytes (Phuc et al., 1991a,b; Okuyama et al., 1992). Therefore, while gonadectomy increases thymic weight in both genders, the absence of estradiol increases the number of mature thymocytes with fewer immature thymocytes. In contrast, androgen removal increases the number of immature thymocytes, thereby demonstrating important differences in the site of action of the two sex steroids (Olsen and Kovacks, 1996).

Many changes to the immune system that occur during pregnancy are caused by estradiol. Estradiol receptors have been identified in both thymocyte populations, the thymal epithelial cells, and the lymphocytes (Weustein et al., 1986; Jakob et al., 1992). In bone marrow, ovariectomy results in an increase in B cells, an effect that can be reversed with estradiol administration (Masuzawa et al., 1994). During pregnancy when estradiol levels are elevated, there is a suppression of B lymphopoesis (Medina et al., 1993). Since the maturation of B cells is influenced by IL-7, it is likely the estradiol effects occur by modulating the actions of IL-7 (Medina et al., 1993). A list of cytokines and the effects of sex steroids are outlined in Table 7-2. Due to the role of cytokines in modulating apoptosis in T cells, estradiol may also modulate apoptosis via its effects on the cytokines.

In goldfish, incubation of macrophages with either estradiol or cortisol inhibited chemotaxis and phagocytosis (Wang and Belosevic, 1995). While estradiol had no effects on nitric oxide production, cortisol was strongly inhibitory, suggesting different mechanisms of action for the two steroid hormones. Estradiol administration *in vivo* to goldfish infected with a haemoflagellate *(Trypanosoma danilewski)* significantly inhibited the proliferation of peripheral lymphocytes (Wang and Belosivic, 1994).

Progestins

Both estradiol and progesterone are important immunosuppressive agents during pregnancy. One of the first lines of defense of the immune system are the mononuclear cells, or macrophages. During pregnancy, the activity of these mononuclear cells is decreased by progesterone, which is produced by

Table 7-2. Effect of sex steroids on cytokine production by immune cells

Cytokine	Origin	Function	Steroid effect
Interleukin-1	Monocytes/Macrophages	Inflammatory response	Decrease by androgens
Interleukin-4	T cells	Differentiation of B cells	Increased by estradiol
TGF-B		Antiproliferation of T cells and thymocytes	Increased by androgen
Interleukin-5	Activated thymocytes	B cell growth	Increased or decreased by androgen
Interferon-y	Thymocytes	Activation, MHC I and II induction.	Increased by estradiol, decreased by androgen.

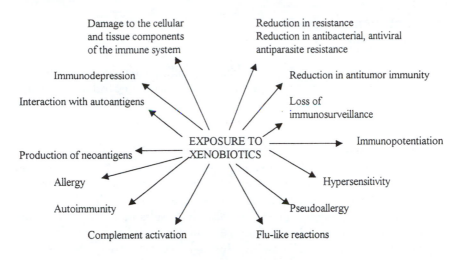

Figure 7-1.
Potential immunotoxic effects of xenobiotics (modified from Wong et al., 1992).

the trophoblast (Feinberg et al., 1992). This immunosuppressive effect of progesterone is a vital step in allowing the fetus to survive and protecting it from the mother's immune cells. It has been observed that during pregnancy, uterine macrophage levels of such cell activators as nitric oxide and TGF-B are suppressed. This inhibition in TGF-B and nitric oxide production appears regulated by progesterone, which is a strong inhibitor of nitric oxide synthetases. Interestingly, however, macrophages do not possess progesterone receptors. In macrophages, glucocorticoids stimulate the production of the I capa B gene. Studies with progesterone also resulted in increased I capa B gene expression, thereby suggesting that progesterone acts on the macrophages via the glucocorticoid receptor pathway.

Neuroendocrine Hormones

The immune system can synthesize neuroendocrine hormones that are produced by the adenohypophysis. Peripheral blood and spleen lymphocytes, as well as thymocytes, can synthesize prolactin (PRL) and the PRL receptor. Activated T cells have an increased number of PRL receptors, which has led to the suggestion that PRL is involved in T cell activation (reviewed by Savino and Dardenne, 1995). Epithelial cells of the thymus can also produce growth hormone (GH), which can stimulate the differentiation and migration of thymocytes. The effects of GH are mediated by insulin growth factor I (IGF-I). IGF-I administration to mice also stimulates the production of B cells in bone marrow, suggesting that IGF-I stimulates both T and B cell lymphopoietic organs (Savino and Dardenne, 1995).

Pollutant Immunotoxicity

General Mechanisms

Various medicines, chemicals, and industrial pollutants, more generally called xenobiotics, are known to alter immune function with, depending on the case, immunodepression or, conversely, immunostimulation (Dean and Murray, 1990; Wong et al., 1992). Figure 7-1 summarizes the possible consequences of these two phenomena.

Classically, chemical immunodepression is manifested as a reduction in resistance to microbial and viral infections, and as increased incidence of some cancers. Also, the interaction of the immune system with xenobiotic detoxification and biotransformation reactions induces a reduction in the elimination of chemicals in immunodeficient individuals. The results of chemical immunostimulation are more complex, and are generally characterized by the induction of a qualitatively abnormal response. Several hypersensitivity reactions have been identified in humans and laboratory animals: an increase in the frequency of allergic phenomena, pseudoallergic reactions, asthma-type pulmonary manifestations, flu-like reactions, autoimmune hemolytic anemias, and other autoimmune diseases (Exon et al., 1987; Luster et al., 1992, 1993, 1994; Revillard, 1994). However, laboratory exposure accompanied by immunologic changes alone is not enough for the xenobiotic involved to be immunotoxic under natural exposure conditions. Thus, the dose levels and exposure pathways, as well as the severity of a particular effect, must be taken into account.

Mode of Action of Immunotoxic Contaminants

The immunotoxic effects of chemicals include effects at the cellular level and on the organs of the immune system (Descotes, 1988; Gleichmann et al., 1989; Luster et al., 1988, 1992, 1993, 1994; Trizio et al., 1988; Wong et al., 1992).

Of the environmental pollutants, dioxins, furans, polycyclic aromatic hydrocarbons (PAHs), polybrominated biphenyls (PBB), polychlorinated biphenyls (PCB), alkylating substances, and organochlorine insecticides are immunodepressive xenobiotics (Bradley and Morahan, 1982; Crocker et al., 1976; Dean et al., 1986; Fournier et al., 1986; Penn, 1987; Krzystyniak et al., 1989; Tryphonas et al., 1989, 1991).

The immunotoxic effect of tetrachlorodibenzo-p-dioxin (TCDD) is mediated through the binding of TCDD to an intracellular protein, in this case the Ah receptor (Poland et al., 1976; Poland and Knutson, 1982). The complex thus formed would interact with specific sites in the nuclear DNA (Cuthill et al., 1988; Whitlock, 1990). PCB immunotoxicity was recently reviewed (Tryphonas, 1995). The mechanisms by which the majority of PCB congeners would exert their toxic potential are comparable to those of the dioxins. However, it is very interesting to note that the immunotoxic potential of PCBs is strongly affected by the level of chlorination of the molecule.

Heavy metals are generally considered as immunotoxic and immunodepressive agents (Lawrence, 1985). They produce effects ranging from the loss of resistance to infection to the development of autoimmune diseases. Also, some of them can act on the immune system at doses that do not produce any other sign of toxicity (Exon, 1984). The effects of these metals have been the subject of several reviews (Wong et al., 1992; Bernier et al.,1995). The cellular and molecular mechanisms by which heavy metal immunotoxicity is expressed vary. Occurring first are processes of cytotoxicity and apoptosis. Heavy metals could also exert their toxic effect by interfering with intercellular contact processes. Thus, cadmium and lead are known to inhibit the interaction of lymphocyte subpopulations (Prozialeck and Niewenhuis, 1991; McCabe and Lawrence, 1991). Heavy metals are also known to affect intracellular signaling mechanisms, calcium mobilization, protein synthesis, or gene expression (Chavez et al., 1985; Cifone et al., 1989, 1990; Noelle and Snow, 1990; Wan et al., 1993; Lachapelle et al., 1993).

Several other environmental contaminants with immunotoxic potential exist. These have been recently reviewed by Wong et al. (1992) and Thomas (1995). However, for most of them, the data available are incomplete. A number of them were studied using doses or routes of administration that are not at all related to natural exposure conditions. This was particularly the case for insecticides and herbicides.

Regardless of the contaminant considered, one should note that there are differences in sensitivity among individuals (age, sex, etc.) and among species. Furthermore, only very limited epidemiologic data involving accidental exposure to these chemicals exist (Revillard, 1994). These factors are real obstacles when extrapolation from experimental animals to humans is attempted. Thus, the magnitude of the risk environmental pollutants pose to human health remains unknown.

Immunologic Biomarkers

Table 7-3 depicts the different levels at which biomarkers for immune system changes can be investigated. Given the structural and functional complexity of the immune system, it is difficult to draw conclusions or to make predictions about a possible dysfunction in the general immune response by investigating one parameter in the cascade of immune reactions.

The essential part of immunotoxicologic work is the evaluation of the immunotoxicity of certain substances on vertebrates and, more particularly, on mammals, through different laboratory tests (Schüürman et al., 1984; Vos and Van Loveren, 1987; Luster et al., 1988, 1992, 1994; Hinton, 1992), and the approach used is most often a sequential approach (Koller and Exon, 1985; Luster et al., 1988; Weeks et al., 1992). It should be noted that tests on some

Table 7-3. Complexity of the immune system

Structure of the immune system	Lymphoid organs (spleen, thymus, etc.) Circulating leucocytes (total or relative number) Plasma protein fraction (total or specific)
Nonspecific immunity	Phagocytosis Inflammatory response Killer cells (NK, NCC)
Specific immunity	Antibody response (circulating antibody levels, number of secreting cells) Cellular response (graft versus host reaction, lymphoproliferative responses to mitogen and antigen stimulation)
Functions regulated by the immune system	Susceptibility to infections Appearance of tumors

types of wildlife (duck, rainbow trout) or sentinel species (earthworms) are included in these approaches.

Such tests are commonly used for the evaluation of potentially immunotoxic new chemicals and for determining the mode of action of these and other chemicals on the immune system. However, it is very difficult and even impossible to apply these tests to organisms under toxic stress in the natural environment. Thus, for the most part, application of these tests has been restricted to the laboratory environment.

Besides laboratory mammals, where determining immunotoxic markers is based more on toxicology than on ecotoxicology, the groups of organisms where data are available on the use of immunologic biomarkers are earthworms, bivalve mollusks, fish, and marine mammals.

Immunologic Biomarkers in Invertebrates

With respect to immunologic biomarkers in invertebrates, oligochaete annelids (earthworms) and mollusks have been investigated to the greatest extent.

Immunologic Biomarkers in Oligochaete Biomarkers

Earthworms are frequently used in standard toxicologic study protocols (Fitzpatrick et al., 1990; Reinecke, 1992; Furst et al., 1993; Callahan et al., 1994). Also, their immune responses are somewhat better known than those of the other invertebrates, and they are a relatively inexpensive animal model. All these characteristics, in addition to their major ecological interest, make them a model of choice for research on immunologic biomarkers.

The effects of different environmental contaminants on the immune system of earthworms have been studied on two model species, *Eisenia foetida* and *Lumbricus terrestris*. Two complementary approaches are used, *in vitro* tests on isolated coelomocytes and *in vivo* tests on whole worms. One of the advantages of the coelomocytes of earthworms is that they can be easily obtained using noninvasive extrusion methods (Eyambe et al., 1991).

Immunological biomarkers in earthworms consist of:

a. Coelomocyte counts and characterization and the potential of differentiation of new coelomocytes (Eyambe et al., 1991; Goven et al., 1993, 1994b). There are different types of coelomocytes that can be characterized by their affinity for certain dyes. Acidophil coelomocytes are responsible for phagocytosis, whereas basophilic coelomocytes are involved in the mechanisms of humoral immunity (Hostetter and Cooper, 1974; Stein et al., 1977; Mohrig et al., 1984)

b. Determination of lysozyme activity using coelomic liquid (Ma, 1982; Lasalle and Lassegues, 1986; Lasalle et al., 1988; Valembois et al., 1986; Goven et al., 1993, 1994b)

c. Phagocytosis (Rodriguez et al., 1989; Eyambe, 1991; Fitzpatrick et al., 1992; Goven et al., 1994a)

d. The production of free radicals of oxygen (Chen et al., 1991; Goven et al., 1994a)

e. The formation of agglutinin-secreting rosettes (Rodriguez et al., 1989)

Exposure of worms to a mixture of PCBs, or Aroclor 1254, is accompanied by qualitative and quantitative changes in the coelomocytes (Goven et al., 1993, 1994b), including a reduction in phagocytotic activity (Fitzpatrick et al., 1992; Ville et al., 1995) and an increase in graft versus host reaction (Cooper and Roch, 1992). Exposure to nonlethal concentrations and to levels normally found in vertebrates living in the natural environment induce inhibition of the formation of agglutinin-secreting rosettes (Rodriguez et al., 1989; Fitzpatrick et al., 1990). Furthermore, exposure to Aroclor 1254 inhibits the natural killer cell activity in the coelomocytes (Suzuki et al., 1995) and increases their susceptibility to bacterial infections (Roch and Cooper, 1991). As for metals, exposure to copper sulfate results in a reduction in lysozyme activity (Goven et al., 1993, 1994b). Various heavy metals induce a reduction in phagocytotic activity following in vivo exposure (Fugère et al., 1995) and a reduction in the production of free radicals of oxygen following in vitro exposure (Chen et al., 1991; Goven et al., 1994a).

It is interesting to note that, as with many other species (Krzystyniak et al., 1989; Flipo et al., 1992; Fournier et al., 1992; Voccia et al., 1994), phagocytosis in earthworms seems to be affected on *in vivo* and/or *in vitro* exposure to different xenobiotics. In view of the sensitivity of this parameter and the critical role phagocytosis plays in the development of several immune responses, examination of this biomarker is the parameter of choice for these organisms.

Immunologic Biomarkers in Mollusks

Most of the existing published data are derived from experiments using bivalves as the experimental model. In these organisms, the immune defense system has a cellular and a humoral component in which hemocytes play the predominant role (Cheng, 1981). The presence of an antigen initiates the migration of the hemocytes. This is followed by phagocytosis and the destruction of the foreign agent through the action of enzymes (Pipe, 1990) and the production of free radicals of oxygen (Pipe, 1992; Noël et al., 1993). This destruction can also be extracellular, following the devagination of the hemocytes. Furthermore, the organism's defense is based on the secretion of agglutinins and cytotoxic molecules (Leippe and Renwrantz, 1988).

The effects of environmental contaminants on the function of the immune system of bivalves has been the subject of a limited number of publications. An increase in the number of circulating hemocytes is frequently observed in bivalves following exposure to a large variety of environmental stressors, whether it is thermal stress (Renwrantz, 1990), or exposure to a pathogen (Anderson et al., 1992b; Oubella et al., 1993), or toxic substance (Cheng, 1988a, b; Renwrantz, 1990; Anderson et al., 1992a; Coles et al., 1994). Others have reported a decrease in the number of circulating hemocytes (Suresh and Mohandas, 1990), but this phenomenon seems to occur at very high exposure levels and is not related to the concentrations normally detected in the environment. Different types of hemocytes can be identified by their size or staining characteristics. In particular, eosinophil hemocytes are large cells responsible for most of the phagocytosis of foreign particles (Pipe, 1990). Studies carried out on *Crassostrea virginica* have demonstrated that the transfer of individuals from a nonpolluted site to a site polluted with PAHs was accompanied by a reduction in the number of large hemocytes and a concomitant increase in the number of small hemocytes (Sami et al., 1992).

With *Mercenaria mercenaria*, exposure to phenol reduces the phagocytotic activity of the hemocytes (Fries and Tripp, 1980), whereas long-term exposures to benzo(a)pyrene, pentachlorophenol, and hexachlorobenzene stimulate this activity slightly (Anderson, 1981). Research on *Crassostrea virginica* and *Mytilus edulis* revealed that heavy metals altered the phagocytic capability of hemocytes. A stimulation was noted with short-term exposure to low concentrations of toxic substances (Cheng and Sullivan, 1984; Cheng, 1988b; Coles et al., 1995). Conversely, an inhibition was signaled with prolonged exposure or a higher concentration (Fries and Tripp, 1980; Cheng and Sullivan, 1984; Cheng, 1988b; Coles et al., 1995).

The production of free oxygen radicals (mainly the superoxide radical) by hemocytes plays an important role in the destruction of foreign bodies in invertebrates (Adema et al., 1991). Although exposures to very high concentrations of heavy metals are likely to inhibit this production (Larson et al., 1989; Anderson et al., 1992a), it seems that the same is not true for concentrations close to those encountered in the natural environment (Anderson et al., 1992a; Coles et al., 1995) which have no effect on this parameter. It is interesting to note that exposure to high concentrations of fluoranthene stimulated the production of the superoxide radical in *Mytilus edulis* (Coles et al., 1994). This apparent contradiction may be due to the specific interaction between the contaminant involved and the production pathway for free radicals of oxygen or the defense mechanisms against oxidative stress (Pipe et al., 1993).

Regarding the effects of chemicals on the release and activity of lysosomal enzymes, the results obtained vary with the type of enzyme and the exposure levels considered. Exposure to a given toxic stress may lead in some cases to an increase in the release and in the activity of certain enzymes (Pickwell and Steinert, 1984; Cheng, 1989) and to a reduction for other enzymes.

Pipe et al. (1995) have attempted to demonstrate possible correlations between different biomarkers of the immune system and the presence of various contaminants in Mediterranean mussels (*Mytilus galloprovincialis*) living in different areas of the Venice lagoon. The biomarkers considered were hemocyte counts (with counts of different types of hemocytes), phagocytosis, the lysosomal enzyme level, and the production of free radicals of oxygen. The contaminants measured included 10 metals and 3 groups of organochlorine molecules, including PCBs. All the parameters measured showed seasonal fluctuations. In some cases, a correlation was observed between an immunologic biomarker and the levels of contaminants measured in mussel tissues.

Thus, eosinophil hemocytes appeared to be more abundant when metal concentrations were low, whereas the total number of circulating hemocytes frequently increased in the sites most contaminated by metals. Also, a tendency towards inhibition of phagocytotic activity was observed in the presence of high concentrations of metals, a phenomenon already demonstrated for PAHs (Sami et al., 1992). Conversely, a certain correlation in the production of free radicals of oxygen with low concentrations of organic pollutants was noted.

Despite these tendencies, the results obtained indicated clearly that many factors other than the contamination level, and particularly the season, are likely to modulate the function of the immune system in organisms living in the natural environment. This obviously complicates the interpretation of the results.

Immunologic Biomarkers in Vertebrates

Immunologic biomarkers in fish

Histologic examination of the lymphoid tissue of exposed animals is an approach that has been frequently applied to fish for the study of the potentially harmful effects of some environmental contaminants (O'Neill, 1981; Zeeman and Brindley, 1981; Wester and Vos, 1994). However, this approach can only be applied terminally. Consequently, research on immunologic biomarkers in fish have focused on the characterization of affects on immune parameters that do not require invasive methods.

The most common immunologic biomarkers used in fish are lymphocyte counts, the lymphocyte's mitotic potential, the leucocrite, phagocytosis, the number of macrophages and their structure, susceptibility to bacterial infections, lysozyme activity, natural cytotoxic activity (NK, NCC), and the level of circulating antibodies.

A certain number of nonspecific tests such as leucocrite determination, the examination of the blood profile, or the lysozyme activity have been used intensively to determine the state of health of the fish (Pickford et al., 1971; McLeay and Gordon, 1977; Esch and Hanzen, 1980; Peters, 1982; Tomasso et al., 1983; Peters and Schwarzer, 1985; Peters and Hong, 1985). The most frequent observation was a change in the blood profile (Smith et al., 1976; Anderson et al., 1989).

In vivo, it is possible to measure the proliferative response of the lymphocytes to mitogenic agents. Regarding exposure to various environmental contaminants, and depending on the type and level of contaminant, the species studied, both stimulation or inhibition of lymphocyte proliferation (Ghanmi et al., 1989; Dunier and Siwicki, 1994b; Albergoni and Viola, 1995b), have been observed.

Functional aspects of macrophages have also been proposed for use as biomarkers for immunotoxic chemicals (Weeks and Warinner, 1984). Weeks et al. (1986, 1987a, b) and Warinner et al. (1988) showed that the exposure of fish to PAHs in the natural environment and in the laboratory resulted in a significant but reversible change in the various functions of the macrophages, including microorganism phagocytosis, chemotaxis, pinocytosis, and melanin accumulation. These functions were either inhibited or stimulated, depending on the species considered. Cossarini-Dunier et al. (1988) reported a stimulation of phagocytosis on exposure of carp to manganese, whereas Dunier and Siwicki (1994a) and Dunier et al. (1994) observed an inhibition of this function following chronic exposure of rainbow trout *(Oncorhynchus mykiss)* to lindane.

Measuring the circulating concentration of specific antibodies to an antigen is another noninvasive biomarker. The most prevalent results observed are immunodepression and immunosuppression. Robohm (1986) demonstrated that at certain concentrations, cadmium stimulated the production of antibodies in *Tautogalabrus adspersus*, while the response was inhibited in *Morone saxatilis*. Anderson et al. (1984) observed a reduction in the number of antibody-secreting cells in rainbow trout *(O. mykiss)* exposed to phenol, formaldehyde, or detergents. Although many metals act as immunosuppressors (Sharma, 1981; O'Neill, 1981; Dunier and Siwicki, 1994b: Voccia et al., 1994; Albergoni and Viola, 1995a), contradictory results (stimulation or inhibition, depending on the concentration and/or the duration of exposure and the species) were obtained with zinc (O'Neill, 1981) and cadmium (Thuvander, 1989).

Increased susceptibility to bacterial infections is a phenomenon reported in different species following exposure to copper (Baker and Knittel, 1983; Ghanmi, 1989; Khangorot and Tripathi, 1991; Rougier et al., 1992, 1994), cadmium (Taffanelli and Summerfelt, 1975) and phenols (Sharma, 1981).

Few studies have focused on demonstrating *in situ* immunotoxic effects, even if some work has shown that the levels of certain contaminants in the environment were likely to alter the immune response (Faisal et al., 1991a; Seeley and Weeks-Perkins, 1991).

Our team carried out an *in vitro* study of the immunomodulatory effects of paper mill effluents on rainbow trout *(O. mykiss)* leucocytes. The results showed different effects, depending on the manufacturing processes used (Voccia et al., 1994). The effluent from a plant with a bisulfite process stimulated phagocytosis and, at low concentrations (8.75%), leucocyte proliferation in response to mitogens. The effluent from a plant with a kraft process inhibited phagocytosis and stimulated leucocyte proliferation. Data derived from *in vivo* experiments are similarly contradictory.

A study carried out on a treatment plant's sludge discharge site in the Firth of Clyde (western Scotland) by Secombes et al. (1995) revealed that in plaice *(Pleuronectes platessa)*, certain immunologic biomarkers (serum lysozyme activity, immunoglobulin concentration, bactericidal activity of renal leucocytes) were negatively correlated with the distance from the discharge point. The observed stimulation of certain immunologic parameters could be due either to the presence of immunostimulant contaminants or to the high microbial concentration near the discharge site (Robohm et al., 1979; Stolen et al., 1985), or possibly to the differences in dietary constituents (Blazer, 1992).

In vertebrates, homeostasis is ensured by the coordinated action of the nervous system and the endocrine system. In fish, the processes of adaptation to the prevailing environmental conditions are accompanied by changes at the cellular and physiological levels (Pickering, 1981; Jobling, 1995). These changes are controlled by three categories of hormones: corticosteroids, catecholamines, and opioids. These three groups of hormones can have direct or indirect effects on the immune system. For example, cortisol, whose serum level increases in response to stress, reduces the number of circulating lymphocytes in Salmonidae (Ralph et al., 1987). The increased production of catecholamines following stress leads to a mobilization of the blood granulocytes and macrophages, which demonstrates increased phagocytotic activity. Stress related to the capture, handling, transport, or storage of fish may greatly hamper the interpretation of the measurement of immunologic biomarkers (Peters et al., 1980; Peters and Schwarzer, 1985; Elsasser and Clem, 1987; Faisal et al., 1989). Furthermore, it should be noted that the effects of contaminants on the immune system of fish are in general reversible and that they disappear when exposure stops, but that this recovery may take several weeks or even months (Weeks et al., 1986; Faisal et al., 1991a, b; Kelly-Reay and Weeks-Perkins, 1994).

Immunologic Biomarkers in Marine Mammals

Marine mammals that inhabit coastal waters are known to accumulate large amounts of persistent contaminants such as metals, pesticides, PAHs, PCBs, or even chlorinated dioxins and furans (PCDD, PCDF). This contamination, promoted by the terminal position that these animals occupy in the trophic systems, is particularly significant in individuals found in the estuary zones, with the St. Lawrence estuary being a classical example (Martineau et al., 1987; Béland et al., 1993). From the many contamination-monitoring studies on these organisms, one can conclude that all marine mammals on the planet are contaminated, although to different degrees (see, for example, Holden, 1975; Tanabe et al., 1983, 1984, 1988; Samiullah, 1990).

Although many studies have reported concentrations of contaminants in the tissues of marine mammals, few have attempted to relate the presence of these contaminants to toxicologic and mainly ecotoxicologic effects (Reijnders, 1980, 1986, 1994). The most prevalent changes reported involve reproduction and the immune system.

A series of epizootic diseases occurring in North America, Europe, and the U.S.S.R. between 1980 and 1990, when thousands of seals and dolphins died (Osterhaus and Vedder, 1988; Dietz et al., 1989; Kennedy, 1990; Domingo et al., 1992; Geraci et al., 1982, 1989; Geraci, 1989; Grachev et al., 1989), have resulted in increased interest in immunotoxicologic studies of marine mammals. In fact, although the main cause of the mass mortality was most often identified as morbilliviruses similar to the canine distemper virus (Osterhaus and Vedder, 1988; Kennedy et al., 1988; Osterhaus et al., 1988, 1989, 1990; Grachev et al., 1989), the very high concentrations of organochlorine compounds found in animals affected by epizootic diseases (Aguilar and Borrell, 1994) have prompted several scientists to ask questions about the role of these contaminants on the course of epizootic diseases. These contaminants may have changed the functioning of the immune system in the most contaminated animals and made them more susceptible to viral infections, with this effect having been demonstrated in laboratory animals for various persistent contaminants such as PCBs, dioxins, furans, pesticides, or heavy metals (see, for example, Vos et al., 1978; Fournier et al., 1986; Krzystyniak et al., 1985, 1989).

The following pathologies are among those that could be related to perturbation of the immune system in marine mammals: stenosis and occlusion of the uterus in Baltic seals (*Phoca hispida*; Helle et al., 1976a, b); osteal lesions (osteoporosis) of the brain case in harbor seals *(Phoca vitulina)* of the Wadden Zee (Stede and Stede, 1990, in Reijnders, 1994) and the Baltic Sea (Mortensen et al., 1992); adrenocortical hyperplasia in ringed seals *(Ph. hispida)* and grey seals *(H. grypus)* of the Baltic Sea (Bergman and Olsson, 1985; Olsson et al., 1992a); and a reduction in the level of vitamin A and thyroid hormones in harbor seals (*Ph. vitulina*; Brouwer et al., 1989).

In addition to the various effects on different systems and organs, a more or less total immunosuppression, linked to the presence of high concentrations of contaminants, was repeatedly suspected in marine mammals. Among other things, immunosuppression due to industrial contaminants was suspected in California sea lions *(Zalophus californianus)* based on morphological characteristics (Britt and Howard, 1983). Immunosuppressive effects of environmental contaminants have also been demonstrated recently in harbor seals (De Swart et al., 1994; Table 7-4). In these experiments, seals were fed contaminated fish derived from the Baltic Sea (Group 1) or the much less contaminated fish from the Atlantic Ocean (Group 2). The NK cell activity and the lymphoproliferative activity of peripheral blood leukocytes to the Con A, PHA, and PWM mitogens were significantly decreased in the Group 1 seals compared to those in Group 2. The lymphoproliferative response to LPS, a B-cell mitogen, was not affected by treatment. Similarly, the WBC levels were higher in the Group 1 seals compared to the levels in Group 2. This increase was due to the significantly higher granulocyte levels reported for Group 1. Collectively these observations suggest that host resistance mechanisms may be impaired.

Recently, a reduction in lymphocyte proliferation in response to mitogens was observed in bottlenosed dolphin *(Tursiops truncatus)*. This reduction correlates with high blood concentrations of PCBs and DDT metabolites (Lahvis et al., 1995). An immunotoxicologic study of the beluga *(Delphinapterus leucas)* of the St. Lawrence (very contaminated) compared to those of the Arctic (hardly contaminated) is currently in progress to examine the effects of relatively long-term exposure to environmental contaminants (Hileman, 1992; Stone, 1992).

A few *in vitro* models were recently developed to evaluate the effects of different contaminants on the cells of marine mammals. Studies on the lymphocytes of arctic beluga (hardly contaminated) have demonstrated a reduction in their proliferative response to mitogens following exposure to concentrations of mercuric chloride similar to those found in the liver of St. Lawrence belugas (De Guise et al, 1996).

Table 7-4. Daily intake (1g/d) of different organochlorine contaminants for seals fed with fish from the Baltic Sea (Group 1) and from the Atlantic Ocean (Group 2) (adapted from Swart et al., 1994)

Contaminants	Group 1 (1g/l)	Group 2 (1g/l)
PCB	1,460	260
PCDD	0.07	0.02
PCDF	0.4	0.03
BHC	42	6
Dieldrin	491	54
JHCH	17	< 5
DDT	497	102

Discussion—Possibilities

Despite an apparent diversity, there are a certain number of analogies (immunocyte involvement, for example) between the immune systems of different groups of animals, including invertebrates and vertebrates. Regardless of the species considered, immunologic biomarkers may reveal harmful effects for exposure levels that are clearly below those that produce acute toxicity. Like other biomarkers, they may detect effects related to a simultaneous exposure to multiple contaminants. However, it is impossible to identify the contaminant responsible for the effect, since no specific immune response of a contaminant has been determined to date.

The responses of immunologic biomarkers are sometimes contradictory since, for a given contaminant, the effects can range from almost total immunodepression to immunostimulation, including no effect. This response will depend, for example, on the species studied, the mode and duration of exposure, the test protocol used, etc. (Sharma, 1981; Koller and Exon, 1985). This variability in responses is the result of species-specific characteristics, as well as the complexity of the immune system, even though the standardization of protocols and the consideration of the other forms of stress limit it to a certain extent (Weeks et al., 1992). Particular attention must be paid to the sources of stress (capture, transport, collection of samples, etc.) likely to change the immune response in organisms or in the samples (tissue, cells, etc.) used for investigating these biomarkers. In any case, the use of immunologic biomarkers must be based on a multimarker approach, where the markers are preferably measured using noninvasive techniques involving general biomarkers (e.g., immunocyte counts and characterization, histology, NK activity, phagocytosis), which allow a rapid diagnosis of dysfunction in the immune system, and other more precise biomarkers, which provide indications on the mechanisms involved (e.g., delayed hypersensitivity, activity of the T lymphocytes) (Weeks et al., 1992). It should be noted that the responses of the second type of biomarker to environmental contaminants have not been extensively studied up to now. Over the longer term, pathogen sensitivity should be integrated into this approach, taking care to select pathogens specific to the species or the group studied, and to standardize the study methods. Furthermore, it would be desirable to have reference contaminants available, producing a positive response (immunosuppression) for each class of organism.

The use of nonspecific immunity biomarkers (e.g., phagocytic cells, NK activity) seems, a priori, more promising for an *in situ* application. In fact, specific immunity biomarkers, although more sensitive, more often require the use of protocols that are not extensively adapted to the contingencies of field work (exposure to a specific antigen at least 1 week before the biomarker is measured, for example).

There is a somewhat urgent need to develop biomarkers to evaluate the effects of contamination of marine environments. In particular, it is critical that we learn the effects of contamination observed in marine mammals from the perspective of their conservation. The possibility of contaminant-related immunosuppression, as was demonstrated experimentally in seals (De Swart et al., 1994), is very interesting to study because of the scope of the possible implications of such a possibility on the health of the animals involved, and even on the fate of entire populations. The evaluation of the potential effects of contaminants on the immune system of marine mammals, whether directly or using animal models or *in vitro* exposures, should therefore be a priority in establishing programs for monitoring the health of aquatic ecosystems.

Conclusions

The immune system is a major target of many environmental contaminants. Its tissue, cell, and molecular components have, for several toxic substances, a greater sensitivity than the components in the other physiological systems, and consequently make immune parameters attractive as biomarkers. Well known and relatively easy to measure in laboratory organisms, these biomarkers must still be validated for their use *in situ*. Recent studies carried out on invertebrates and vertebrates have shown the interest of a multimarker approach, while demonstrating the involvement of immunomodulating factors other than environmental contaminants, which make the response of the immunologic biomarkers more complex (Pipe et al., 1995; Secombes et al., 1995). Discovering a link between these biomarkers and possible ecotoxicologic effects is a difficult task, even if a certain amount of data seem to indicate a correlation between some immunomodulatory effects of the environmental contaminants and the health status of the populations exposed, and even their dynamics. Moreover, the recent evidence showing the contribution of endocrine modulators in establishing immunotoxicological sensitivity point out the need of an integrated approach in ecotoxicological surveys, as contaminants can act as hormones or antihormones.

References

Adema, C.M., Van-Deutekom-Mulder, E.C., Van Der Knaap, W.P.W., Meuleman, E.A., and Sminia, T. (1991). Generation of oxygen radicals in hemocytes of the snail *Lymnaea stagnalis* in relation to the rate of phagocytosis. *Develop. & Comp. Immunol.* 15(1–2): 17–26.

Aguilar, A., and Borrell, A. (1994). Abnormally high polychlorinated biphenyl levels in striped dolphins *(Stenella coeruleoalba)* affected by the 1990–1992 mediterranean epizootic. *Sc. Tot. Environ.* 154: 237–247.

Albergoni, V., and Viola, A. (1995). Effects of cadmium on lymphocytes proliferation and macrophage activation in catfish, *Ictalurus melas. Fish Shellfish Immunol.* 5: 301–311.

—— and —— (1995). Effects of cadmium on catfish, *Ictalurus melas.* Humoral immune response. *Fish & Shellfish Immunol.* 5(2): 89–95.

—— and —— (1995). Effects of cadmium on lymphocyte proliferation and macrophage activation in catfish, *Ictalurus melas. Fish & Shellfish Immunol.* 5(4): 301–311.

Anderson, D.P., Dixon, O.W., and Van Ginkel, F.W. (1984). Suppression of bath immunization in rainbow trout by contaminant bath pretreatments. In *Chemical Regulation of Immunity in Vetenary Medicine.* (M. Kende, J. Gainer, and M. Chirigos, eds.), pp. 289–293. Liss, New York.

Anderson, C.R., Campbell, G., and Payne, M. (1989). Metabolic origins of 5-hydroxytryptamine in enteric neurons in a teleostean fish *(Platycephalus bassensis)*, a toad *(bufo marinus)* and the guinea-pig. *Comparative Biochem. Physiol. Comp. Pharmacol. & Toxicol.* 92(2): 253–258.

——, Paynter, K.T., and Burreson, E.M. (1992)b. Increased reactive oxygen intermediate production by hemocytes withdraw from *Crassostrea virginica* infected with *Perkinsus marinus. Biol. Bull.* 183: 476–481.

Anderson, R.S., Oliver, L.M., and Jacobs, D. (1992). Immunotoxicity of cadmium for the eastern oyster [*Crassostrea virginica* (Gmelin, 1791)]: Effects on hemocyte chemiluminescence. *J. Shellfish Res.* 11(1): 31–35

Baker, R.J., and Knittel, M.D. (1983). Susceptibility of chinook salmon, *Oncorhynchus tshawytscha* (walbaum) and rainbow trout, *Salmo gairdneri*, to infection with *Vibrio anguillarum* following sublethal copper exposure. *J. Fish Dis.* 6: 267–275.

Béland, P., De Guise, S., Girard, C., Lagacé, A., Martineau, D., Michaud, R., Muir, D.C.G., Norstrom, R.J., Pelletier, E., Ray, S. and Shugart, L.R. (1993). Toxic compounds and health and reproductive effects on St. Lawrence beluga whales. *J. Great Lakes Res.* 19: 766–775.

Bergman, A., and Olsson, M. (1985). Pathology of baltic grey seal *(Halichoerus gryphus)* and ringed seal *(Pusa hispida)* females with special reference to adreno-cortical hyperplasia: Is environmental pollution the cause of a widely distributed disease syndrome? In *Proceedings of the Symposium on Seals in the Baltic and Eurasiar Lakes,* 5–8 June (1985). Savonlinna, Finland, International Council for the Exploitation of the Sea, Copenhagen. CM. 985/N:21/Ref E., 27; pp.1–27.

Berndtsson, R., Hogland, W., Larson, M., Enell, M., Wennbe, L., and Forsberg, J. (1989). The effects of storm water on Lake Vaxjosjon (Sweden). *Vatten* 45(2): 167–173.

Bernier, J., Brousseau, P., Tryphonas, H., Krzystyniak, K., and Fournier, M. (1995). Immunotoxity of selected heavy metals pertinent to great lakes contamination. *Env. Health Perspec.* 103: 23–34.

Bigger, C.H. (1980). *Biol. Bull.* 159: 117–134.

Borgmann, U., and Ralph, K.M. (1985). Feeding, growth and particle-size-conversion efficiency in white sucker larvae and young common shiners. *Environ. Biol. Fishes* 14(4): 269–279.

———— and ———— (1986). Effects of cadmium, 2,4-dichlorophenol and pentachlorophenol on feeding, growth, and particle-size-conversion efficiency of white sucker *(Catostomus commersoni)* larvae and young common shiners *(Notropis cornutus)*. *Arch. Environ. Contam. & Toxicol.* 15(5): 473–480.

————, ————, Norwood, W.P. (1989). Toxicity test procedures for *Hyalella azteca*, and chronic toxicity of cadmium and pentachlorophenol to *Hyalella azteca*, *Gammarus fasciatus*, and *Daphnia magna*. *Arch. Environ. Contam. & Toxicol.* 18(5): 756–764.

Bossart, G.D., and Dierauf, L.A. (1990). Marine mammal clinical laboratory medicine. In CRC *Handbook of Marine Mammal Medicine: Health, Disease and Rehabilitation.* (L.A. Dierauf, ed.), pp. 1–52. CRC, Boca Raton.

Britt, J.O., and Howard, E.B. (1983). Tissue residues of selected environmental contaminants in marine mammals. In *Pathobiology of Marine Mammal Diseases. Vol. II.* (E.B. Howard, ed.), pp. 80–94. CRC, Boca Raton.

Brousseau, P., Fugere, N., Bernier, J., Coderre, D., Nadeau, D., Poirier, G., and Fournier, M. (1997). Evaluation of earthworm exposure to contaminated soil by cytometric assay of coelomocyte phagocytosis in *Lumbricus terrestris* (oligochaeta). *Soil Biol. Biochem.* 29: 681–684.

Brouwer, A., Reijinders, P.J.H., and Koeman, J.H. (1989). Polychlorinated biphenyl (pcb)-contaminated fish induces vitamin a and thyroid hormone deficiency in the common seal *(Phoca vitulina)*. *Aquat. Toxicol.* 15: 99–106.

Callahan, C.A., Shirazi, M.A., and Neuhauser, E.F. (1994). Comparative toxicity of chemicals to earthworms. *Environ. Toxicol. Chem.* 13(2): 291–298.

Calzori, A. (1898). *Arch. Ital. Biol. Torino* 307: 71

Chavez, E., Briones, R., Michel, B., Bravo, C., and Jay, D. (1985). Evidence for the involvement of dithiol groups in mitochondrial calcium transport: Studies with cadmium. *Arch. Biochem. Biophys.* 242: 493.

Chen, S.C., Fitzpatrick, L.C., Goven, A.J., Venables, B.J., and Cooper, E.L. (1991). Nitroblue tetrazolium dye reduction by earthworm *(Lumbricus terrestris)* coelomocytes: an enzyme assay for nonspecific immunotoxicity of xenobiotics. *Environ. Toxicol. Chem.* 10(8): 1037–1044.

————, ————, ————, ————, and ———— (1991). Suppression of nitroblue tetrozolium dye reduction in lumbricus terrestris coelomocytes harvested from earthworms exposed to refuse derived fuel fly ash. *Environ. Toxicol. Chem.* 10: 1037–1043.

Cheng, T.C. (1988a). *In vivo* effects of heavy metals on cellular defense mechanisms of *Rassostrea virginica:* Total and differential haemocyte counts. *J. Invertebr. Pathol.* 51: 207–214.

———— (1988b). *In vivo* effects of heavy metals on cellular defense mechanisms of *Rassostrea virginica:* phagocytic and endocytic indices. *J. Invertebr. Pathol.* 51: 215–220.

————, Sullivan, J.T. (1984). Effects of heavy metals on phagocytosis by mollusca hemocytes. *Mar. Envir. Res.* 14: 305–315.

Cifone, M.G., Alesse, E., Di Eugenio, R., Napolitano, T., Morrone, S., Paolini, R., Santoni, G., and Santoni, A. (1989b). *In vivo* cadmium treatment alters natural killer activity and large granular lymphocyte number in the rat. *Immunopharmacology* 18(3): 149.

————, ————, Procopio, A., Paolini, R., Morrone, S., Di Eugenio, R., Santoni, G., and Santoni, A. (1989a). Effects of cadmium on lymphocyte activation. *Biochimica et Biophysica Acta* 1011(1): 25.

————, Rocopio, A., Napolitano, T., Alesse, E., Santoni, G., and Santoni, A. (1990). Cadmium inhibits spontaneous (NK), antibody-mediated (ADCC) and Il-2-stimulated cytotoxic functions of natural killer cells. *Immunopharmacology* 20(2): 73.

Coles, J.A., Fairly, S.R., and Pipe, R.K. (1994). The effects of fluoranthene on the immunocopetence of the common marine mussel, *Mytilus edulis. J. Aquat. Toxicol.* 30:367–379.

————, ————, and ———— (1995). Alteration of the immune response of the common marine mussel, *Mytilus edulis,* resulting from exposure to cadmium. *Dis. Aquat. Org.* 22:59–65.

Cooper, E.L. (1968). Transplantation immunity in annelids. I. Rejection of xenografts exchanged between *Lumbricus terrestris* and *Eisenia foetida. Transplantation* 6: 322–337.

———— (1969). Specific tissue graft rejection in earthworms. *Science* 66: 1414.

———— Roch P. (1992). The capacities of earthworms to heal wounds and to destroy allografts are modified by polychlorinated biphenyls (PCB). *J. Invert. Pathol.* 60: 59–63.

Cossarini-Dunier, M., Demael, A., Lepot, D., and Guerin, V. (1988). Effect of manganese ions on the immune response of carp *(Cyprinus carpio)* against *Yersinia ruckeri. Dev. Comp. Immunol.* 12: 573–579.

———— and Hattenberger, A.M. (1988). Effect of pesticides (atrazine and lindane) on the replication of spring viremia of carp virus *in vitro. Ann. Recherches Veterinaires* 19(3): 209–211.

Crocker, J.F.S., Ozere, R.L., Safe, S.H., Digout, S.C., Rozee, K.R., and Hutzinger, O. (1976). Lethal interaction of ubiquitous insecticide carriers with virus. *Science* 192: 1351–1393.

Cuthill S., Wilhelmsson, A., Mason, G.G.F., Gillner, M., Poellinger, L., and Gustafsson, J.-A. (1988). The dioxin receptor: A comparison with the glucocorticoid receptor. *Toxicol. Appl. Pharmacol.* 94: 141–149.

Davis, G.E., and Anderson, T.W. (1989). Population estimates of four kelp forest fishes and an evaluation of three in situ assessment techniques. *Bull. Marine Sci.* 44(3): 138–1151.

De Guise, S., Bernier, J., Martineau, D., Béland, P., and Fournier, M. (1996). *In vitro* exposure of beluga whale lymphocytes to selected heavy metals. *Environ. Toxicol. Chem.* 15: 1357–1364.

De Swart, R.L., Ross, P.S., Vedder, L.J., Timmerman, H.H., Heisterkamp, S.H., Van Loveren, H., Vos, J.G., Reltnders, P.J.H., and Osterhaus, A.D.M.E. (1994). Impairment of immune function in harbour seals *(Phoca vitulina)* feeding on fish from polluted waters. *Ambio.* 23: 155–159.

Dean, J.H., and Murray, M.J. (1990). Toxic responses of the immune system. In *Toxicology: The Basic Science of Poisons, Vol. 4.* (C.D. Klaassen, M.O. Amdur, and J. Doull, eds.), pp. 282–333. Macmillan, New York.

————, Ward, E.C., and Murray, M.J. (1986). Immunosuppression following 7:12-dimethylbenz(a)anthracene exposure in B6C3F1 mice. II. Altered cell-mediated immunity and tumor resistance. *Int. J. Immunopharmac.* 8: 189–198.

Descotes, J. (1988). Immunotoxicity of chemicals. In *Immunotoxicology of Drugs and Chemicals*, pp. 297–444. Elsevier, Amsterdam.

Dietz, R., Heide-Jorgensen, M.D., and Harkonen, T. (1989). Mass deaths of harbor seals *(Phoca vitulina)* in Europe. *Ambio.* 18: 258–264.

Domingo, M., Visa, J., Pumarola, M., Marco, A.J., Ferrer, L., Rabanal, R., and Kennedy, S. (1992). Pathologic and immunocytochemical studies of morbillivirus infection in striped dolphins *(Stenella coeruleoalba)*. *Vet. Pathol.* 29: 1–10.

Dunier, M., Siwicki, A.K., Scholtens, J., Dal-Molin, S., Vergnet, C., and Studnicka, M. (1994). Effects of lindane exposure on rainbow trout *(Oncorhynchus mykiss)* immunity. *Ecotoxicol. Environ. Safety* 27(3): 324–334.

Elsasser, C.F., and Clem, L.W. (1987). Cortisol induced hematologic and immunologic changes in channel catfish *Ictalarus punctatus. Comp. Biochem. Physiol.* 87A: 405408.

Esch, G.W., and Hazen, T.C. (1980). Stress and body condition in a population of largemouth bass: Implication for red-sore-disease. *Trans. Am. Fish. Soc.* 109: 532–536.

Exon, J.H. (1984). The immunotoxicity of selected environmental chemicals, pesticides, and heavy metals. In *Chemical Regulation in Veterinary Medicine*, p. 355. Liss, New York.

————, Kerkvliet, N.I., and Talcot, P.A. (1987). Immunotoxicity of carcinogenic pesticides and related chemicals. *J. Environ. Sci. Health* Part C. *Environ. Carcinog.* Rev. 5: 73–120

Eyambe, G.S., Goven, A.J., Fitzpatrick, L.C., Venables, B.J., and Cooper, E.L.

(1991). A non-invasive technique for sequential collection of earthworm *(Lumbricus terrestris)* leukocytes during subchronic immunotoxicity studies. *Lab. Animals* 25: 61–67.

Faisal, M., Chiapelli, F., Ahmed, I.I., Cooper, E.L., and Weiner, H. (1989). Evidence of aberration of the natural cytotoxic cell activity in *Fundulus heteroclitus* (Pisces: Cyprinodontidae) from the Elizabeth River, Virginia. *Vet. Immunol. Immunopathol.* 29: 339–351.

————, Marzouk, M.S.M., Smith, C.L., and Huggett, R.J. (1991). Mitogen induced proliferative responses of lymphocytes from spot *(Leiostomus xanthurus)* exposed to polycyclic aromatic hydrocarbon contaminated environments. *Immunopharmacol. Immunotoxicol.* 13(3): 311–328.

————, Weeks, B.A., Vogelbein, W.K., and Huggett, R.J. (1991). Evidence of aberration of the natural cytotoxic cell activity in *Fundulus heteroclitus* (Pisces: Cyprinodontidae) from the Elizabeth River, Virginia (USA). *Vet. Immunol. Immunopathol.* 29(3–4): 339–352.

Fitzpatrick, L.C., Goven, A.J., Benables, B.J. and Cooper, E.L. (1990). Earthworm immunoassays for evaluating biological effects of exposure to hazardous materials. In *In Situ Evaluation of Biological Hazards of Environmental Pollutants.* (S.S. Sandhu, ed.), pp.119–129. Plenum, New York.

————, Sassani, R., Venables, B.J., Goven, A.J. (1992). Comparative toxicity of polychlorinated biphenyls to earthworms *Eisenia foetida* and *Lumbricus terrestris. Environ. Pollution* 77: 65–69.

Flipo, D., Bernier, J., Girard, D., Krzystyniak, K., and Fournier, M. (1992). Simultaneous effects of selected pesticides on humoral immune response in mice. *Internat. J. Immunopharmacol.* 14: 747–752.

Fournier, M., Bernier, J., Flipo, D., and Krzystyniak, K. (1986). Evaluation of pesticide effects on humoral response to sheep erythrocytes and mouse hepatitis virus 3 by immunosorbent analysis. *Pest. Biochem. Physiol.* 26: 353–364.

————, Friborg, J., Girard, D., Mansour, S., and Krzystyniak K. (1992). Limited immunotoxic potential of technical formulation of the herbicide atrazine, Aatrex, in mice. *Toxicol. Lett.* 60: 263–274.

Franceschi, C., Cossarizza, A., Monti, D., and Ottaviani, E. (1991). Cytotoxicity and immunocyte markers in cells from the freshwater snail *Planorbarius corneus (Gastropoda Pulmonata):* Implications for the evolution of natural killer cells. *Eur. J. Immunol.* 21: 489–493.

Frie, R.V., Anderson, J.K., and Larson, M.J. (1989). Age verification of walleyes from Lake of the Woods, Minnesota (USA). *J. Great Lakes Res.* 15(2): 298–305

Fries, C.R., and Tripp, M.R. (1980). Depression of phagocytosis in *Mercenaria* following chemical stress. *Dev. Comp. Immunol.* 4: 233–244.

Furst, A., Chien, Y., and Chien, P.K. (1993). Worms as a substitute for rodents in toxicology: Acute toxicity of three nickel compounds. *Toxicol. Methods* 3(1): 19–23.

Geraci, J.R. (1989). Clinical investigation of the 1987-88 mass mortality of bottlenose dolphins along the US central and south Atlantic coasts. Final Report, NMFS, U.S. Navy, Off. Naval Res., 63 pp.

———, Anderson, D.M., Timperi, R.J., St. Aubin, D.J., Early, G.A., Prescott, J.H., and Mayo, C.M. (1989). Humpback whales *(Megaptera novaeangliae)* fatally poisoned by dinoflagellate toxin. *Can. J. Fish. Aquat. Sci.* 46: 1895–1898.

———, St. Aubin, D.J., Barker, I.K., Webster, R.G., Hinshaw, V.S., Bean, W.J., Ruhnke, H.R., Prescott, J.H., Early, G., Baker, A.S., Madoff, S., and Schooley, R.T. (1982). Mass mortality of harbor seals: Pneumonia associated with influenza A virus. *Science* 215: 1129–1131.

Ghanmi Z., Rouabhia, M., Othmane, O., and Deschaux, P.A. (1989). Effect of metal ions on cyprinid fish immune response: *In vitro* effects of Zn^{2+} and Mn^{2+} on the mitogenic response of carp pronephros lymphocytes. *Ecotox. Environ. Safety* 17: 183–489.

Gleichmann, E., Kimber, I., and Purchase, I.F.H. (1989). Immunotoxicology: Suppressive and stimulatory effects of drugs and environmental chemicals on the immune system. *Arch. Toxicol.* 63: 257–273.

Goven, A.J., Chen, S.C., Fitzpatrick, L.C., and Venables, B.J. (1994). Lysozyme activity in earthworm *(Lumbricus terrestris)* coelomic fluid and coelomo-cytes: Enzyme assay for immunotoxicity of xenobiotics. *Environ. Toxicol. Chemistry,* 13(4): 607–613.

———, ———, ———, and ——— (1994a). Lysozyme activity in earthworm *(Lumbricus terrestris)* coleomic fluid and coleomocytes: Enzyme assay for immunotoxicity of xenobiotics. *Environ. Toxicol. Chem.* 13: 607–613.

———, Eyambe, G.S., Fitzpatrick, L.C., Venables, B.J., Cooper, E.L. (1993). Cellular biomarkers for measuring toxicity of xenobiotics: Effects of poly-chlorinated biphenyls on earthworm *Lumbricus terrestris* coelomocytes. *Environ. Toxicol. Chem.* 12(5): 863–870.

———, ———, ———, ———, and ——— (1993). Cellular biomakers for measuring toxicity of xenobiotics: Effects of polychlorinated biphenyls on earthworm *Lumbricus terrestris* coelomocytes. *Environ. Toxicol. Chem.* 12: 863–870.

———, Fitzpatrick, L.C., and Venables, B.J. (1994b). Chemical toxicity and host defense in earthworms: An invertebrate model. *Annals N.Y. Acad. Sci.* 712: 280–300.

Grachev, M.A., Kumarev, V.P., Mamaev, L.V., Zorin, V.L., Baranova, L.V., Denikjna, N.N., Belikov, S.I., Petrov, E.A., Kolesnik, V.S., Kolesnik, R.S., Dorofeev, V.M., Beim, A.M., Kudelin, V.N., Nagieva, F.G., and Sidorov, V.N.

(1989). Distemper virus in Baikal seals. *Nature*, 338: 209.

Grass, P.P., Poisson, R.A., and Tuzet, O. (1970). In *Zoologie I, 2d ed*. Masson et Cie, Paris.

Grossman, C.J. (1985). Interactions between the gonadal steroids and the immune system. *Science* 227: 257–261

Helle, E., Olsson, M., and Jensen, S. (1976a). DDT and PCB levels and reproduction in ringed seal from the Bothnian Bay. *Ambio*. 5: 188–189.

———, ———, and ——— (1976b). PCB levels correlated with pathological changes in seal uteri. *Ambio*. 5: 135–137.

Hileman, B. (1992). Effects of organohalogens on marine mammals to be investigated. *Chem. Eng. News* 70: 23–24.

Hinton, D.M. (1992). Testing guidelines for evaluation of the immunotoxic potential of direct food additives. *Crit. Rev. Food Sci. Nutrition* 32: 173–190.

Holden, A.V. (1975). The accumulation of oceanic contaminants in marine mammals. *Rapp. P.-V. Rèun. Cons. Int. Explor. Mer.* 169: 353–361.

Humphreys, T., and Reinherz, E.L. (1994). Invertebrate immune recognition, natural immunity and the evolution of positive selection. *Immunol. Today* 15: 316–320.

Jessop, B.M., and Anderson, W.E. (1989). Effects of heterogeneity in the spatial and temporal pattern of juvenile alewife *(Alosa pseudoharengus)* and blueback herring *(Alosa aestivalis)* density on estimation of an index of abundance. *Can. J. Fisheries Aquatic Sci.* 46(9): 1564–1574.

Jobling, M., Arnesen, A.M., Baardvik, B.M., Christiansen, J.S., and Jorgensen, E.H. (1995). Monitoring voluntary feed intake under practical conditions, methods and applications. *J. App. Ichthyol.* 11(3–4): 248–262.

Karp, R.D. (1990). Cell-mediated immunity in invertebrates: The enigma of the insect. *Biosci.* 40: 732–737.

Karp, R.D., and Hildemann, W.H. (1976). Specific allograft reactivity in the sea star *Dermasteria imbricata*. *Transplantation* 22: 434–439.

Kelly-Reay, K., and Weeks-Perkins, B.A. (1994). Changes in humoral and cell mediated immune response and in skin and respiratory surfaces of catfish, Sacchabranchus fossiles, following copper exposure. *Ecotoxicol. Environ. Saf.* 22: 2191–2208.

Kennedy, S. (1990). A review of the 1988 European seal morbillivirus epizootic. *Vet. Rec.* 12: 563–567.

———, Smyth, J.A., McCullough, S.J., Allan, G.M., McNeilly, F., and McQuaid, S. (1988). Confirmation of the cause of recent seal deaths. *Nature* 335: 404.

Khangarot, B.S., and Tripathi, D.M. (1991). Changes in humoral and cell-mediated immune responses and in skin and respiratory surfaces of catfish, *Saccobranchus fossilis*, following copper exposure. *Ecotoxicol. Environ. Safety* 22(3): 291–308.

Koller, L.D., and Exon, J.H. (1985). The rat as a model for immunotoxicity assessment. In *Immunotoxicology and Immunopharmacology*. (J.H. Dean, M.I. Luster, A.E. Munson, and H. Amos, eds.), pp. 99–111. Raven, New York.

Krzystyniak, K., Flipo, D., Mansour, S., and Fournier, M. (1989). Suppression of avidin processing by mouse macrophage after sublethal exposure to dieldrin. *Immunopharmacol.* 18: 157–166.

———, Hugo, P., Flipo, D., and Fournier, M. (1985). Increased susceptibility to mouse hepatitis virus 3 of peritoneal macrophages exposed to dieldrin. *Toxicol. Appl. Pharmacol.* 80: 397–408.

Lachapelle, M., Guertin, F., Marion, M., Fournier, M., and Denizeau, F. (1993). Mercuric chloride affects protein secretion in rat primary hepatocyte cultures: A biochemical, ultrastructural and gel immunocytochemical study. *J. Tox. Env. Health* 38: 343–354.

Lahvis, G.P., Wells, R.S., Kuehi, D.W., Stewart, J.L., Rhinehart, H.L., and Via, C.S. (1995). Decreased lymphocytes response in free-ranging bottlenose dolphins *(Tursiops truncatus)* are associated with increased concentrations of PCBs and DDT in peripheral blood. *Environ. Health Perspec.* 103: 67–72.

———, ———, Casper, D., and Via, C.S. (1993). *In vitro* lymphocyte response of bottlenose dolphins *(Tursiops truncatus):* Mitogen-induced proliferation. *Mar. Env. Res.* 35: 115–119.

Lasalle, F., and Lassegues, M. (1986). Humoral defense in earthworm lysozyme and yellow pigment. *Soc. Fran. Immunol. Immunobiol.* 173: 231–233.

———, ———, and Roch, P.H. (1988). Protein analysis of earthworm coelomic fluid. 4. Evidence, activity induction and purificaiton of *Elsenia foetida andrei* lysozyme. *Comp. Biochem. Physiol.* 91B: 1887–1892.

Laulan, A., Morel, A., Lestage, J., Delaage, M., and Chateaureynaud-Duprat, P. (1985). Evidence of synthesis by *Lumbricus terrestris* of specific substances in response to an immunization with a synthetic hapten. *Immunology* 56: 751–757.

Lawrence, D.A. (1985). Immunotoxicity of heavy metals. In *Immunotoxicology and Immunopharmacology*. (J.H. Dean, M.I. Luster, A.E. Munson, and H. Amos, eds.), p. 341. Raven, New York.

Leslie, J.K., Metcalfe, J.L., and Ralph, K.M. (1986). Early development of the common shiner, *Notropis cornutus*, in southern Ontario (Canada). *Can. Tech. Rep. Fisheries Aquatic Sci.* 0(1455): I–IV: 1–37.

Lubbock, R. (1980). Clone specific cellular recognition in a sea anemone. *Proceeding of National Academy of Sciences, USA* 77: 6667–6669.

Luster, M.I., Munson, A.E., Thomas, P.T., Holsapple, M.P., Fenters, J.D., White, K.L., Lauer, L.D., Germolec, D.R., Rosenthal, G.J., and Dean, J.H. (1988). Development of a testing battery to assess chemical-induced immunotoxicity: National Toxicology Program's guidelines for immunotoxicity evaluation in mice. *Fundam. Appl. Toxicol.* 10: 2–9.

————, Portier, C., Pait, G.G., and Germolec, D.R. (1994). Use of animal studies in risk assessment for immunotoxicology. *Toxicol.* 92: 229–243.

————, ————, ————, Rosenthal, G., Germolec, G., Corsini, E., Blaylock, B., Pollock, P., Kouchi, Y., Craig, W., White, K.L., Munson A.E., and Comment, C. (1993). Risk assessment in immunotoxicology. II. Relationships between immune and host resistance tests. *Fund. Appl. Toxicol.* 21: 71–82.

————, ————, ————, White, K.L., Gennings C., Munson A.E., and Rosenthal, G. (1992). Risk assessment in immunotoxicology. I. Sensitivity and predictability of immune tests. *Fund. Appl. Toxicol.* 18: 200–210.

Ma, W.C. (1982). The influence of soil properties and worm-related factors on the concentration of heavy metals in earthworms. *Pedobiologia* 24: 109–119.

Martineau, D., Bland, P., Desjardins, C. and Lagac, A. (1987). Levels of organochlorine chemicals in tissues of beluga whales *(Delphinapterus leucas)* from the St. Lawrence Estuary, Québec, Canada. *Arch. Environ. Contam. Toxicol.* 16: 137–147.

McCabe, M.J., and Lawrence, D.A. (1991). Lead, a major environmental pollutant is immunomodulatory by its differential effects on $CD_{4+}T$ subset. *Toxicol. App. Pharmacol.* 111: 13.

McLeay, D.J., and Gordon, M.R. (1977). Leucocrit: A simple hematological technique for measuring acute stress in salmonid fish, including stressful concentration of pulpmill effluent. *J. Fish Res. Bd. Can.* 34: 2164–2175.

Mullainadhan, P., and Renwrantz, L. (1986). Lectin-dependent recognition of foreign cells by hemocytes of the mussel, *Mytilus edulis. Immunobiol.* 171(3): 263–273.

Noelle, R.J., and Snow, E.C. (1990). Cognate interactions between helper T cells and B cells. *Immunol. Today* 11: 361.

O'Neil, J.G. (1981). Effects of intraperitoneal lead and cadmium on the humoral immune response of *Salmo trutta. Bull. Environ. Contam. Toxicol.* 27: 42–48.

———— (1981). The humoral immune response of *Salmo trutta* l. and *Cyprinus carpio* exposed to heavy metals. *J. Fish Biol.* 19: 297–306.

Olsson, M., Andersson, O., Bergman, A., Blomkvist, G., Frank, A., and Rappe, C. (1992). Contaminants and diseases in seals from Swedish waters. *Ambio.* 21: 561–562.

Osterhaus, A.D.M.E., and Vedder, E.J. (1988). Identification of virus causing recent seal deaths. *Nature* 335: 20.

————, Groen J., De Vries P., Uytdehaag F.G.C.M., Klingeborn B. and Zarnke R. (1988). Canine distemper virus in seals. *Nature* 335: 403–404.

————, ————, Spijkers, H.E.M., Broeders, H.W.J., Uytdehaag, F.G.C.M., De Vries, P., Teppema, J.S., Visser, I.K.G., Van De Bildt, M.W.G., and Vedder,

E.J. (1990). Mass mortality in seals caused by a newly discovered morbillivirus. *Vet. Microbiol.* 23: 343–350.

————, ————, Uytdehaag, F.G.C.M., Visser, I.K.G., Van De Bildt, M.W.G., Bergman, A., and Klingeborn, B. (1989). Distemper virus in Baikal seals. *Nature* 338: 209–210.

Oubella, R., Maes, P., Paillard, C., and Auffret, M. (1993). Experimentally induced variation in hemocyte density for *Ruditapes philippinarum* and *R. decussatus* (Mollusca, Bivalvia). *Dis. Aquat. Org.* 15: 193–197.

Penn, I. (1987). The neoplastic consequences of immunodeppression. In *Immunotoxicology.* (A. Berlin, J. Dean, M.H. Draper, E.M.B. Smith, F. Spreafico, eds.), pp. 69–82. Martinus Nijhoff, Dordrecht.

Peters, G.H. (1982). The effect of stress on the stomach of the European eel, *Anguilla anguilla. L. J. Fish Biol.* 21: 497–512.

———— and Schwarzer, R. (1985). Changes in the hemopoietic tissues of rainbow trout under the influence of stress. *Dis. Aquat. Org.* 1: 1–10.

————, Delventhal, H., and Klinger, H. (1980). Physiological and morphological effects of social stress in the eel (*Anguilla anguilla* L.). *Arch. Fisch. Wiss.* 30: 157–180.

———— and Schwarzer, R. (1985). Changes in hemopoietic tissue of rainbow trout *(Salmo gairdneri)* under influence of stress. *Dis. Aquatic Organisms* 1(1): 1–10.

Pickering, A.D., ed. (1981). *Stress and Fish.* Academic, London.

Pickford, G.E., Srivastava, A.K., Slicher, A.M., and Pang, P.K.T. (1971). The stress response in the abundance of circulating leucocytes in the killifish *Fundulus heteroclitus.* III. The role of the adrenal cortex and a concluding discussion of the leucocyte stress syndrome. *J. Exp. Zoll.* 177: 109–117.

Pipe, R.K. (1990). Hydrolytic enzymes associated with the granular haemocytes of the marine mussel *Mytilus edulis. Histochem. J.* 22(11): 595–603.

———— (1992). Generation of reactive oxygen metabolites by the haemocytes of the mussel *Mytilus edulis. Develop. and Comp. Immunol.* 16(2–3): 111–122.

———— and Coles, J.A. (1995). Environmental contaminants influencing immune function in marine bivalve molluscs. *Fish & Shellfish Immunol.* 5(8): 581–595.

Poland, A., and Knutson, J.C. (1982). 2,3,7,8-Tetrachlorodibenzo-*p*-dioxin and related halogenated aromatic hydrocarbons: Examination of the mechanism of toxicity. *Ann. Rev. Pharmacol. Toxicol.* 22: 517–554.

————, Glover, E., and Kende, A.S. (1976). Stereospecific, high affinity binding of 2,3,7,8-tetrachlorodibenzo-*p*-dioxin by hepatic cytosol: Evidence that the binding species is the receptor for induction or aryl hydrocarbon hydroxylase. *J. Biol. Chem.* 251: 4936.

Porchet-Hennere, E., Dugimont, T., and Fischer, A. (1992). Natural killer cells in a lower invertebrate, *Nereis diversicolor. European J. Cell. Biol.* 58: 99–107.

Prozialeck, W.C., and Niewenhuis, R.J. (1991). Cadmium disrupts calcium-dependent cell-cell junctions and alters the pattern of E-cadherin immunofluorescence in LLC-PK-1 cells. *Biochem. Biophys. Res. Comm.* 181(3): 1118.

Reijnders, P.J.H. (1980). Organochlorine and heavy metal residues in harbour seals from the Wadden Sea and their possible effects on reproduction. *Neth. J. Sea Res.* 14: 30–65.

Reijnders, P.J.H. (1986). Reproductive failure in common seals feeding on fish from polluted coastal waters. *Nature* 324: 456–457.

Reinecke, A.J. (1992). A review of ecotoxicological test methods using earth-worms. In *Phylogeny of Earthworms.* (P.W. Greig-Smith, M.B. Becker, P.J. Edwards, and F. Heimbach, eds.), pp. 7–19. Intercept, Andover, U.K.

Renwrantz, L. (1990). Internal defence system of *Mytilus edulis.* In *Studies in Neuroscience: Neurobiology of Mytilus edulis.* (G.B. Stefano, ed.), pp. 256–275. Manchester University Press, Manchester.

Revillard, J.P. (1994). Immunotoxicité des xénobiotiques. In *Immunologie.* pp. 289–296. De Boeck-Wesmael, Bruxelles.

Robohm, R.A. (1986). Paradoxical effects of cadmium exposure on antibacter-ial antibody responses in two fish species: Inhibition in cunners *(Tautogolabrus adspersus)* and enhancement in stripped bass *(Morone saxatilis). Vet. Immunol. Immunopathol.* 12: 251–262.

———, Brown, C., and Murchelano, R.A. (1979). Comparison of antibodies in marine fish from clean and polluted waters of the New York bight: Relative levels against 36 bacteria. *Appl. Environ. Microbiol.* 38: 248–257.

Roch, P., and Cooper, E.L. (1991). Cellular but not humoral antibacterial activ-ity of earthworms is inhibited by Aroclor 1254. *Ecotoxicol. Environ. Safety,* 22(3): 283–290.

Rodriguez-Grau, J., Venables, B.J., Fitzpatrick, L.C., Goven, A.J., and Cooper, E.L. (1989). Suppression of secretory rosette formation by PCBs in *Lumbricus terrestris:* An earthworm immunoassay for humoral immunotoxicity of xenobiotics. *Environ. Toxicol. Chem.* 8: 1201–1207.

Roitt, I.M., Brostoff, J., and Male, D.K. (1993). *Immunology, 3d ed.* Gower, London.

Rougier, F., Menudier, A., Troutaud, D., Bosgiraud, C., N'Doye, A., Nicolas, J.A., and Deschaux, P. (1992). *In vivo* effect of zinc and copper on the develop-ment of listeriosis in zebrafish *Brachidanio rerio. J. Fish Dis.* 15: 454–456.

———, Troutaud, D., N'Doye, A., and Deschaux, P. (1994). Non-specific immune response of zebrafish *Brachydanio rerio* (Hamilton-Buchanan) fol-lowing copper and zinc exposure. *Fish & Shellfish Immunol.* 4(2): 115–127.

Sami, S., Faisal, M., and Huggett, R.J. (1992). Alterations in cytometry characteristics of hemocytes from the American oyster *Crassostrea virginica* exposed to a polycyclic aromatic hydrocarbon (PAH) contaminated environment. *Mar. Biol.* 113: 247–252.

Schuurman, H.J., Kuper, C.F., and Vos J.G. (1994). Histopathology of the immune system as a tool to assess immunotoxicity. *Toxicol.* 86: 187–212.

Secombes, C.J., White, A., Fletcher, T.C., Stagg, R., and Houlihan, D.F. (1995). Immune parameters of plaice, *Pleuronectes platessa* L., along a sewage sludge gradient in the Firth of Clyde, Scotland. *Ecotoxicol.* 4(5): 329–340.

Seeley, K.R., and Weeks-Perkins, B.A. (1991). Altered phagocytic activity of macrophages in oyster toadfish from a highly polluted subestuary. *J. Aquatic Animal Health* 3(3): 224–227.

Shardlow, T.F., Hoyt, T.G., and Anderson, A.D. (1988). Review of 1986 south coast (British Columbia, Canada) salmon troll fisheries. *Can. Tech. Rep. Fisheries Aquatic Sci.* 0(1560): I-VIIII, 1–99.

Sharma, R.P. (1981). *Immunologic Considerations in Toxicolology, Vol. II*, 145 pp. CRC, Boca Raton.

Sheldon, W.M., Jr., and Blazer, V.S. (1991). Influence of dietary lipid and temperature on bactericidal activity of channel catfish macrophages. *J. Aquatic Animal Health* 3(2): 87–93.

Siwicki, A.K. (1994). Dunier-M effects of lindane exposure on rainbow trout *(Oncorhynchus mykiss)* immunity. *Ecotoxicol. Environ. Safety* 27(3): 316–323.

Smith, B.P., Hejmancik, E., and Camp, B.J. (1976). Acute effects of cadmium on *Ictalarus punctatus* (catfish). *Bull. Environ. Contam. Toxicol.* 15: 271–275.

Stein, E.A., Wojdani, A., and Cooper, E.L. (1982). Agglutinins in the earthworm *Lumbricus terrestris*: Naturally occurring and induced. *Develop. Comp. Immunol.* 6: 407–421.

Stolen, J.S., Gahn, T., Kasper, V., and Nagle, J.J. (1985). Natural and adaptive immunity in marine teleosts to bacterial isolates from sewage sludge. In *Fish Immunology.* (M.J. Manning and M.F. Tatner, eds.), pp. 207–220. Academic, London.

Stone, R. (1992). Swimming against the PCB tide. *Science* 255: 798–799.

Suresh, K., and Mohandas, A. (1990). Effect of sublethal concentrations of copper on hemocyte number in bivalves. *J. Invertebrate Pathol.* 55(3): 325–331.

Suzuki, M.M., Cooper, E.L., Eyambe, G.S., Goven, A.J., Fitzpartick, L.C., and Venables, B.J. (1995). Polychlorinated biphenyls (PCBs) depress allogeneic natural cytotoxicity by earthworm coelomocytes. *Environ. Toxicol. Chem.* 14(10): 1697–1700.

Swart, R.D.L., Ross, P.S., Vedder, L.J., Timmerman, H.H., Heisterkamp, S., VanLoveren, H., Vos, J.V., Reijnders, P.J.H., and Osterhaus, A.D.M.E. (1994).

Impairment of immune function in harbor seals *(Phoca vitulina)* feeding on fish from polluted waters. *Ambio.* 23(2): 155–159.

Taffanelli, R., and Summerfelt, R.C. (1975). Cadmium induced histopathological changes in goldfish. In *Pathology of Fishes.* (W.E. Rebelin and G. Migaki, eds.). University of Wisconsin Press, Madison.

Tanabe, S., Mori, T., Tatsukawa, R., and Miyazaki, N. (1983). Global pollution of marine mammals by PCBs, DDTs and HCHs (BHCs). *Chemosphere* 12: 1269–1275.

———, Tanaka, H., and Tatsukawa, R. (1984). Polychlorobiphenyls, SDDT, and hexachlorocyclohexane isomers in the western north Pacific ecosystem. *Arch. Environ. Contam. Toxicol.* 13: 731–738.

———, Watanabe, S., Kan, H., and Tatsukawa, R. (1988). Capacity and mode of metabolism of PCB in small cetaceans. *Mar. Mamm. Sci.* 4: 103–124.

Thomas, I.G., and Ratcliffe, N.A. (1982). Integumental grafting and immunorecognition in insects. *Develop. Comp. Immunol.* 6: 643–654.

Thuvander, A. (1987). Cadmium exposure of rainbow trout, *Salmo gairdneri* Richardson: Effects on immune functions. *J. Fish Biol.* 35: 521–529.

Tomasso, J.R., Simco, B.A., and Davis, K.B. (1983). Circulating corticosteroid and leucocyte dynamics in channel catfish during net confinement. *Tex. J. Sci.* 35: 83–88.

Trizio, D., Basketter, D.A., Botham, P.A., Graepel, P.H., Lambre, C., Magda, S.J., Pal, T.M., Riley, A.J., Ronneberger, H., Van Sittert, N.J., and Bontinck, W.J. (1988). Identification of immunotoxic effects of chemicals and assessment of their relevance to man. *Fd. Chem. Toxic.* 26: 527–539.

Tryphonas, H. (1995). Great Lakes health effects: Immunotoxicity of PCBs (Aroclors). *Env. Health Perspec.* 103: 35–46.

———, Hayward, S., O'Grady, L., Loo, J.C.K., Arnold, D.L., Bryce, F., and Zawidzka, Z.Z. (1989). Immunotoxicity studies of PCB (Aroclor 1254) in the adult rhesus *(Macaca mulatta)* monkey: Preliminary report. *Int. J. Immunopharmacol.* 11: 199–206.

———, Luster, M.I., White, Jr., K.L., Naylor, P.H., Erdos, R., Burlesson, G.R., Germolec, D., Hodgen, M., Hayward, S., and Arnold, D.L. (1991). Effects of PCB (Aroclor 1254) on non-specific immune parameters in rhesus *(Macaca mulatta)* monkey. *Int. J. Immunopharmacol.* 13: 639–648.

Valembois, P., Roch, P., and Lassegues, M. (1986). Antibacterial molecules in annelids. In *Immunity in Invertebrates.* (M. Brehelin ed.), pp. 74–93. Springer-Verlag, Berlin.

Voccia, I., Krzystyniak, K., Dunier, M., Flipo, D., and Fournier, M. (1994). *In vitro* mercury-related cytotoxicity and functional impairment of the immune cells of rainbow trout *(Oncorhyncus mykiss)*. *Aquatic Toxicol.* 29: 27–48.

Vos, J.G., Kreeftenberg, J.G., Engel, H.W.B., Minderhoud, A., and Van Noorle Jansen, L.M. (1978). Studies on 2,3,7,8-tetrachlorodibenzo-*p*-dioxin–induced immune suppression and decreased resistance to infection: Endotoxin hypersensitivity, serum zinc concentrations and effect of thymosin treatment. *Toxicol.* 9: 75–86.

——— and Van Loveren, H. (1987). Immunotoxicity testing in the rat. In *Advances in Modern Environmental Toxicology, Vol. XIII: Environmental Chemical Exposures and Immune System Integrity.* (E.J. Burger, R.J. Tardiff, and J.A. Bellanti, eds.), pp. 167–180. Princeton Scientific Publishing, Princeton.

Wan, S., Lachapelle, M., Marion, M., Fournier, M., and Denizeau, F. (1993). Mechanisms of cadmium cytotoxicity effects on albumin and glutathione metabolism in primary cultures of exposed hepatocytes toxicol. *J. Tox. Env. Health* 38: 381–392.

Weeks, B.A., and Warinner, J.E. (1984). Effects of toxic chemicals on macrophage phagocytosis in two estuarine fishes. *Mar. Environ. Res.* 14: 327–335.

———, Anderson, D.P., Dufour, A., Fairbrother, A., Goven, A.J., Lahvis, G.P., and Peters, G. (1992). Immunological biomarkers to assess environmental stress. In *Biomarkers.* (R.J. Huggett, R.A. Kimerie, P.M. Mehrie, and H.L. Bergman, eds.), pp. 211–234. Levis, Boca Raton.

———, Keisler, A.S., Myrvik, Q.N., and Warinner, J.E. (1987). Differential uptake of neutral red by macrophages from three species of estuarine fish. *Develop. Comp. Immunol.* 11(1): 117–124.

———, ———, Warinner, J.E., and Mathews, E.S. (1987). Preliminary evaluation of macrophage pinocytosis as a technique to monitor fish health. *Marine Environ. Res.* 22(3): 205–214.

——— and Warinner, J.E. (1986). Functional evaluation of macrophages in fish from a polluted estuary. *Vet. Immunol. Immunopathol.* 12: 313–320.

———, ———, Mason, P.L., and McGinnis, D.S. (1986). Influence of toxic chemicals on the chemotactic response of fish macrophages. *J. Fish Biol.* 28(6): 653–658.

Weeks-Perkins, B.A., and Ellis, A.E. (1995). Chemotactic responses of Atlantic salmon *(Salmo salar)* macrophages to virulent and attenuated strains of *Aeromonas salmonicida. Fish & Shellfish Immunol.* 5(4): 313–323.

Wester, P.W., and Vos, J.G. (1994). Toxicological pathology in laboratory fish: An evaluation with two species and various environmental contaminants. *Ecotoxicol.* 2: 21–44.

Whitlock, J.P. (1990). Genetic and molecular aspects of 2,3,7,8-tetra-chlorodibenzo-*p*-dioxin action. *Ann. Rev. Pharmacol. Toxicol.* 30: 251–277.

Wojdani, A., Stein, E.A., Lemmi, C.A., and Cooper, E.L. (1982). Agglutinins and proteins in the earthworm, *Lumbricus terrestris*, before and after injection of erythrocytes, carbohydrates and other materials. *Develop. Comp. Immunol.* 6: 613–624.

Wong, S., Fournier, M., Coderre, D., Banska, W., and Krzystyniak, K. (1992). Environmental immunotoxicology. In *Animal Biomarkers as Pollution Indicators.* (D. Peakall, ed.), pp. 167–189. Chapman and Hall, London.

Zeeman, M.G., and Brindley, W.A. (1981). Effects of toxic agents upon fish immune systems: A review. In *Immunologic Considerations in Toxicology, Vol. II.* (R.P. Sharma, ed.). CRC, Boca Raton.

Chapter 8

ETHOTOXICOLOGY: AN EVOLUTIONARY APPROACH TO BEHAVORIAL TOXICOLOGY

Stefano Parmigiani,[1] Paola Palanza,[1] and Frederick S. vom Saal[2]

[1] Dipartimento di Biologia
   Evolutiva e Funzionale
   Parma University
   43100 Parma, Italy

[2] Division of Biological Sciences
   University of Missouri-Columbia
   Columbia, MO 65211, U.S.A.

Introduction

It is now recognized that a variety of anthropogenic environmental contaminants can interfere with development of the brain and reproductive system in wildlife and humans by interfering with neural and endocrine signals and other cellular functions, such as enzyme activity. These environmental chemicals are referred to as endocrine disruptors (Colborn et al., 1993). The primary focus of the research we will describe is on behavioral effects of endocrine-disrupting chemicals that mimic the action of estrogen, emphasizing consequences of exposure to low, environmentally relevant doses during critical periods in brain and organ development.

Ethotoxicology

Fetal exposure to endocrine disruptors can potentially cause impaired behavioral responsiveness to environmental demands expressed as reduced social adaptation. In our studies we have applied an evolutionary approach of the study of behavior, which is the core of ethology, to the field of toxicology. We refer to this new approach as ethotoxicology (Parmigiani et al., 1998a).

The traditional approach to the study of behavior in toxicology has involved focusing on behavioral test batteries designed to investigate effects of chemicals on learning and memory, sensory function, activity, and neuromuscular function. This experimental approach has used animals as tools to detect alterations in neural mechanisms. Thus, the issue of whether the social and environmental situations in which animals are tested are appropriate for the animal and thus ecologically relevant has not been considered to be important. In contrast, ecological relevance is a central aspect of the ethotoxicological approach to the study of behavior. In the ethotoxicological approach, animal behavior is examined in situations that approximate, as much as possible, the environment and context in which a given behavior was selected. This also creates the greatest probability that the behavior will be observed in its proper form and function (Parmigiani et al., 1999). Ethological analysis thus relies on the study of animal behavior in an evolutionary perspective and takes into account the adaptive significance of the behavior and the selective pressures that had acted on the behavior. Ethology has borrowed concepts from comparative anatomy, such as homology, to understand the evolution of behavior. This is accomplished by the comparative analysis of behavior in related species (Hall, 1994; Gilbert et al., 1996). These principles are the foundation of the ethotoxicological approach we propose for the study of behavioral effects of environmental chemicals.

The Hormonal Control of Brain and Behavioral Development

Across a wide variety of vertebrate species including humans, estrogens, androgens, and other steroid hormones influence aggressive and parental behaviors, as well as other sociosexual behaviors in both males and females. Depending on the species, testosterone can influence the development and expression of behavior in females as well as in males, and estradiol can influence the development and expression of behavior in males as well as in females (vom Saal, 1989; vom Saal et al., 1992). The sociosexual behaviors that are influenced by these hormones are critical with regard to competition for mates and resources and, thus, reproductive success. These behaviors and related physiological systems have been proposed to have evolved via the action of sexual selection. Animals that share a common phylogeny show similar morphological, physiological, and behavioral traits with common adaptive functions that were therefore likely subjected to the same selection pressures. Selective pressures appear to have operated during vertebrate evolution such that all vertebrates share a set of homologous neuroendocrine control mechanisms mediating sociosexual behaviors and reproductive functions (Nelson, 1995). Of great importance with regard to the emerging field of ethotoxicology, the underlying mechanisms (but not the effects) of the action of hormones such as estradiol are fundamentally identical across vertebrates (Katzenellenbogen et al., 1979; Pakdel et al., 1989; vom Saal, 1995).

The general assumption is that evolution has operated on developmental processes such that fitness is maximized, and the sociosexual behaviors of a particular species are adapted to the animals' environment. Therefore, the perturbation of systems that differentiate under endocrine control may result not only in the disruption of organ function but also of the individual's species specific, genetically preprogrammed behaviors, such as fighting, courtship, and other behaviors influenced by sexual selection mechanisms. Thus, the behavioral ecology of a given species may be compromised by exposure to endocrine disruptors during critical periods in brain development. The effects on social behavior may be dramatic. If animals wi05thin a population show changes in sociosexual behaviors, marked disturbance in social structure would be expected to occur. Disturbance in social structure can lead to unpredictable changes in population dynamics, and one consequence could be a population crash (vom Saal, 1984; Cowell, 1998).

What Is the Evidence for Effects of Endocrine Disruption in Wildlife and Human Populations?

A substantial literature has documented that man-made, endocrine-disrupting chemicals are altering development, leading to altered behavior and reproductive capacity in wildlife (Colborn et al., 1993). For example, reproductive system abnormalities in wildlife have been related to endocrine disruptors in fish, alligators and turtles, and birds and mammals (Colborn et al., 1993, 1998). Recent studies are now confirming that the same chemicals implicated in the adverse effects observed in wildlife are also related to detrimental effects in humans (Jacobson and Jacobson, 1996; Lonky et al., 1996; Brouwer et al., 1998; Porterfield, 1998). There are also trends in genital abnormalities in men, such as a 50-year decline in semen quality (Carleson et al., 1993; Swan, 1997), a 20-year steady increase in genital tract malformations such as cryptorchidism (undescended testes) (Radcliffe, 1986), hypospadias (malformed penis and urethra) (Paulozzi et al., 1997), and testicular and prostate cancer (Sharpe and Skakkebaek, 1993; Hass and Sakr, 1997). As yet, studies have not been conducted to determine whether there is a relationship between any of these trends and endocrine disruptors. However, environmental factors are thought to at least contribute to, or possibly even entirely account for, these trends in genital tract abnormalities observed at birth. The basis for this prediction is that testicular abnormalities, such as undescended testes, testicular cancer, and a decrease in testicular sperm production (or testicular and epididymal function; Hess et al.,

1997) occur together (Sharpe and Skakkebaek, 1993). Also, a decrease in testicular sperm production, and developmental changes and cancer in the epididymides, seminal vesicles, and prostate, as well as abnormalities such as cryptorchidism, have been experimentally produced in laboratory animals with levels of environmental estrogens similar to amounts to which humans are commonly exposed (McLachlan et al., 1975; Grocock et al., 1988; Sharpe et al., 1996; Nagel et al., 1997; vom Saal et al., 1998).

Estrogenic Endocrine Disruptors: Evolutionary Conservation of Estrogen Receptors

The emphasis of this review will be on endocrine-disrupting chemicals that can interact with estrogen receptors in cells, although there are endocrine-disrupting chemicals that can interfere with androgen receptors (Kelce et al., 1995) and thyroid hormone function (Brouwer et al., 1998; Porterfield et al., 1998), and operate via other mechanisms (Colborn et al., 1998). The hormone estradiol is identical in all vertebrates; all steroid hormones are identified by their precise cholesterol-based structure, unlike protein hormones, which can vary in amino acid sequences but have the same name in different species. In addition, the region of the classical estrogen receptor (ERα) that binds estradiol in fish is fundamentally the same as that in birds and in women (Katzenellenbogen et al, 1979; Pakdel et al., 1989; White et al., 1994). Species and tissue distribution, and binding characteristics of the recently discovered ERβ are being investigated (Paech et al., 1997; Kuiper et al., 1997).

The extreme conservation of the alpha form of the estrogen receptor over hundreds of millions of years of vertebrate evolution has profound implications with regard to estrogenic endocrine disruptors (vom Saal, 1995). A chemical that can bind to the estrogen receptor in one vertebrate is predicted to bind to estrogen receptors in any other vertebrate, and of course this includes humans. This is not intended to suggest that the outcome of binding to the receptor will be the same in different species. In addition, this does not imply that in different tissues within a species, or even within a tissue at different times in life, will the effects of binding of a chemical to estrogen receptors be the same. The outcome of binding of any estrogenic chemical to the estrogen receptor depends on the conformational change induced in the receptor, the interaction of the receptor with tissue-specific proteins associated with the transcriptional apparatus, and the specific genes associated with estrogen response elements (EREs) to which the transcriptional regulating complex of ligand, receptor, and associated proteins binds, thus regulating the process of transcription (Katzenellenbogen et al., 1996). The central issue is that there will be some physiological consequence of the event of binding of a chemical to estrogen receptors. It is the fact that some change in cell function will occur as a consequence of having estrogen receptors occupied by an environmental estrogenic endocrine disruptor that is of such importance when this occurs during critical periods in development. A change in what would have been the normal course of development will occur, and it is only the specific nature of the change (disruption) that will differ from species to species, from tissue to tissue, and as a function of the time in development that exposure occurs. The fact that there will be a consequence of receptor binding of a chemical in any vertebrate that can be predicted by studying virtually any animal model is critical with regard to recognizing the threat to the health of all species posed by endocrine-disrupting chemicals.

The Fragile Fetus:
Effects of Endocrine Disruptors on Fetal Development

The major threat associated with endocrine-disrupting chemicals is with exposure during fetal life, when interfering with normal endocrine function disrupts the course of differentiation of tissues. This is why the term endocrine disruptor was chosen for chemicals that have this effect (see Consensus Statement in Colborn and Clement, 1992, p. 2). Endocrine signals coordinate cell differentiation and organogenesis, and many of these signaling molecules, such as estradiol, regulate the course of development at much lower concentrations than had been appreciated (vom Saal et al., 1997). Alteration of the developmental program can occur as a result of changing, even slightly, the concentrations of endocrine-signaling molecules that are available to bind to receptors in target cells (due to exposure to chemicals that either mimic or antagonize hormones). This disruption of the normal developmental program can lead to irreversible changes in the functioning of organ systems throughout the remainder of life. This can occur without altering the genetic code in cells by mutations. Instead, disruption can involve altering the processes involved in turning on and off specific genes, as well as setting the rate of activity of genes that are turned on during the developmental period of cell differentiation (one example is by differential methylation of nucleotides in promoters for specific genes leading to differences in rates of transcription; Pakarinen et al., 1989; Li et al., 1997).

Exposure to estrogenic, androgenic, or thyroid hormone-disrupting chemicals in the environment during critical developmental periods in fetal life has the potential to produce permanent changes in the structure and functioning of the brain, leading to changes in behavior (vom Saal et al., 1992; Colborn et al., 1998). The timing of exposure is critical. During the period when the central nervous system is undergoing rapid change and before homeostatic (protective) mechanisms have developed, estrogenic endocrine-disrupting chemicals, at environmentally relevant concentrations within the range of exposure of human and wildlife populations, can lead to irreversible alterations in brain development. This can occur during development at exposure levels that might produce little effect in an adult (Bern, 1992; vom Saal, 1995; vom Saal et al., 1995).

The House Mouse as a Model Animal
for Studying Endocrine Disruptors

Our premise is that a model animal, such as the house mouse *(Mus musculus domesticus)*, can be used with the ethotoxicological approach to understand more general principles of actions of endocrine disruptors on animals in different taxa. Thus, our research findings can also plausibly be applied to predict effects of chemicals in other species, including humans, that cannot be subjected to experimental studies (Parmigiani et al., 1998b; vom Saal et al. 1998). This is possible since the development of physiological systems, at the mechanistic level, is highly conserved (similar to that for estrogen receptors) throughout vertebrate evolution. The high degree of conservation of developmental mechanisms among metazoans has only recently been revealed by the tools of molecular genetics (Gilbert et al., 1996).

We have been investigating the effects of exposure to estrogenic chemicals during fetal life on the subsequent sociosexual behaviors of both male and female house mice. We use Swiss albino outbred stocks of mice (CD-1 and CF-1) in our studies because they are still very similar (especially the male) to wild mice in their social behaviors (Parmigiani et al., 1989). The house mouse is widely distributed throughout the world and thus has been subjected to varied ecological pressures (Bronson, 1979). For this reason, the house mouse is one of the most versatile mammalian species. House mice live in environments as diverse as fields, cold stores, warehouses, hayricks, Pacific atolls, and islands close to Antarctica. Although there are reports of feral populations of this species living totally apart from humans, house mice live most commonly as commensals of man, and mice are thus likely to be exposed to many of the same environmental pollutants as humans (Berry, 1989).

House mice exhibit different social organizations, based on prevailing local conditions of food distribution, available cover, and animal density. In particular, the social organization in different populations of house mice reflects variations in female and male social aggression. House mice thus show flexibility in social behavior and social structure, as well as rapid adaptation (Brain and Parmigiani, 1990). This feature of house mice makes them an interesting model animal for use in studying the impact of environmental endocrine disruptors on behavior and social dynamics, and the relationship between variation in the social structure of populations and pollutants in the environment. House mice also provide the opportunity to examine interindividual variability in behavioral strategies in response to developmental exposure to endocrine disruptors, due to our knowledge of sources of individual differences in endogenous hormones (vom Saal, 1989).

Development of Sociosexual Behaviors and Accessory Reproductive Organs in Mice Prenatally Exposed to Estrogenic Endocrine Disruptors

Based on our evolutionary perspective of reproductive function and behavior, our interest has been in determining the degree to which endogenous hormones, as well as environmental toxicants that act as endocrine disruptors, can perturb development, thus impacting reproduction and sociosexual behaviors. One primary concern is the long-term effects of endocrine disruptors on behavioral interactions within species and with their environments. In social species, such as house mice, an important consequence of intrasex aggression is the regulation of the density of animals, leading to an appropriate spacing. Since sex steroids play a critical role in regulating the development of the neural areas mediating aggression, as well as the expression of aggression in adulthood (in species that have the genetic predisposition for aggressiveness), environmental chemicals that interfere with the normal actions of sex steroids have the potential to alter levels of aggressiveness and other territorial behaviors, such as urine-marking of the environment, in exposed animals. An assumption of our ethotoxicological investigations is that environmental chemicals that alter aggressiveness or other sociosexual behaviors will lead to changes in social interactions, which will be reflected by changes in population dynamics (vom Saal, 1984; Parmigiani et al., 1994).

Urine marking and intermale aggression are mediated by similar neuroendocrine and genetic mechanisms (Parmigiani et al., 1989; Parmigiani and Palanza, 1991; Ferrai et al., 1996). Urine marking is influenced by a male's social status and testosterone levels, both in terms of qualitative differences (urine composition) and quantitative differences (rate of urine deposition). Male mouse urine contains olfactory cues (pheromones) that affect both the behavior and the physiology of other mice. For example, male urine elicits intermale aggressive behavior, is attractive to females, contains primer pheromones that accelerate puberty, induces estrus and blocks pregnancy in unfamiliar females, and also stimulates interfemale aggression (Parmigiani et al., 1989; Palanza et al., 1994). Males urine-mark to advertise their dominance status, and it is likely that this plays a role in territorial defense against potential intruders. The rate of urine marking in response to social stimuli can thus be a useful indicator of a male's social rank and territorial defense potential, which provides one explanation for the use by females of pheromonal cues in male urine to assess the reproductive fitness of males (Bronson, 1979).

We are also interested in anatomophysiological correlates of sociosexual behaviors. It is well known that in mice, preputial glands produce pheromones that are involved in social communication between males and females (Caroom and Bronson, 1971) and influence aggressiveness between males (Mugford and Nowell, 1972; Ingersoll et al., 1986). Preputial gland secretions pass through ducts that empty into the prepuce, which is specially adapted in mice for depositing urine marks (Maruniak et al., 1975). The placing of these pheromones into a male mouse's environment is thus via urine-marking behavior, which is influenced by dominance status. Based on these findings, we not only analyzed territorial (urine-marking) behavior and aggression but also sex accessory glands, such as the preputial glands, associated with the formation and release of pheromones. We also examined the prostate, which is involved in producing seminal fluid. The development of these accessory reproductive organs is regulated by sex hormones. Small changes in circulating estradiol during fetal life alters prostate development in male mice (Nonneman et al., 1992; vom Saal et al., 1997). Thus, it is hypothesized that estrogenic endocrine disruptors would alter the normal development of these organs that influence social interactions and fertility, in addition to altering adult sociosexual behaviors.

Insecticides with Estrogenic Activity

DDT and methoxychlor are insecticides that have the capacity to disrupt the endocrine system (Colborn et al., 1993). o,p'DDT is a contaminant in commercial DDT, and both o,p'DDT and methoxychlor are estrogenic endocrine disruptors (Fry and Toone, 1981; Johnson et al., 1988; Gray et al., 1989; vom Saal et al., 1995). In contrast, the *in vivo* metabolite of p,p'DDT is p,p'DDE, which binds to androgen receptors and acts as an androgen antagonist in rodents (Kelce et al., 1995). We have examined the effects of exposure of male mice to o,p'DDT and methoxychlor during fetal life on territorial urine-marking behavior, the development of reflexes during neonatal life, adult urine marking and intermale aggression, and accessory reproductive organs and testes.

We conducted an experiment in which pregnant CF-1 mice were fed with different doses of the drug diethylstilbestrol (DES), which served as a positive control for estrogen action, and two insecticides: o,p'DDT and methoxychlor (vom Saal et al., 1995). Each chemical was dissolved in tocopherol-stripped corn oil and fed (using a pipetter to deliver an accurate volume) to pregnant female mice. Pregnant mice were allowed to drink the oil solution (which they readily consumed) after placing the oil into the mouth rather than being subjected to gavage. The objective was to minimize stress that would be associated with placing a tube down the animals' throat into their stomachs. We fed each pregnant mouse a chemical 1 time per day from gestation day 11 to 17 (mating = day 0) during the time that the initial development of the brain and reproductive organs occurs in fetuses (vom Saal et al., 1992). The male off-spring were examined in adulthood for urine-marking behavior (number of urine marks deposited) when placed into a novel environment.

We found that a low dose of DES (20 ng/kg maternal body weight/day; 20 parts per trillion, ppt) significantly increased urine-marking behavior relative to control males (produced by mothers that were unhandled and fed just the oil vehicle), and as the dose increased to 2 µg/kg (2 parts per billion, ppb), rates of urine marking increased. Interestingly, the result showed an inverted-U dose response, in that males whose mothers were fed a 200 µg/kg/day dose of DES showed significantly lower rates of urine marking than did males produced by mothers fed 2 µg/kg (vom Saal et al., 1995). This type of inverted-U dose–response function has also been observed in other studies of the effects of fetal exposure to estrogens during development (vom Saal et al., 1997).

We observed similar effects on urine-marking behavior in response to maternal ingestion of o,p'DDT at a dose between 1,000 to 10,000 times higher than DES. Methoxychlor showed a similar effect at a dose between 10,000 to 100,000 times higher than DES. It appears that we did not administer a high enough dose of these insecticides to see an inhibition of urine marking at very high doses (vom Saal et al., 1995). The doses of these chemicals that resulted in changes in behavior are within the range of human exposure and are thus environmentally relevant (ATSDR, 1993, 1994). In summary, using much lower (ppt to ppb) doses of estrogenic chemicals than have previously been examined, prenatal exposure to DES, o,p'DDT, and methoxychlor produced alterations in urine-marking behavior. This suggested that a possible concomitant increase at low doses in intermale aggression might also be observed in male mice exposed prenatally to estrogenic chemicals.

We recently confirmed this prediction in a study in which 20 and 200 ng/kg/day doses of DES and 20 and 200 µg/kg/day doses of o,p'DDT were administered using the same procedures described above, but in this experiment we used another stock of outbred Swiss mice (CD-1; Palanza et al., 1999). Pregnant female mice were fed a chemical as described above, and the male offspring in each group were examined in adulthood for aggressive behavior (biting and chasing the opponent) during a 10-min test with an intruder who was placed into the home cage of the experimental animal. Relative to control (unhandled and vehicle-fed) animals, the proportion of males exhibiting aggression was significantly increased by exposure to the 20 and 200 ng/kg/day DES doses. Males prenatally exposed to 20 and 200 µg/kg/day dose of o,p'DDT were also more likely than controls to attack same-sex intruders. Males exposed to the 20 µg/kg/day dose of DDT were examined for preputial gland and testis weight after completion of the behavior experiment. Relative to controls, there was a significant increase in preputial gland size and a significant decrease in testis weight in DDT-exposed males. We previously have shown that prenatal exposure to a small increase in estradiol, or in response to low doses of estrogenic chemicals, such as DES, results in an increase in adult prostate weight (vom Saal et al., 1997). In addition, the monomer used to make polycarbonate plastic, bisphenol A, was found to increase prostate weight and decrease daily sperm production in male offspring of pregnant CF-1 mice administered 2 or 20 μg/kg/day as described above (Nagel et al., 1997; vom Saal et al., 1998).

The Merging of Ethopharmacology and Ethotoxicology for Understanding the Neurochemical Effects of Endocrine Disruptors

Steroid hormones are involved during fetal life in setting the number of the brain receptors for steroids and neurotransmitters. These anatomical and physi-ological data parallel alterations in sociosexual behaviors. The finding of alterations in aggressive behaviors by endocrine disruptors due to exposure during fetal life suggests that these compounds might interfere with the normal development of serotonergic and GABA-ergic receptors systems, thus impairing an individual's coping strategies in response to stressful situations. This raises the possibility of using ethopharmacological techniques to understand the impact of these compounds on neurotransmitter systems.

Ethopharmacology combines the evolutionary perspective of ethological analysis with psychoactive drug administration to understand mechanisms and adaptive functions of behavior. This evolutionary approach to behavioral phar-macology has been recently applied to mice (Palanza et al.,1994; Parmigiani et al., 1998b). The modification of behavior and the potential alteration of func-tion observed in drug-treated animals helps to clarify proximate (neurochem-ical substrates) and ultimate (adaptive significance) causation of a behavior and increases the understanding of drug action on brain and behavior.

For example, a serotonergic drug, Fluprazine (a phenylpiperazine derivative with antiaggressive properties), has been used as a tool to probe the neuro-chemical mechanisms underlying different forms of competitive aggression (intrasexual attack and infanticide) in male and female mice. As previously stated, these behaviors are very important in sexual selection mechanisms, in shaping social structure, and in population dynamics in the house mouse. Fluprazine (1, 2, and 5 mg/kg) dose-dependently inhibited attack by resident territorial males and females on same-sex conspecific intruders, and also inhib-ited the killing of genetically unrelated pups. However, predatory attack on an insect larva was unaltered by any dose of Fluprazine. These data showed that neurochemical substrates underlying intrasexual attack and infanticide may be similar to each other but different from those that regulate predatory attack both in male and female mice. This finding supports the hypothesis that male and female intrasex aggression and infanticide are competitive forms of aggres-sion and share similar neural and motivational control and, hence, function, rather than infanticide being an expression of intaspecific predation or canni-balism (Parmigiani and Palanza, 1991; Palanza et al., 1994; Ferrai et al., 1996, Parmigiani et al., 1998). What is important in these studies is the fact that drugs can change the action of neuromodulators in a way that alters the function of a given behavior in the context that it was selected for, thus undermining the individual response and fitness in that particular context. This evolutionary approach to behavioral pharmacology allows a better understanding not only of drug action on neurochemical substrates; the approach also allows us to more fully understand the ultimate causation of behavior, that is, its adaptive signifi-cance (Parmigiani et al., 1998b). In this perspective, the merging of ethotoxico-logical and ethopharmacological approaches offers a powerful way of identifying the neurochemical mechanisms of behaviors altered by environ-mental chemicals. The administration of psychoactive drugs to animals prena-tally exposed to endocrine disruptors may reveal possible neurochemical alterations and consequent impairment of adaptive function of the behaviors.

The focus on proximate mechanisms of behavior involves the study of indi-viduals within populations. This has been the focus of toxicological research involving laboratory animals. Ethotoxicological analysis involves examination of alterations of proximate mechanisms due to exposure to environmental chemi-cals, and thus their effects on an individual's behavior. But this information is used to provide the basis for understanding changes at the population level, which has been the focus of ecologists interested in the effects of pollutants on populations (for example, emigration and immigration rates and distribution of animals of different ages and sex).

Summary

Our findings show that exposure during fetal life to very low doses of endocrine-disrupting estrogenic chemicals, within the range of exposure to these chemicals by humans and wildlife, can alter the development of sociosex-ual behaviors as well as the size and functioning of reproductive organs. It is generally assumed that natural selection operates to create a phenotype that is optimum for a particular environment. If exposure to endocrine disruptors changes that phenotype, leading to a less than optimum set of traits, such as an altered level of aggressiveness for that environment, a negative impact on those individuals in the population is likely to occur and changes in population dynamics will likely follow.

While sociosexual behaviors of a particular species are adapted to specific environmental conditions (Krebs and Davies, 1981), there are many factors that give rise to individual differences in these behaviors (vom Saal, 1989). Consequently, there is an evolved range of social behaviors that occurs among animals within any population due to variation in genotype, hormone levels, experience, etc. Shifts in the proportion of animals within a population that show specific traits, such as increased aggressiveness, can influence social structure and population dynamics (vom Saal, 1984; Brain and Parmigiani, 1990; Cowell et al., 1998).

Natural and sexual selection operate at the level of phenotype, not directly on genes. Thus, even though environmental factors may not alter genes via mutations, they do alter phenotype and thus the object of selection. It is well recognized in toxicology that not all individuals are equally sensitive to environmental chemicals, and there has long been a concern with sensitive subpopulations that show a greater response to chemical exposure than other members of a population. Thus, when exposed to endocrine disruptors during development, some individuals will be shifted in phenotype to a greater degree than will other individuals. One consequence will be a decrease in the range of phenotypes in the population and possibly a decrease in population variance. Our hypothesis is that this will also lead to a "feed forward" process that leads to a change in genotype within the population, that is, a change in the course of evolution. The genetically predisposed individuals that show the greatest change in phenotype in response to endocrine disruptors could have had, in an uncontaminated environment, the optimum phenotype for the environment. Subsequent to developmental exposure to endocrine disruptors, this phenotype could be lost. In fact, the phenotype resulting from chemical exposure of these sensitive individuals may now be the least optimum for the environment. By this mechanism, pollutants operating on phenotype may eventually lead to a change in gene frequency in the population.

Our ethotoxicological approach reveals the importance of evolutionary processes for the understanding of the potential effects both on individuals and on populations due to exposure to endocrine disruptors. This requires multidisciplinary studies in which analysis of the effects of these chemicals is undertaken using molecular, cellular, physiological, behavioral, and ecological methods in order to understand the full range of their effects.

References

Adkins-Regan, E. (1983). Sex steroids and the differentiation and activation of avian reproductive behavior. In *Hormones and Behaviour in Higher Vertebrates* (J. Balthazart, E. Prove, and R. Giles, eds.), pp. 218–228. Springer-Verlag, Berlin.

ATSDR (1993). Toxicological Profile for DDT. Agency for Toxic Substances and Disease Registry, U.S. Department of Health and Human Services, Atlanta.

——— (1994). Toxicological Profile for Methoxychlor. Agency for Toxic Substances and Disease Registry, U.S. Department of Health and Human Services, Atlanta.

Bern, H.A. (1992). "The fragile fetus." In *Chemically Induced Alterations in Sexual and Functional Development: The Wildlife/Human Connection* (T. Colborn and C. Clement, eds.), pp. 9–16. Princeton Scientific Publishing, Princeton.

Berry, R.J. (1989). Genes, behaviour and fitness in mice: Concepts and confusions. In *House Mouse Aggression*. (P. Brain, D. Mainardi, and S. Parmigiani, eds.), pp. 23–48. Harwood Academic Publishers, Chur.

Brain, P.F., and Parmigiani, S. (1990). Variation in aggressiveness and social structure in the house mouse populations. *Biol. J. Linnean Soc.* 41: 257–269.

Bronson, F.H. (1979). The reproductive ecology of the house mouse. *Q. Rev. Biol.* 54: 246–299.

Brouwer, A., Morse, D.C., Lans, M.C., Schuur, A.G., Murk, A.J., Klasson-Wehler, E., Bergman, A., and Visser, T.J. (1998). Interactions of persistent environmental organohalogens with the thyroid hormone system: Mechanisms and possible consequences for animal and human health. *Tox. Ind. Health* 14: 59–84.

Carlsen, E., Giwereman, A., Keding, N., and Skakkebaek, N. (1995). Declining semen quality and increasing incidence of testicular cancer: Is there a common cause? *Environ. Health Perspec.* 103 (Suppl. 7): 137–139.

Caroom, D., and Bronson, F.H. (1971). Responsiveness of female mice to preputial attractant: Effects of sexual experience and ovarian hormones. *Physiol. Behav.* 7: 659–662.

Colborn, T., and Clement, C. (1992). Chemically-induced alterations in sexual and functional development: The wildlife/human connection. In *Advances in Modern Environmental Toxicology* (M.A. Mehlman, ed.), p. 403. Princeton Scientific Publishing, Princeton.

————, Smolen, M.J., and Rolland, R. (1998). Environmental neurotoxic effects: The search for new protocols in functional teratology. *Environ. Health Perspec.* 14: 9–23.

————, vom Saal, F.S., and Soto, A.M. (1993). Developmental effects of endocrine-disrupting chemicals in wildlife and humans. *Environ. Health Perspec.* 101: 378–384.

Cowell, L.G., Crowder, L.B., and Kepler, T.B. (1998). Density-dependent prenatal androgen exposure as an endogenous mechanism for the generation of cycles in small mammal populations. *J. Theor. Biol.* 190: 93–106.

Ferrai, P.F., Palanza, P., Rodgers, J.R., Mainardi, M., and Parmigiani, S. (1996). Comparing different forms of male and female aggression in wild and laboratory mice: An ethopharmacological study. *Physiol. Behav.* 60: 549–554.

Fry, D.M., and Toone, C.K. (1981). DDT-induced feminization of gull embryos. *Science* 213: 922–924.

Gilbert, S.E.A. (1996). Resynthesizing evolutionary and developmental biology. *Develop. Biol.* 173: 357–372.

Gray, L.E., Ostby, J., Ferrell, J., Rehnberg, G., Linder, R., Cooper, R., Goldman, J., Slott, V., and Laskey, J. (1989). A dose-reponse analysis of methoxychlor-induced alterations of reproductive development and function in the rat. *Fund. Appl. Toxicol.* 12: 92–108.

Grocock, C.A., Charlton, H.M., and Pike, M.C. (1988). Role of the fetal pituitary in cryptorchidism induced by exogenous maternal oestrogen during preganacy in mice. *J. Reprod. Fertil.* 83: 295–300.

Hall, B.K., ed. (1994). *Homology: The Hierarchical Basis of Comparative Biology.* Academic, New York.

Hass, G.P., and Sakr, W.A. (1997). Epidemiology of prostate cancer. *CA Cancer J. Clin.* 47: 273–287.

Hess, R.A., Bunick, D., Ki-Ho, L., Bahr, J., Taylor, J.A., Korach, K.S., and Lubahn, D.B. (1997). A role for oestrogens in the male reproductive system. *Nature* 390: 509–512.

Ingersoll, D.W., Morley, K.T., Benvenga, M., and Hands, C. (1986). An accessory sex gland aggression-promoting chemosignal in male mice. *Behav. Neurosci.* 100: 187–191.

Jacobson, J.L., and Jacobson, S.W. (1996). Intellectual impairment in children exposed to polychlorinated biphenyls in utero. *N. Engl. J. Med.* 335: 783.

Johnson, D.C., Kogo, H., Sen, M., and Dey, S.K. (1988). Multiple estrogenic action of o,p'-DDT: Initiation and maintenance of pregnancy in the rat. *Toxicol.* 53: 79–87.

Katzenellenbogen, B.S., Katzenellenbogen, J.A., and Mordecai, D. (1979). Zearalenones: Characterization of the estrogenic potencies and receptor interactions of a series of fungal-resorcylic acid lactones. *Endocrinol.* 105: 33–40.

Katzenellenbogen, J.A., O'Malley, B.W., and Katzenellenbogen, B.S. (1996). Tripartite steroid hormone receptor pharmacology: Interaction with multiple effector sites as a basis for the cell- and promoter-specific action of these hormones. 10: 191–231.

Kelce, W.R., Stone, C.R., Laws, S.C., Gray, L.E., Kemppainen, J.A., and Wilson, E.M. (1995). Persistent DDT metabolite p,p'-DDE is a potent androgen receptor antagonist. *Nature* 375: 581–585.

Krebs, J.R., and Davies, N.B. (1981). *An Introduction to Behavioral Ecology.* Sinauer, Sunderland.

Kuiper, G.G.J.M., Carlsson, B., Grandien, K., Enmark, E., Häggblad, J., Nilsson, S., and Gustafsson, J.-A. (1997). Comparison of the ligand binding specificity and transcript tissue distribution of estrogen receptors alpha and beta. *Endocrinol.* 138: 863–870.

Li, S., Washburn, K.A., Moore, R., Uno, T., Teng, C., Newbold, R.R., McLachlan, J.A., and Megishi, M. (1997). Developmental exposure to diethylstilbestrol elicits demythelation of estrogen-responsive lactoferrin gene in mouse uterus. *Cancer Res.* 57: 4356–4359.

Lonky, E., Reihman, J., Darvill, T., Mather, J., and Daly, H. (1996). Neonatal behavioral assessment scale performance in humans influenced by maternal consumption of environmentally contaminated Lake Ontario fish. *J. Great Lakes Res.* 22: 198–212.

Maruniak, J.A., Desjardins, C., and Bronson, F.H. (1975). Adaptations for urinary marking in rodents: Prepuce length and morphology. *J. Reprod. Fertil.* 44: 567–570.

McLachlan, J., Newbold, R., and Bullock, B. (1975). Reproductive tract lesions in male mice exposed prenatally to diethylstilbestrol. *Science* 190: 991–992.

Mugford, R.A., and Nowell, N.W. (1972). The dose-response to testosterone propionate of preputial glands, pheromones and aggression in mice. *Horm. Behav.* 3: 39–46.

Nagel, S.C., vom Saal, F.S., Thayer, K.A., Dhar, M.G., Boechler, M., and Welshons, W.V. (1997). Relative binding affinity-serum modified access (RBA-SMA) assay predicts the relative in vivo bioactivity of the xenoestrogens bisphenol A and octylphenol. *Environ. Health Perspec.* 105: 70–76.

Nelson, R.J. (1995). An introduction to behavioral endocrinology. Sinauer, Sunderland.

Nonneman, D., Ganjam, V., Welshons, W., and vom Saal, F. (1992). Intrauterine position effects on steroid metabolism and steroid receptors of reproductive organs in male mice. *Biol. Reprod.* 47: 723–729.

Paech, K., Webb, P., Kuiper, G.G.J.M., Nilsson, S., Gustafsson, J.-A., Kushner, P.J., and Scanlan, T.S. (1997). Differential ligand activation of estrogen receptors by ERα and ERβ at AP1 sites. *Science* 277: 1508–1510.

Pakarinen, P., and Huhtaniemi, I. (1989). Gonadal and sex steroid feedback regulation of gonadotrophin mRNA levels and secretion in neonatal male and female rats. *J. Mol. Endocrinol.* 3: 139–144.

Pakdel, F., Le Guellec, C., Vaillant, C., Gaelle Le Roux, M., and Valotaire, Y. (1989). Identification and estrogen induction of two estrogen receptors (ER) messenger ribonucleic acids in the rainbow trout liver: Sequence homology with other ERs. *Mol. Endocrinol.* 3: 44–51.

Palanza, P., Parmigiani, S., Huifer Liu, H., and vom Saal, F. (1999). Prenatal exposure to low doses of the estrogenic chemicals diethylstilbestrol and o,p'-DDT alters agressive behavior of male and female house mice. *Pharmacol. Biochem. Behav.* In press.

———, vom Saal, F.S. and Parmigiani, S. (1994). Male urinary cues stimulate intrasexual aggression and urine-marking in wild female mice, *Mus musculus domesticus. Anim. Behav.* 48: 245–247.

Parmigiani, S., and Palanza, P. (1991). Fluprazine inhibits intermale attack and infanticide, but not predation, in male mice. *Neurosci. Biobehav. Rev.* 15: 511–513.

————, Brain, P.F., and Palanza, P. (1989). Ethoexperimental analysis of different forms of intraspecific aggression in the house mouse. In *Ethoexperimental Approaches to the Study of Behavior.* (R. Blanchard, P. Brain, D. Blanchard, and S. Parmigiani, eds.), pp. 418–431. Kluwer, Dordrecht.

————, Palanza P. , and vom Saal, F.S. (1998a). Etho-toxicology: An evolutionary approach to the study of enironmental endocrine-disrupting chemicals. *Tox. Ind. Health* 14: 333–339.

————, ————, Mainardi D., and Brain, P.F. (1994). Infanticide and protection of young in house mice *(Mus domesticus)*: Female and male strategies. In *Infanticide and Parental Care.* (S. Parmigiani and F. vom Saal eds.), pp. 341–363. Harwood Academic Publishers, Chur.

————, Ferrari P.F., and Palanza P. (1999) An evolutionary approach to behavioral phramacology: Using drugs to understand proximal mechanisms of different forms of aggression in mice. *Neurosci. Biobehav. Rev.,* In press.

Paulozzi, L.J., Erickson, J.D., and Jackson, R.J. (1997). Hypospadias trends in two US surveillance systems. *Pediatrics* 100: 831–834.

Porterfield, S., and Hendry, L.B. (1998). Impact of PCBs on thyroid hormone directed brain development. *Tox. Ind. Health* 14: 103–120.

Radcliffe, John (1986). Cryptorchidism: An apparent substantial increase since 1960. *Br. Med. J.* 293: 1401–1404.

Sharpe, R.M., Fisher, J.S., Millar, M.M., Jobling, S., and Sumpter, J.P. (1996). Gestational and lactational exposure of rats to xenoestrogens results in reduced testicular size and sperm production. *Environ. Health Perspec.* 103: 1136–1143.

————, and Skakkebaek, N.E. (1993). Are oestrogens involved in falling sperm count and disorders of the male reproductive tract? *Lancet* 341: 1392–1395.

Swan, S.H., Elkin, E.P., and Fenster, L. (1997). Have sperm densities declined? A reanalysis of global trend data. *Environ. Health Perspec.* 105: 1228–1232.

vom Saal, F.S. (1989). Sexual differentiation in litter bearing mammals: Influence of sex of adjacent fetuses in utero. *J. Anim. Sci.* 67: 1824–1840.

———— (1984). The intrauterine position phenomenon: Effects on physiology, aggressive behavior and population dynamics in house mice. In *Biological Perspectives on Aggression.* (K. Flannelly, R. Blanchard, and D. Blanchard, eds.), pp. 135–179. Liss, New York.

———— (1995). Environmental estrogenic chemicals: Their impact on embryonic development. *Hume Ecol. Risk Assess.* 1: 3–15.

————, Cooke, P.S., Palanza, P., Thayer, K.A., Nagel, S., Parmigiani, S., and

Welshons, W.V. (1998). A physiologically based approach to the study of bisphenol A and other estrogenic chemicals on the size of reproductive organs, daily sperm production, and behavior. *Toxicol. Ind. Health* 14: 239–260.

———, Montano, M.M., and Wang, H.S. (1992). Sexual differentiation in mammals. In *Chemically Induced Alterations in Sexual and Functional Development: The Wildlife/Human Connection*. (T. Colborn and C. Clement, eds.), pp. 17–83. Princeton Scientific Publishing, Princeton.

———, Nagel, S.C., Palanza, P., Boechler, M., Parmigiani, S., and Welshons, W.V. (1995). Estrogenic pesticides: Binding relative to estradiol in MCF-7 cells and effects of exposure during fetal life on subsequent territorial behavior in male mice. *Toxicol. Lett.* 77: 343–350.

———, Timms, B.G., Montano, M.M., Palanza, P., Thayer, K.A., Nagel, S.C., Dhar, M.D., Ganjam, V.K., Parmigiani, S., and Welshons, W.V. (1997). Prostate enlargement in mice due to fetal exposure to low doses of estradiol or diethylstilbestrol and opposite effects at high doses. *Proc. Natl. Acad. Sci.* 94: 2056–2061.

White, R., Jobling, S., Hoare, S.A., Sumpter, J.P., and Parker, M.G. (1994). Environmentally persistent alkylphenolic compounds are estrogenic. *Endocrinol.* 135: 175–182.

Chapter 9

EMBRYONIC AND NEONATAL EXPOSURE TO ENDOCRINE-ALTERING CONTAMINANTS: EFFECTS ON MAMMALIAN FEMALE REPRODUCTION

Taisen Iguchi

Department of Biology
Yokohama City University
Kanazawa-ku, Yokohama 236, Japan

Introduction

Perinatal estrogen exposure serves as a model for assessing the potential toxicity of a broad range of chemicals that have estrogenic activity—that is, actions mimicking that of estrogen (for reviews see Bern, 1992a, b; Iguchi, 1992). These chemicals, which are widespread in the environment, include some chlorinated organic compounds, polycyclic aromatic hydrocarbons, herbicides, and pharmaceutical agents (for reviews see Colborn and Clement, 1992; Colborn et al., 1993; Guillette, 1995). It has been hypothesized that such chemicals may contribute directly or indirectly to the increasing rate of breast cancer (Davis and Bradlow, 1995) and disorders of the male reproductive tract (Peterson et al., 1992). Several of these chemicals, including dichlorodiphenyl-trichloroethane (DDT), 2,3,7,8-tetrachlorodibenzo-p-dioxin (TCDD, dioxin), and methoxychlor (an estrogenic pesticide currently used as a substitute for DDT) have reproductive effects similar to those of DES after *in utero* exposure in rodents (Cooke and Eroschenko, 1990; Roman et al., 1995).

In female embryonic mice, perinatal sex-hormone exposure induces anovulatory sterility (caused by permanent alterations of hypothalamo-hypophysial-ovarian system) and lesions in mammary gland, ovary, oviduct, uterus, cervix, and vagina (Raynaud, 1961; Takasugi et al., 1962; Dunn and Green, 1963; Takasugi and Bern, 1964). These early studies of mice emphasized the possible relevance of the mouse findings to the development of cancer in humans. After these initial studies, Herbst et al. (1970, 1971, 1972, 1975) demonstrated a close correlation between occurrence of vaginal clear-cell adenocarcinoma and early intrauterine exposure to a synthetic estrogen, diethylstilbestrol (DES), in young women (for review see Herbst and Bern, 1981). Mice exposed perinatally to estrogens, including DES, provide a model for exploration of the consequences of DES exposure in the human, because mouse genital tract development at birth is similar to that of the human fetus at the end of the first trimester. The mouse model has demonstrated the chronic effects of perinatal sex hormone exposure on the female reproductive tract (for reviews see Bern et al., 1976; Takasugi, 1976, 1979; Forsberg, 1979; Herbst and Bern, 1981; Walker, 1984; Mori and Nagasawa, 1988; Kincle, 1990; Bern, 1992a, b; Iguchi, 1992; Iguchi and Ohta, 1996; Iguchi and Bern, 1996; Newbold and McLachlan, 1996). Perinatal treatment of female mice with natural and synthetic steroids—including DES, androgens, phytoestrogens, and tamoxifen, an antiestrogen—induces various abnormalities in the reproductive tract, including ovary-independent vaginal keratinization, adenosis, and tumors; uterine hypoplasia, epithelial metaplasia, and tumors; oviducal tumors; and polynuclear oocytes, polyovular follicles, and polyfollicular ovaries (for reviews see Iguchi, 1992; Iguchi and Ohta, 1996). The growth response of neonatally DES-exposed reproductive organs to estrogen is reduced (Mair et al., 1985; Iguchi and Takasugi, 1987), as are estrogen receptor levels (Bern et al., 1987), epidermal growth factor (EGF) receptor levels (Iguchi et al., 1993), along with other hormone receptor levels (for reviews see Bern, 1992a, b). This chapter emphasizes the utility of the neonatal mouse model in indicating permanent changes, both overt and cryptic, in a variety of female reproductive structures, with particular attention to newer information of estrogenic compounds on estrogen receptor (ER) expression, oncogene expression, reproductive abnormalities, fertilizability, decidual reaction, complications of pregnancy, and skeletal and muscular tissues.

Estrogen Receptor Interactions

Estrogen action is mediated through a specific receptor in estrogen target tissues (see Chap. 4). These receptors are present early in embryonic development; indeed, Hou and Gorski (1993) and Gorski and Hou (1995) demonstrated that ER mRNA is present in mouse oocytes and fertilized eggs; message concentration declined at the 2-cell stage and reached its lowest level at the 5- to 8-cell stage. Progesterone receptor (PR) mRNA was not detectable until the blastocyst stage. ER mRNA was not detectable at the morula stage but reappeared at the blastocyst stage. Greco et al. (1991) demonstrated ER by immunoblotting on fetal day 10 in extracts of whole mouse embryos and by immunostaining in female reproductive organs at fetal day 15.

The uterus of postnatal mice does not express ER protein in epithelial cells until at least the third postnatal day (Korach et al., 1988; Yamashita et al., 1989; Bigsby et al., 1990; Greco et al., 1991; Sato et al., 1992, 1996). However, a single injection of 17β-estradiol (E2) or DES on the day of birth (day 0) induces uterine epithelial cell ER protein expression in a dose-dependent fashion (Yamashita et al., 1990; Sato et al., 1992). Sato et al. (1996) showed that ER mRNA and ER protein in uterine epithelial cells were induced 4 and 12 hr later, respectively, by a single injection of DES on day 0. The potency of 50 μg E2 in inducing ER is equivalent to that of 0.3 μg DES. Neonatal injection of 3 μg DES induced ER and also significantly stimulated cell division of the uterine epithelium (Sato et al., 1996). Exposure of more than 0.1 μg DES or 10 μg E2 to newborn mice induced abnormalities in reproductive tracts in later life (Takasugi, 1976; Iguchi, 1992). The doses of DES and E2 that induce ER in neonatal mouse uterine epithelial cells are correlated with those that induce persistent changes in reproductive tracts in mice. ER was increased in uterine epithelial cells by testosterone given on day 0, but not by 5α-dihydrotestosterone (5α-DHT) (Sato et al., 1996), suggesting that testosterone acted as an ER inducer directly or after conversion into estrogen (Ryan et al., 1972; Iguchi et al., 1988). In the rat uterus, ER immunoreaction appeared in epithelial cells at day 5 and in stromal cells at day 1, and two daily injections of E2 induced ER in epithelial cells. PR immunoreaction was detected at day 5 in rat uterine epithelial cells and day 12 in stromal cells, which was not affected by ovariectomy or E2 administration (Ohta et al., 1996). ER was also present in periosteum of pubic bones and bone cells of all parts of pubis at 10 to 60 days (Uesugi et al., 1992) and in anococcygeus muscle cells until 60 days in mice (Iguchi et al., 1995), which was also permanently affected by neonatal exposure to estrogenic compounds.

DES is not the only synthetic steroid that affects development of the female reproductive system. Tamoxifen is a nonsteroidal triphenylethylene derivative that binds to ER (Katzenellenbogen et al., 1983; Ignar-Trowbridge et al., 1991), acting as an estrogen agonist or antagonist (Gronemeyer et al., 1992). Neonatal tamoxifen exposure (100 μg) induces abnormalities in the uterus, ovary, and pelvic bones in mice (Iguchi, 1992). A single injection of tamoxifen on day 0 induces ER protein expression in uterine epithelial cells 24 hr later, with the staining intensity increasing from 24 to 48 hr after the injection. However, the expression level of ER mRNA in uterine epithelial cells remains low for 24 hr after the injection of tamoxifen. Tamoxifen also induces ER in uterine epithelial cells, although it is weaker in action than DES (Sato et al., 1996). The time course of nuclear binding of E2 and tamoxifen in the rat uterus are different: Binding of E2 is maximized 2 hr after injection, whereas binding of tamoxifen is maximized at 8 hr after injection (Ennis and Stumpf, 1988). Binding affinities of tamoxifen to ER and DES to ER are 1/50 and 1/1.25 of that of E2 (Korach et al., 1978; Miller et al., 1986). The differences in time course of nuclear binding and/or affinity for ER between E2 and tamoxifen may be ascribed to the difference in reaction time and/or amount of ER induced by DES and tamoxifen

In situ hybridization studies reveal that both DES and tamoxifen induce ER mRNA in uterine and vaginal epithelial cells 12 hr after injection in 50-day-old ovariectomized adult mice. The injection of DES or tamoxifen significantly stimulated the cell division in uterine and vaginal epithelial cells (Sato et al., 1996). Shupnik et al. (1989) and Medlock et al. (1991) showed that ER mRNA level in the uterus of ovariectomized adult rats is reduced by E2. In mice, both DES and tamoxifen act as ER inducer in the uterus and vagina in both neonatal and ovariectomized adult mice, although ER mRNA induction time of tamoxifen is slower than that of DES in neonatal mice (Sato et al., 1996).

Scrocchi and Jones (1991) reported that ER mRNA expression was not detected by northern blot analysis in the vagina of mice exposed neonatally to E2. However, Kamiya et al. (1996) showed that the concentration of ER mRNA of the uterus of neonatally DES-exposed, ovariectomized adult mice was significantly higher than that of the DES-unexposed, ovariectomized controls. In the vagina of DES-exposed mice, however, ER mRNA was much lower in concentration than in the controls. These results coincide with the previous findings that ER was reduced by neonatal DES exposure (Bern et al., 1987), and that a small percentage of ER-immunoreactive cells was detected only in the basal layer in the vagina of DES-exposed mice (Sato et al., 1992). Tamoxifen also raises the uterine activity levels of protooncogenes in rats (Kirklad et al., 1993; Nephew et al., 1993) and in mice (Nishimura et al., 1993). Tamoxifen exhibits an estrogenic activity in the induction of ER and cell division at levels similar to that of DES in adult ovariectomized mice. However, tamoxifen shows developmentally dependent responses, as the responsiveness of the reproductive tracts to tamoxifen is different between newborn and adult mice (Kamiya et al., 1996).

One hypothesis explaining this developmentally dependent response is that there are altered forms of steroid hormone receptors that are associated with hormone independence and/or resistance. Support for this hypothesis was obtained from human breast cancer cells, which show low E2 binding due to a deletion of the estrogen-binding domain (Murphy and Dotzlaw, 1989). Similarily, Xing and Shapiro (1993) showed that a mutant ER containing two copies of an amphipathic helix exhibited a 10-fold increase in constitutive estrogen-independent transcription. Kamiya et al. (1996), however, demonstrated that no major variants were detected in the estrogen-binding domain of ER mRNA from both the uterus and vagina of neonatally DES-exposed mice.

Oncogene Expression

Stimulation of cell cycle progression by mitogens requires the coordinated expression of several genes in a strictly controlled temporal order (Pardee, 1989). Substantial differences exist in the pattern of gene activation by different mitogens, depending on their nature and on the cell type (Baserga, 1990). Previous studies on the rat uterus demonstrated that the activation of *c-jun* and *c-fos* rapidly reached a peak within a few hours after estrogen injection and declined with time after the peak (Weisz and Bresciani, 1988; Weisz et al., 1990; Webb et al., 1993; Khan and Stancel, 1994). In the mouse vagina, Scrocchi and Jones (1991) demonstrated that the expression of *c-fos* mRNA was increased 3 hr after estrogen injection. Kamiya et al. (1996) showed that postpuberal E2 injection resulted in a rapid similar transient increase in the expressions of *c-jun* and *c-fos* mRNAs in the uterus and vagina of ovariectomized mice, reached a peak 3 hr and 1 hr after the E2 injection, respectively, and declined with time thereafter. In the rat uterus, estrogen stimulated 3 to 5 times greater expression of *c-myc* gene (Weisz and Bresciani, 1988; Persico et al., 1990). However, no change in the expression of *c-myc* occurred in the mouse uterus, and it was even lowered in the vagina following E2 injection (Kamiya et al., 1996). Activation of *c-jun* and *c-fos* genes in the rat uterus was demonstrated to be a primary response to the ER complex (Weisz and Bresciani, 1988; Weisz et al., 1990).

The uterus and vagina of neonatally DES-exposed mice expressed high levels of *c-jun* and *c-fos* mRNAs, higher than those of ovariectomized control mice given no postpuberal E2 injection (Kamiya et al., 1996), which coincide with results showing a high mitotic activity in uterus and vagina of mice treated neonatally with DES (Iguchi et al., 1985a). In the vagina, however, the expression of *c-jun* and *c-fos* mRNAs in the DES-exposed mice was greater than in the control mice receiving an E2 injection. The expressions of *c-jun* and *c-fos* mRNAs in the uterus and vagina of the DES-exposed mice were not changed by the later E2 injection, supporting the previous results that sensitivity to E2 is low in neonatally "estrogenized" mouse reproductive tracts (Iguchi, 1992). The products of *c-jun* and *c-fos* genes can associate as homo- or heterodimers to form AP-1, a transacting transcription factor complex serving as a mediator of growth factors, transforming oncogenes and tumor promoters (Angel and Karin, 1991). In addition, deregulated expression of these genes can lead to neoplastic transformation of rat fibroblasts (Schuermann et al., 1989). Microinjection of anti-fos antibody or transfection of *c-fos* antisense RNA inhibits DNA synthesis or cell proliferation in cultured fibroblasts (Holt et al., 1986; Riabowol et al., 1988). These findings suggest that neonatal exposure of estrogen to mice causes deregulated expression of *c-jun* and *c-fos* mRNAs in the postnatal vagina, resulting in estrogen-independent persistent proliferation and cornification and hyperplastic down-growths in the vaginal epithelium. Tamoxifen also acts as an estrogen agonist in induction of ER and cell division, as well as in the induction of protooncogenes in mice.

Growth Factor Expression

Cunha et al. (1985, 1992) suggested that estrogen-induced female genital tract epithelial cell proliferation is mediated by the stroma. Estrogens caused a rapid increase of epidermal growth factor (EGF) in the uterus (DiAugustine et al., 1988) and EGF-stimulated proliferation of uterine epithelial cells *in vivo*

(Ignar-Trowbridge et al., 1992) and *in vitro* (Tomooka et al., 1986; Uchima et al., 1991). Estrogens also increase transforming growth factor-α (TGF-α) in the mouse uterus (Nelson et al., 1992). Therefore, expression of ER in uterine epithelial cells of neonatal mice could be mediated by estrogen-induced EGF (McLachlan and Newbold, 1996). However, a single injection of 5 μg EGF did not induce ER expression or cell division (Sato et al., 1996). Neonatal estrogen exposure induced the premature and permanent induction of growth factor genes like EGF and lactoferrin (Teng et al., 1989) that are normally under steroid hormone control. Both mRNA and the protein for EGF and lactoferrin have been demonstrated to be persistent and are observed even in the adult animals exposed neonatally to DES (Nelson et al., 1994). We have also demonstrated that ovary-independent persistent expression of EGF and TGF-α mRNAs in neonatally DES-exposed mouse vagina, but not in uterus (Sato et al., unpublished data). Several other genes, such as insulin-like growth factor-I (IGF-I), IGF-II, and keratinocyte growth factor were up-regulated after ovariectomy in the control mouse uterus. Expressions of these genes up-regulated in the neonatally DES-exposed uterus were decreased after ovariectomy, indicating that regulation of gene expression is altered by neonatal DES exposure (Sato et al., 1997). Tumor necrosis factor-α (TNF-α) and Fas ligand may be associated with apoptotic cell death of mouse reproductive tracts after ovariecctomy, since mRNAs of these factors increased after ovariectomy in control mouse uterus and vagina (Sato et al., 1996). However, no increase in expression of these genes was found in neonatally DES-exposed uterus and vagina (Sato et al., 1997). Therefore, ovary-independent persistent proliferation of reproductive organs in neonatally DES-exposed mice is partly explained by the down-regulation of these death factors. The up-regulated expression of estrogen-inducible genes, EGF and/or TGF-α, as well as down-regulated expression of death factors, TNF-α and/or Fas ligand, could play roles in the development of lesions and in the etiology of peneoplastic and neoplastic lesions that are manifested by developmental exposure to natural and xeno-estrogens.

Steroidogenesis

Neonatal exposure to estrogens induces anovulatory sterility in female rats and mice, with the rodents having no corpora lutea in the ovary (Takasugi, 1976). Prenatal DES exposure from days 9 to 16 of gestation did not result in consistent effects on corpora lutea in ovaries from the adult offspring, but some females exposed to the higher doses failed to exhibit corpora lutea (McLachlan et al., 1982). Ovaries from 8-week-old NMRI mice exposed neonatally to DES showed no corpora lutea, were composed predominantly of interstitial tissue, and had abnormal steroidogenesis (Tenenbaum and Forsberg, 1985). Haney et al. (1984) reported that ovaries of 3- to 14-month-old mice exposed prenatally to DES (100 μg/kg body weight on days 9–16 of gestation) showed increased *in vitro* production of E2, progesterone, and testosterone. Homogenates of ovaries from neonatally DES-exposed, 8-week-old NMRI mice also showed increased synthesis of progesterone, androstenedione, and E2, whereas the synthesis of 17α-hydroxyprogesterone and testosterone was reduced (Tenenbaum and Forsberg, 1985; Halling and Forsberg, 1990; Halling, 1992a). In the ovarian homogenates of neonatally DES-exposed (5 mg for 5 days) C57BL mice, production of 17α-hydroxyprogesterone and E2, but not progesterone and testosterone, were significantly higher than in the controls (Ohta et al., 1995). Though the difference in steroidogenesis of ovaries maybe ascribable to the difference in mouse strains used, these results indicate that perinatal DES exposure directly alters steroidogenesis in mouse ovaries. In female mice treated neonatally with tamoxifen, steroid metabolism in ovarian homogenates was significantly lower than in the controls and DES-exposed mice. However, levels of 17α-hydroxyprogesterone, androstenedione, testosterone, and E2 were higher in the ovarian homogenates than in the controls (Ohta et al., 1995). These findings suggest a possibility that the altered steroidogenesis causes a larger number of oocyte deaths, as reported previously in ovaries of neonatally tamoxifen-exposed mice (Iguchi et al., 1986a). Perinatal DES also alters the hypothalamo-hypophyseal ovarian axis; the plasma concentration of testosterone, but not of progesterone, was significantly lower in DES-exposed females than in the controls (Halling and Forsberg, 1989). Therefore, ovarian steroidogenesis is impaired directly by neonatal exposure to DES or tamoxifen, as well as indirectly through alterations of the hypothalamo-hypophyseal system.

After transplantation of DES-exposed ovaries to ovariectomized control females, the steroid pattern changed to that typical of control ovaries (Tenenbaum and Forsberg, 1985). Control ovaries transplanted to DES-treated females had a steroid pattern similar to that of DES-exposed ovaries. As has been known for many years owing to the occurrence of ovary-dependent vaginal changes abolished by ovariectomy following neonatal estrogen exposure (Takasugi, 1959, 1976; Iguchi et al., 1976; Bern and Talamantes, 1981), neonatal DES exposure alters the hypothalamo-hypophysial system in the developing female mice (Halling, 1992a). Ovarian changes after neonatal DES exposure appear to be secondary to the abnormal endocrine milieu in which the ovary develops (Halling, 1992b).

Polyovular Follicles

Neonatal exposure of DES or E2 causes an increased occurrence of polynuclear oocytes (Iguchi, 1985, 1992) as well as polyovular follicles (follicles with 2–23 oocytes in ovaries of immature mice; Forsberg et al., 1985; Iguchi, 1985; Iguchi et al., 1986b). Spontaneous occurrence of polyovular follicles has been reported in intact mice, although the incidence is low. Polyovular follicle incidence (percentage of polyovular follicles per ovary) and polyovular follicle frequency (percentage of mice with polyovular follicles) were significantly greater in 34-day-old BALB/cCrgl mice receiving five daily injections of DES, progesterone, 17α-hydroxyprogesterone caproate, or testosterone as neonates, but not in mice receiving nonaromatizable 5α-DHT. In DES-treated mice, polyovular follicle incidence was 120 to 340 times higher than in the controls. In 30-day-old C57BL mice treated neonatally for 5 days with a daily dose of testosterone, E2 or DES, polyovular follicle incidence also increased by 2 to 50 times. However, 5α-or 5β-DHT failed to increase polyovular follicle incidence. Polyovular follicle incidence increased gradually from 10 to 30 days of age only when neonatal DES treatment was begun between days 0 and 3 (Iguchi et al., 1986b), although polyovular follicle occurrence has been reported as a consequence of prolonged DES implantation in adult squirrel monkeys (Graham and Bradley, 1971). Natural and synthetic estrogens induced a higher polyovular follicle incidence than did aromatizable androgen (Iguchi et al., 1986b). Simultaneous injections of an aromatase inhibitor lowered the induction of polyovular follicles by testosterone, indicating that testosterone enhanced polyovular follicle formation as a result of its conversion to estrogen (Iguchi et al., 1988). The minimum daily dose of DES during neonatal life necessary to induce polyovular follicle was 10^{-2} to 10^{-3} µg in mice (Iguchi, 1985). Polyovular follicles also occurred in the ovary of 30-day-old offspring of mother ICR mice given four daily injections of DES from days 15 to 18 of gestation. Polyovular follicle incidence in prenatally DES-exposed offspring was increased 33 to 112 times over controls. Polyovular follicles appeared as early as 5 days in DES-exposed mice, and the incidence increased linearly until 30 days of age (Iguchi and Takasugi, 1986). Polyovular follicles were induced by neonatal DES exposure in all strains of mice examined (C57BL, C3H, BALB/c, ICR, and SHN) and in T strain rats. There was a negative correlation in polyovular follicle incidence and incidence of onset of luteinization between DES-exposed mice and DES-unexposed control mice at 30 days of age (Iguchi et al., 1987). A high incidence of polyovular follicles was also found in newborn mouse ovaries transplanted for 30 days into ovariectomized adult hosts given DES injections. When neonatal ovaries were cultured in a serum-free medium containing DES for 5 days and then transplanted into ovariectomized hosts, polyovular follicles were found in the grafts (Iguchi et al., 1990). Polyovular follicles also occur in humans and may be stimulated by an activated gonadotropin-estrogen axis (Dandekar et al., 1988). The significance of polyovular follicles may lie in the prenatal utilization of the limited number of female germ cells and hence to "aging" of the ovary at an earlier age (see Finch, 1990, p. 165).

Guillette et al. (1994) reported that female alligators from Lake Apopka—contaminated by an extensive spill of xenoestrogens, dicofol, and DDT or its metabolites—exhibited abnormal ovarian morphology with large numbers of polyovular follicles and polynuclear oocytes. This result shows a close connection between wildlife and laboratory rodents exposed to estrogenic compounds in early life, and the presence of polyovular follicles can be used as a marker of early exposure to xenoestrogens in alligators as well as in mice.

In ovaries of tamoxifen-exposed mice, luteinization was never observed (Forsberg, 1985; Taguchi and Nishizuka, 1985). Ovaries of tamoxifen-exposed 150-day-old mice contained small follicles whose oocytes frequently underwent degeneration. Frequency of oocyte death rose with an increase in tamoxifen dose given to the neonates (Iguchi et al., 1986a). Follicular growth was markedly suppressed in ovaries of tamoxifen-exposed mice, being similar to the ovary of hypophysectomized rats (cf. Selye et al., 1933). In addition, responsiveness of ovaries to prepubertally injected gonadotropin (hCG) was strikingly reduced in tamoxifen-exposed mice (Iguchi et al., 1986a). These findings imply that the hypothalamo-hypophysio-ovarian system was impaired by neonatal tamoxifen exposure as with neonatal estrogen exposure (Takasugi, 1976; Gorski et al., 1977). Specific binding of follicle-stimulating hormone to ovaries of 40-day-old C57BL mice exposed neonatally to 5 µg DES and 100 µg tamoxifen for 5 days from the day of birth are significantly lower than that in the controls (Iguchi, 1992). Tamoxifen (100 µg/day) injections starting within 5 days after birth caused a high incidence of polyovular follicles in the ovary of C57BL mice (Irisawa and Iguchi, 1990).

Reduction of Fertilizability

Polyovular follicles induced by neonatal exposure have been ovulated by injections of pregnant mare's serum gonadotropin (PMSG) and human chorionic gonadotropin (hCG), indicating that neonatal ovaries exposed to estrogen *in vivo* and *in vitro* are capable of responding to gonadotropins, resulting in a reduction of polyovular follicle incidence (Iguchi et al., 1990). The number of granulosa cells per tubal ovum from neonatally DES-exposed mice given PMSG and hCG was significantly greater than in control mice. Granulosa cells of DES-exposed mice have larger gap junctions (Iguchi et al., 1991). A total of 77% of oocytes from uniovular follicles of control mice developed up to 8-cell stage embryos following *in vitro* insemination; 66% of those from similar follicles of DES-exposed mice developed to the same stage. By contrast, only 47% of oocytes from polyovular follicles of DES-exposed mice showed division up to the 8-cell stage 72 hr after insemination, indicating a significantly lower fertilization rate compared to the oocytes from uniovular follicles of control and DES-exposed mice (Iguchi et al., 1991).

Wordinger and Derrenbacker (1989) demonstrated that a single injection of DES on day 15 of gestation decreased the time between the stages of follicular development, resulting in a greater number of developmentally advanced stages of follicles during neonatal ovarian development. Wordinger et al. (1989) also demonstrated that ovaries of mice exposed prenatally to DES are capable of responding to gonadotropins but show reduced preblastocyst developmental potential. Menczer et al. (1986) demonstrated that 22 of 40 women exposed to DES *in utero* had primary infertility and showed a higher rate of anatomical structural defects. After treatment with ovulation-stimulating drugs, spontaneous abortion and tubal pregnancy were frequent in both fertile and infertile groups of DES-exposed women. About 30% of the infertile women had mild hyperprolactinemia.

Halling and Forsberg (1990) showed that ova from neonatally DES-exposed mouse ovaries grafted into control females gave rise to normal living offspring in fertility tests with untreated males and that the litter size was similar to that seen after grafting control ovaries into control females. Grafting of control ovaries to DES-exposed females never resulted in pregnancy. DES-exposed females were treated with hormones to induce ovulation and artificially inseminated (Halling and Forsberg, 1991). Most zygotes died in the oviduct of DES-exposed females, only a few surviving to the 4-cell stage. When the early embryos were cultured *in vitro*, several survived to the implantation stage, but survival was lower than in similar experiments using control females, suggesting that both an egg factor and an oviducal factor were responsible for early embryonic death in DES-exposed females (Halling and Forsberg, 1991).

Walker and Kurth (1995) reported that 10 ovarian adenomas and 10 uterine adenocarcinomas in 143 offspring, which were transferred to mice exposed prenatally to DES, when blastocysts stage. Among 92 offspring from blastocyst transfers between mice exposed prenatally to vehicle, there was one ovarian adenoma and one uterine adenocarcinoma. These results indicate that the prenatal exposure of the host to DES produced a maternal environment that increased the incidence of ovarian and uterine tumors. In the reverse type of transfer, blastocysts from female mice exposed prenatally to DES were transferred into mice exposed to vehicle only prenatally; 6 ovarian adenomas and 16 uterine adenocarcinomas were developed in 99 offspring, indicating that DES also has a multigenerational effect transmitted through the blastocyst, which is consistent with fetal germ cell mutation from DES. In DES-exposed daughters and sons, no evidence for transgenerational effects have yet been reported (Giusti et al., 1995; Mittendorf, 1995).

Oviducal Effects on Embryos

Newbold et al. (1983) reported that prenatal DES exposure resulted in uncoiled and shorter oviducts. In addition, the demarcation between the oviduct and uterus was not readily apparent. The prenatally DES-exposed oviduct showed proliferation of columnar epithelium lining the lumen with gland formation extending into the underlying stroma, absence of or reduced fimbrial tissue, increased thickness of the muscular wall, and inflammatory cell infiltration. As a functional test of uterotubal junction integrity, Coomassie Blue dye was injected into the uterus. The control uterotubal junction confined the fluid to the uterus; however, in 80% to 100% of the animals exposed prenatally to DES, independent of the extent of gross abnormality, the dye readily flowed into the oviduct and filled the ovarian bursa (Newbold et al., 1983). It is apparent that perinatal exposure of female mice to DES causes changes in function of the reproductive tract later in life, which may contribute to infertility. Women exposed to DES *in utero* had structural changes of the genital tract, such as transverse ridges, cervical collars, hoods, coxcombs, hypoplastic cervices, and pseudopolyps (see review in Rotmensch et al., 1988). Kaufman et al. (1980, 1986) reported on 267 exposed women who underwent hysterosalpingograms, 69% of whom demonstrated such abnormalities as T-shaped uterus, small uterine cavity, and constriction rings.

Halling and Forsberg (1992b) transferred 2-cell embryos flushed from the oviduct into the oviduct of 8-week-old neonatally DES-exposed females or to females treated with E2 for 2 days before and 2 days after transfer. Two days after transfer, a significantly lower number of embryos were recovered from oviducts of neonatally DES-exposed females than were seen in control females; a still lower number occurred in E2-treated females. The recovered embryos were cultured *in vitro* for 4 days to test trophoblast outgrowth. The incidence of embryos reaching this stage after development in the oviduct of DES-exposed females was only half that for embryos passing through control oviducts or E2-treated oviducts. These results suggested that the adult oviducal environment in neonatally DES-exposed females significantly decreased early embryo developmental potential. They concluded that the oviducal factor(s) harmful to the embryo is related to persistent and increased levels of circulating estrogen in neonatally DES-exposed females (Halling and Forsberg, 1992b). However, it is also possible that hormones and growth factors needed for early embryonic development were deficient in the oviducts of DES-exposed mothers.

Embryos at the 2-cell stage from control females were transferred to the oviducts of neonatally DES-exposed females. Slightly fewer embryos were recovered from oviducts of DES-exposed females compared with control oviducts (81% vs. 92%) (Halling et al., 1993). When the recovered embryos were cultured *in vitro*, 64% of the embryos from control oviducts reached the blastocyst stage in contrast to only 24% of those from oviducts of DES-exposed mice; fewer of the latter showed trophoblastic outgrowth (76% vs. 93%). Oviducal transport and uterine attachment were studied by introducing blue-stained dextran microspheres into the oviduct of DES-exposed mice. The transport of microspheres through the oviduct in DES-exposed mice was accelerated, and retention of spheres within the uterus was reduced; presumably embryos would have been exposed to an inappropriate oviducal and uterine environment if similar effects occur with embryos (Halling et al., 1993).

Alterations of Decidual Reaction

In many mammals, the endometrium differentiates into a decidua in response to blastocyst implantation. The main event in this differentiation is the transformation of stromal cells into polyploid decidual cells (Finn, 1971). Similar endometrial differentiation occurs in experimentally induced pseudopregnant rodents by artificial stimuli mimicking blastocyst stimulation (Finn, 1971; Ohta, 1995). The decidual response has been used in the study of hormonal control of implantation and employed as a criterion for luteal function (Gibori et al., 1984). Neonatal exposure of estrogen or aromatizable androgen to female rats during a critical period results in anovulatory sterility (Takewaki, 1962; Barraclough, 1966; Gorski, 1971; Ohta, 1995). The capacity of the endometrium to differentiate into deciduoma in response to endometrial stimulation is markedly reduced in the uterus of neonatally estrogen and androgen-exposed rats receiving an appropriate regimen of progesterone and estrogen injections (Ohta, 1995). The vagina of neonatally tamoxifen-exposed rats ovariectomized at ages of 10 and 60 days failed to respond to estrogen priming, showing no estrous smears (Ohta et al., 1989). Therefore, continued vaginal diestrus in neonatally tamoxifen-exposed rats may be accounted for by changed sensitivity of the ovary to gonadotropin and/or of the vagina to sex hormones. The endometrium of neonatally tamoxifen-exposed rats had a reduced capacity for the decidual reaction (Ohta et al., 1989), suggesting that tamoxifen administered in neonatal life elicits a sustained, low uterine responsiveness to the deciduogenic stimulus. Ohta (1982, 1985) has demonstrated that endogenous estrogen is not needed for the capacity of the uterus to form deciduoma within 10 days after birth. Thus, exposure of estrogen, androgen, and tamoxifen to the developing uterus results in the lowered uterine responsiveness to the deciduogenic stimulus in rats (Ohta, 1995).

Complications of Pregnancy

Bibbo et al. (1977) suggested that DES-exposed women were less likely to become pregnant than unexposed controls. Herbst et al. (1980) showed that 42 of 226 DES-exposed women had infertility lasting more than 12 months, and that 19 of the 42 women had never become pregnant, in comparison with 21 of 203 unexposed women. An increased risk of pregnancy complications in DES-exposed women has been indicated. The major consequences are ectopic pregnancy, premature labor, and midtrimester loss (Barnes et al., 1980; Cousins et al., 1980; Herbst et al., 1981; Mangan et al., 1982). DES-exposed women with cervical malformations and hypoplasia developed incompetent cervices during the second trimester of pregnancy (Goldstein, 1978). A greater risk of unfavorable pregnancy outcome is reported in DES-exposed women than in controls (Barnes et al., 1980; for review see Giusti et al., 1995). The incidence of cervical incompetence is increased due to the high prevalence of cervical anomalies. Nonviable births were more common among DES-exposed women with cervical ridges (Herbst et al., 1980). Ectopic pregnancy occurred in 7% of 150 DES-exposed women; none occurred in 181 controls (Herbst et al., 1981). However, prenatal DES exposure was not related to diagnosis or symptoms of early menopause (Hornsby et al., 1995).

Female offspring from pregnant CD-1 mice injected with DES at 17 days of gestation were mated with untreated males for 4 months. Among 74 mated mice, 34 became pregnant and 11 of those pregnancies ended in abortion or stillbirth. In addition, two fetuses showed compressed heads, one of which seemed blocked from delivery by a vaginal adenocarcinoma; two uterine tumors were noted, one of which was a teratocarcinoma (Walker, 1983). In contrast, 16 of 17 control mice became pregnant. These results showing reduction of frequency of successful pregnancies in the DES-exposed mice are similar to those reported for DES-exposed women discussed above.

Mammary Gland Abnormalities

Bibbo et al. (1978) reported 32 incident cases of breast cancers among 693 mothers exposed to DES, compared with 21 cases of breast cancers among 668 controls. Vessey et al. (1983) reported no difference in the rate of breast cancer between the DES-treated and control groups. Colton et al. (1993), however, showed a small but statistically significant increase in the relative risk for breast cancer in DES-exposed mothers in more than 30 years of follow-up studies.

The long-term effects of perinatal treatments with sex hormones on normal and neoplastic mammary gland growth in mice were reviewed by Mori et al. (1980), Bern and Talamantes (1981), and Nagasawa and Mori (1988). Numerous anomalies in mammary gland development in male and female offspring were reported following treatment of pregnant mice and rats on day 14 of gestation with E2 (Raynaud, 1961). Tomooka and Bern (1982) demonstrated that neonatal injections of DES inhibited mammary gland growth on the 6th day, whereas the treatment stimulated growth after 4 weeks. When E2 administration was begun between 1 to 5 days of age, mammary abnormalities such as dilated ducts, hyperplastic alveolar nodule-like lesions, and aberrant secretory states were encountered. The incidence of abnormalities declined markedly when the treatment was begun after day 5, indicating the presence of a critical period for the responsiveness to estrogen (Bern et al., 1983). Neonatal treatment with estrogen, progesterone, or androgen induced an increase in mammary tumor development in several strains of mice bearing mammary tumor virus (Mori et al., 1980) and mammary tumor virus-unexposed female BALB/c mice (Jones and Bern, 1979; Mori, 1986). Nagasawa et al. (1980) showed that spontaneous mammary tumorigenesis of SLN mice given DES on day 12 of gestation was significantly lower than that of mice receiving progesterone or oil on the same day. These results indicate that long-term effects of perinatal exposure to DES or steroids on mammary tumorigenesis at advanced ages are largely dependent on perinatal age of the subjects and the facts that sex hormones and related substances after maturity can modulate these perinatal hormone effects (Nagasawa and Mori, 1988).

Rothschild et al. (1987) showed that prenatal DES exposure alone, or neonatal DES treatment alone, or a combination of the two, yielded significantly higher mammary tumor incidence and decreased tumor latency in ACI rats. Boylan and Calhoon (1979, 1981, 1983) found that the combination of neonatal exposure to DES and postnatal treatment with carcinogen resulted in a significant increase in the number of mammary tumors per rat compared to rats treated with carcinogen alone. However, earlier studies when estrogen treatment was begun slightly later indicated a decreased sensitivity to a carcinogen (Shellabarger and Soo, 1973). In female hamsters, prenatal DES exposure resulted in a higher incidence of mammary tumors from a carcinogen than did carcinogen alone (Rustia and Shubik, 1979). The increased sensitivity to a carcinogen as a result of perinatal exposure to DES obtained in experimental animals raises some concern regarding the possible occurrence of a similar phenomenon in the breast tissue of humans exposed to DES or other abnormal hormonal states such as oral contraception and phytoestrogens *in utero*. However, as yet there are no epidemiological data pointing to increased breast cancer in DES-exposed human offspring.

Perinatal Effects of Phytoestrogens

The possible effects of dietary environmental estrogens in humans, laboratory rodents, and wildlife animals during critical periods of reproductive development are a matter of considerable concern (Iguchi, 1992; Sheehan, 1995). The predominant dietary estrogens are naturally occurring phytoestrogens, such as coumestrol isolated from ladino clover, zearalenone isolated from Fusarium, and genistein from soybeans (see Sheehan, 1995). Epidemiological studies indicate that foods containing phytoestrogens may reduce the risk of certain hormonally related cancers, particularly breast and prostate, and cardiovascular disease (Don and Muir, 1985; Tominaga, 1985; Setchell and Adlercreutz, 1988; Lee et al., 1991; Messina et al., 1994). Soybeans containing more than 10 isoflavones, such as genistein and daidzein, are important components of Asian diets that have been associated with a reduced risk of breast and prostate cancer (Coward et al., 1993). Anticarcinogenic activity of phytoestrogens suggests an antiestrogenic action (Setchell and Adlercreutz, 1988). However, Setchell et al. (1987) have postulated that the soy phytoestrogens were responsible for observed infertility and liver disease in captive cheetahs, since hormonal estrogens are known to be associated with hepatotoxicity (Zimmer and Maddrey, 1987). The effects of phytoestrogens on livestock reproduction have been well documented (Kaldas and Hughers, 1989). Cassidy et al. (1994) reported that soy protein (60 g containing 45 mg isoflavones) given daily for 1 month significantly increased follicular phase length and/or delayed menstruation, and that midcycle surges of luteinizing hormone and follicle-stimulating hormone were significantly suppressed during dietary intervention with soy protein in women. Neonatal treatment of female rats with coumestrol induced persistent vaginal cornification, reduced ovarian weight (Leavitt and Meismer, 1968), premature uterine gland development, increased uterine weight as an early response, and lowered uterine weight and severe suppression in the estrogen receptor level at later ages (Medlock et al., 1995). Prenatal exposure to genistein induced alterations of anogenital distance, birth weight, and puberty onset in rats (Levy et al., 1995). Lamartiniere et al. (1995) demonstrated that neonatal genistein exposure chemoprevention provided against chemically induced mammary cancer in rats. Neonatal injections of coumestrol resulted in precocious vaginal cornification, cervicovaginal hyperplasia, down-growths, adenosis, cysts, uterine squamous metaplasia, and ovary-independent persistent vaginal cornification in mice (Burroughs et al., 1985, 1990a, b; Burroughs, 1995). Neonatal zearalenone exposure of female rats resulted in uterine expansion (Sheehan et al., 1984). Neonatal exposure of female mice to zearalenone caused ovary-independent genital tract alterations, ovarian dysfunction, dense collagen deposition in the uterine stroma, absence of uterine glands, squamous metaplasia, and dysplastic lesions in the vagina. However, no such alterations were found in ovariectomized mice exposed neonatally to zearalenone (Williams et al., 1989). Cline et al. (1996) showed that soybean estrogens, 26.6 mg free genistein per monkey per day, did not induce keratinization of vaginal epithelium. Recently, Sato, Burroughs, Sheehan, and Iguchi (unpublished data) found that a single injection of 100 μg coumestrol and genistein, but not daidzein, to newborn mice induced ER expression in uterine epithelial cells 24 hr after the injection, as has been shown by E2 and DES (Sato et al., 1992, 1996a), indicating that these phytoestrogens act as ER inducers in newborn mouse uteri and cause reproductive abnormalities in later life through ER.

β-Sitosterol, a phytoestrogen present in high concentration in plant oils, legumes, and wood (Pollack and Kritchevsky, 1981), shares the four-ring structure common to cholesterol and E2 and stimulates uterine growth in mice, sheep, and rabbits (Elghamry and Hansel, 1969; Ghanuddi et al., 1978; El Samannoudy et al., 1980; Malini and Vanithakumari, 1993; Rosenblum et al., 1993). This compound binds to rat hepatic cytosolic estrogen receptors (Rosenblum et al., 1993). β-Sitosterol enters the aquatic environment in high concentrations in the effluent discharged from bleached kraft pulp mill. Injection of β-sitosterol induces a significant decrease in plasma levels of the reproductive steroids due to decrease in the gonadal biosynthetic capacity through effects on cholesterol availability or the activity of the side chain cleavage enzyme P450scc in goldfish (MacLatchy and van der Kraak, 1995).

Perinatal Effects of Xenoestrogens

Bulger and Kupfer (1985) reviewed estrogenic activity of chlorinated hydrocarbon pesticides, DDT and its metabolites, methoxychlor and chlordecone, and other xenobiotics (polychlorinated biphenyls, arochlor, etc.) on the adult rodent uterus and male reproductive tract. A wide variety of chemicals—including herbicides, fungicides, insecticides, nematocides, and industrial chemicals—show estrogenic actions and antiangrogenic actions in animals (Colborn et al., 1993; Kelce et al., 1995a, b). Abnormal sexual development in reptiles (Guillette et al., 1994, 1995; Guillette and Crain, 1995) or birds (Fly, 1995), as well as feminized responses and vitellogenin expression in male fish (Sumpter and Jobling, 1995), have suggested an association with environmental chemicals functioning as estrogens. Methoxychlor, a substitute for DDT because of its reduced toxicity, has a short life in mammals and is biodegradable (Bulger and Kupfer, 1985). DDT, methoxychlor, and other organochlorine insecticides were detected in human milk (Tuinstra, 1971; Miller et al., 1979; Jani et al., 1988). Fifteen-day-old suckling female pups of lactating mouse dams receiving E2 or methoxychlor showed stimulated vagina and uterus, indicating that E2 or methoxychlor were secreted in milk and remained biologically active in the suckling mice (Appel and Eroshenko, 1992). Neonatal treatment of female mice with methoxychlor for 14 days stimulated the reproductive organs of female mice showed precocious vaginal opening and persistent vaginal cornification, increased reproductive weights and vaginal and uterine epithelial hypertrophy (Walters et al., 1993), and increased uterine protein synthesis as in E2-treated mice (Eroschenko and Rourke, 1992). Prenatal exposure, days 11 to 17 of gestation, of o,p′-DDT and methoxychlor as well as DES, increased urine marking in a novel territory of male mouse offspring (vom Saal et al., 1995). However, Kelce et al. (1995b) reported that the DDT metabolite p,p′-DDE is a potent androgen receptor antagonist. A single injection of PCBs to lactating maternal rats induced an increase in both maternal and neonatal hepatic cytochrome P-450, cytochrome b5, and cytochrome-c-(P-450) reductase, and increased expression of the protooncogenes *c-Ha-ras* and *c-raf* in the liver of mother and the neonates (Borlack et al., 1996).

The environmental pollutant TCDD induces severe reproductive defects in male rats exposed *in utero* and during lactation (Mably et al., 1992a, b, c), as well as thymic atrophy, immune suppression, and carcinogenesis by binding aryl hydrocarbon receptor, which acts as a nuclear ligand-induced transcription factor (Safe, 1995). In female offspring of Holtzman rats given 1 μg/kg TCDD on gestational day 15, serum estrogen levels were significantly lower; ER mRNA levels were higher in the hypothalamus, uterus, and ovary but lower in the pituitary; and ER DNA-binding activity in the uterus was higher than the controls (Chaffin et al., 1996), suggesting that TCDD acts systemically. In culture of osteoblasts from fetal rat calvaria, 10 nM TCDD dramatically suppressed postconfluent bone nodule formation, also alkaline phosphatase and osteocalcin production (Gierthy et al., 1994), suggesting that TCDD alters effects of estrogen and/or vitamin D on bones. They, therefore, suggested the possibility that chronic exposure to low environmental levels of TCDD and related compounds could result in altered fetal bone development or increased magnitude of osteoporosis in the adult.

Brown and Lamartiniere (1995) reported that puberal treatment of Sprague-Dawley rats with 50 ng DES, 50 μg genistein, and 50 μg o,p'-DDT, but not 25 μg arochlor and 2.5 ng TCDD for a week, enhanced epithelial cell proliferation and gland development in the mammary gland.

Nonylphenol is an alkylphenolic compound used in the preparation of lubricating oil additives, plasticizers, resins, detergents, and surface-active agents. This compound has been shown to possess estrogen-like properties in human estrogen-sensitive MCF-7 breast cancer cells and in rat endometrial cells in induction of cell proliferation (Soto et al., 1991). Colerangle and Roy (1996) showed that puberal treatment of Noble rats with nonylphenol (0.01 and 7.1 mg/24 hr by osmotic minipump) for 11 days induced proliferation of mammary epithelial cells and alteration of cell-cycle kinetics and increased the conversion of mammary epithelial cells from G0 to G1 and S-phase cells compared to that of controls. When compared to the effect of DES, 105- to 106-fold higher concentration of nonylphenol was required to produce the same biological effects as DES.

Abnormalities in Skeletal Tissue and Muscle

In addition to the perinatal effects discussed above, nongenital abnormalities of the immune system, central nervous system, and hypothalamohypophysial complex and behaviors have also been reported in mice exposed perinatally to natural and synthetic hormones (Bern and Talamantes, 1981, Fig. 10-1). A breakdown of symphysial bone and cartilage replacement by connective tissue occurs (for review, see Iguchi, 1992). Estrogen feminizes the bone structure of the pelvis and pubic symphysis in many mammals (Gardner, 1936; Iguchi et al., 1995; Fukazawa et al., 1996). The shape of the innominate bone is transformed to the male type under the influence of early postnatal androgen (Iguchi et al., 1989; Iguchi, 1992; Uesugi et al., 1992). Neonatal treatment with tamoxifen induces elongation of the pubic ligament and retards the growth of the ilium and pubis in mice by changing the activities of osteoclasts and osteoblasts. Tamoxifen acts directly on the neonatal mouse pubis as an antiestrogen to inhibit its ossification.

Neonatal DES exposure induced persistent reduction of calcium and phosphorus in pelvis and femur in aged mice (Migliacchio et al., 1992, 1995; Iguchi et al., 1995; Fukazawa et al., 1996). These results indicate that neonatal DES and tamoxifen exposure result in permanent changes in bone tissue in mice.

The anococcygeus muscle is a paired smooth muscle in the perineal area, which shows sexual dimorphism; the cross-sectional area of the muscle in male mice is significantly larger than that in females (Fukazawa et al., 1992). Neonatal exposure to DES significantly reduced the muscle area in male mice but strikingly increased the muscle area in females. The muscle area of neonatally DES-exposed female mice was significantly larger than the controls, and ovariectomy did not alter this, indicating that DES had an irreversible stimulatory effect on the muscle of neonatal female mice (Iguchi et al., 1995). These examples suggest that more attention should be paid to abnormalities in nongenital organs exposed to various estrogenic agents during embryonic, fetal and early postnatal development in mammals including humans, as well as in nonmammalian vertebrates.

Recently, *Xenopus* embryos were developed in water containing 10^{-10} to 10^{-5} M E2, 17α-estradiol, DES, or 10^{-5} M progesterone or 5α-DHT from developmental stage 3. Survival rates of the embryos developed in water containing 10^{-5} M E2 and DES decreased remarkably after stage 27, and all embryos were dead by stages 42 and 32, respectively. Embryos treated with 10^{-5} M E2 showed malformations and suppressed organogenesis, including crooked vertebrae at stage 38: The head was smaller and the abdomen was larger than those in the controls. Similar effects were observed in embryos developed in 10^{-5} M DES, but not in 17α-estradiol, progesterone, or 5α-DHT. In 10^{-5} M E2 treatment, abnormalities were induced only when the treatment was started before stage 39. ER 4 was expressed in adult liver, unfertilized and fertilized eggs, and embryos, indicating that 10^{-5} M E2 and DES induced embryo death and malformations, and ER may be involved in the induction of the developmental defects in *Xenopus* embryos (Nishimura et al., 1997).

The Threat of Environmental Estrogens

Recently, much evidence has accumulated showing that environmental phytoestrogens and various xenobiotic agents (pesticides, herbicides, polychlorinated compounds, and alkylphenols) may affect human and animal populations, including wildlife. The action of such agents during embryonic and fetal development demands extensive attention (Colborn and Clement, 1992; Soto et al., 1992; Colborn et al., 1993; Guillette, 1995). The issue involved is not only the possible occurrence of birth defects but also the possible long-term effects that in humans may not manifest themselves until adolescence, and even much later in life. As these effects may result in structural, reproductive, endocrinological, metabolic, immunological, neurological, behavioral, dysplastic, and neoplastic changes, the search for the consequences on the offspring of exposure during intrauterine life must be stringent and diversified (Takasugi and Bern, 1988). Analyses of transgenerational effects of xenobiotic agents are needed in order to allow potential dangers to human and wildlife population to be estimated and confronted.

Acknowledgments

This work was supported by a Grant-in-Aid for Scientific Research from the Ministry of Education, Science and Culture of Japan, a research grant from Kihara Science Foundation, and a grant in Support of the Promotion of Research at Yokohama City University.

References

Angel, P., and Karin, M. (1991). The role of Jun, Fos and the AP-1 complex in cell-proliferation and transformation. *Biochim. Biophys. Acta* 1072: 129-157.

Appel, R.J., and Eroschenko, V.P. (1992). Passage of methoxychlor in milk and reproductive organs of nursing female mice. 1. Light and scanning electron microscopic observations. *Reprod. Toxicol.* 6: 223-231.

Barnes, A.B., Colton, T., Gunderson, J., Noller, K.L., Tilley, B.C., Strama, T., Townsend, D.E., Hatab, P., and O'Brien, P.C. (1980). Fertility and outcome of pregnancy of women exposed *in utero* to diethylstilbestrol. *N. Engl. J. Med.* 302: 609–613.

Barraclough, C.A. (1966). Modification in reproductive function after exposure to hormones during the prenatal and early postnatal period. In *Neuroendocrinology.* (L. Martini, and W.F. Ganong, eds.), pp. 61–99. Academic, New York.

Baserga, R. (1990). The cell cycle: Myths and realities. *Cancer Res.* 50: 6769–6771.

Bern, H.A., Jones, L.A., Mills, K.T., Kohrman, A., and Mori, T. (1976). Use of the neonatal mouse in studying long-term effects of early exposure to hormones and other agents. *J. Toxicol. Environ. Health* (Suppl.) 1: 103–116.

———— and Talamantes, F.J., Jr. (1981). Neonatal mouse models and their relation to disease in the human female. In *Developmental Effects of Diethylstilbestrol (DES) in Pregnancy.* (A.L. Herbst, and H.A. Bern, eds.), pp. 129–147. Thieme Stratton, New York.

————, Mills, K.T., and Jones, L.A. (1983). Critical period for neonatal estrogen exposure in occurrence of mammary gland abnormalities in adult mice.

Proc. Soc. Exp. Biol. Med. 172: 239–242.

——, Edery, M., Mills, K.T., Kohrman, A.F., Mori,T., and Larson, L. (1987). Long-term alterations in histology and steroid receptor levels of the genital tract and mammary gland following neonatal exposure of female BALB/cCrgl mice to various doses of diethylstilbestrol. *Cancer Res.* 47: 4165–4172.

—— (1992a). The fragile fetus. In *Chemically-induced Alterations in Sexual and Functional Development: The Wildlife/Human Connection.* (T. Colborn and C. Clement, eds.), pp. 9–15. Princeton Scientific Publications, Princeton.

—— (1992b). Diethylstilbestrol (DES) syndrome: Present status of animal and human studies. In *Hormonal Carcinogenesis.* (J.J. Li., S. Nandi, and S.A. Li, eds.), pp. 1–8. Springer-Verlag, New York.

Bibbo: M., Gill, W.B., Azizi, F., Blough, R., Fang, V.S., Rosenfield, R.L., Schumacher, G.F.B., Sleeper, K., Sonek, M.G., and Wied, G.L. (1977). Follow-up study of male and female offspring of DES-exposed mothers. *Obstet. Gynecol.* 49: 1–8.

——, Haenszel, W.M., Wied, G.L., Hubby, M., and Herbst, A.L. (1978). A twenty-five-year follow-up study of women exposed to diethylstilbestrol during pregnancy. *N. Engl. J. Med.* 298: 763–767.

Bigsby, R.M., Aixin, L., Luo: K., and Cunha, G.R. (1990). Strain differences in the ontogeny of estrogen receptors in murine uterine epithelium. *Endocrinology* 126: 2592–2596.

Borlak, J.T., Scott, A., Henderson, J., Jenkins, H.J., and Wolf, C.R. (1996). Transfer of PCBs via lactation simultaneously induces the expression of P-450 isoenzymes and the protooncogenes c-Ha-ras and c-raf in neonates. *Biochem. Pharmacol.* 51: 517–529.

Boylan, E.S., and Calhoon, R.E. (1979). Mammary tumorigenesis in the rat following prenatal exposure to diethylstilbestrol and postnatal treatment with 7,12-dimethylbenz(a)anthracene. *J. Toxicol. Environ. Health* 5: 1059–1071.

—— and ——(1981). Prenatal exposure to diethylstilbestrol: Ovarian-independent growth of mammary tumors induced by 7,12–dimethylbenz(a)anthracene. *J. Natl. Cancer Inst.* 66: 649–652.

—— and ——(1983). Transplacental action of diethylstilbestrol on mammary carcinogenesis in female rats given one or two doses of 7,12-dimethylbenz(a)anthracene. *Cancer Res.* 43: 4879–4884.

Brown, N.M., and Lamartiniere, C.A. (1995). Xenoestrogens alter mammary gland differentiation and cell proliferation in the rat. *Environ. Health Perspect.* 103: 708–713.

Bulger, W.H., and Kupfer, D. (1985). Estrogenic activity of pesticides and other xenobiotics on the uterus and male reproductive tract. In *Endocrine Toxicology.* (J.A. Thomas, K.S. Korach, and J.A. McLachlan, eds.), pp. 1-33. Raven, New York.

Burroughs, C.D., Bern, H.A., and Stokstad, E.L.R. (1985). Prolonged vaginal cornification and other changes in mice treated neonatally with coumestrol, a plant estrogen. *J. Toxicol. Environ. Health.* 15: 51–61.

———, Mills, K.T., and Bern, H.A. (1990a). Reproductive abnormalities in female mice exposed neonatally to various doses of coumestrol. *J. Toxicol. Environ. Health* 10: 105–122.

———, ———, and ———(1990b). Long-term genital tract changes in female mice treated neonatally with coumestrol. *Reprod. Toxicol.* 4: 127–135.

——— (1995). Long-term reproductive tract alterations in female mice treated neonatally with coumestrol. *Proc. Soc. Exp. Biol. Med.* 208: 78–81.

Cassidy, A., Bingham, S., and Setchell, K.D.R. (1994). Biological effects of a diet of soy protein rich in isoflavones on the menstrual cycle of pre-menopausal women. *Am. J. Clin. Nutr.* 60: 333–340.

Chaffin, C.L., Peterson, R.E., and Hutz, R.J. (1996). In utero and lactational exposure of female Holtzman rats to 2,3,7,8-tetrachlorodibezo-*p*-dioxin: Modulation of the estrogen signal. *Biol. Reprod.* 55: 62–67.

Cline, J.M., Paschold, J.C., Anthony, M.S., Obasanijo: I.O., and Adams, M.R. (1996). Effects of hormonal therapies and dietary soy phytoestrogens on vaginal cytology in surgically postmenopausal macaques. *Fertil. Steril.* 65: 1031–1035.

Colborn, T., and Clement, C., eds. (1992). *Chemically-induced Alterations in Sexual and Functional Development: The Wildlife/Human Connection*, pp. 403. Princeton Scientific Publications, Princeton.

———, vom Saal, F.S., and Soto, A.M. (1993). Developmental effects of endocrine-disrupting chemicals in wildlife and humans. *Environ. Health Perspec.* 101: 378–384.

Colerangle, J.B., and Roy, D. (1996). Exposure of environmental estrogenic compound nonylphenol to Noble rats alters cell-cycle kinetics in the mammary gland. *Endocrine* 4: 115–122.

Colton, T., Greenberg, E.R., Noller, K., Reseguie, L., Van Bennekom, C., and Heeren, T. (1993). Breast cancer in mothers prescribed diethylstilbestrol in pregnancy: Further follow-up. *JAMA* 269: 2096–2100.

Cooke, P.S., and Eroshenko, V.P. (1990). Inhibitory effects of technical grade methoxychlor on development of neonatal male mouse reproductive organs. *Biol. Reprod.* 42: 585–596.

Cousins, L., Karp, W., Lacey, C., and Lucas, W.E. (1980). Reproductive outcome of women exposed to diethylstilbestrol *in utero. Obstet. Gynecol.* 56: 70–76.

Coward, L., Barnes, N.C., Setchell, K.D.R., and Barnes, S. (1993). Genistein, daidzein and their beta-glycoside conjugates-antitumor isoflavones in soybean foods from American and Asian diets. *J. Agr. Food Chem.* 41:

1961–1967.

Cunha, G.R., Bigsby, R.M., Cooke, P.S., and Sugimura, Y. (1985). Stromal-epithelial interactions in the determination of hormonal responsiveness. In *Estrogens in the Environment II. Influences on Development*. (J.A. McLachlan, ed.), pp. 273–287. Elsevier, New York.

———— and Young, P. (1992). Role of stroma in oestrogen-induced epithelial proliferation. *Epithelial Cell Biol*. 1: 18–31.

Dandekar, P.V., Martin, M.C., and Glass, R.H. (1988). Polyovular follicles associated with human *in vitro* fertilization. *Fertil. Steril*. 49: 483–486.

Davis, D.L., and Bradlow, H.L. (1995). Can environmental estrogens cause breast cancer? *Sci. Amer*. Oct. 144–149.

DiAugustine, R.P., Petrusz, P., Bell, G.I., Brown, C.F., Korach, K.S., McLachlan, J.A., and Teng, C.T. (1988). Influence of estrogens on mouse uterine epidermal growth factor precursor protein and messenger ribonucleic acid. *Endocrinology* 122: 2355–2363.

Don, A.S., and Muir, C.S. (1985). Prostatic cancer: Some epidemiological factors. *Bull. Cancer* 72: 381–390.

Dunn, T.B., and Green, A.W. (1963). Cysts of the epididymis, cancer of the cervix, granular cell myoblastoma, and other lesions after estrogen injection in newborn mice. *J. Natl. Cancer Inst*. 31: 425–455.

Elghamry, M.I., and Hansel, R. (1969). Activity and isolated phytoestrogen of shrub palmetto fruits (*Serenoa repens* Small), a new estrogenic plant. *Experientia* 25: 828–829.

El Samannoudy, F.A., Shareha, A.M., Ghannudi, S.A., Gillaly, G.A., and El Mougy, S.A. (1980). Adverse effects of phytoestrogens-7: Effect of b-sitosterol treatment on follicular development, ovarian structure and uterus in the immature female sheep. *Cell. Mol. Biol*. 26: 255–266.

Ennis, B.W., and Stumpf, W.E. (1988). Binding of estrogen and antiestrogen in uterine cell nuclei: *In vivo* autoradiographic studies. *J. Steroid Biochem*. 31: 405–409.

Eroschenko, V.P., and Cooke, P.S. (1990). Morphological and biochemical alterations in reproductive tracts of neonatal female mice treated with the pesticide methoxychlor. *Biol. Reprod*. 42: 573–583.

———— and Rourke, A.W. (1992). Stimulatory influences of technical grade methoxychlor and estradiol on protein synthesis in the uterus of the immature mouse. *J. Occupa. Med. Med. Toxicol*. 1: 307–315.

Finn, C.A. (1971). The biology of decidual cells. *Adv. Reprod. Physiol*. 15: 1–26.

Finch, C.E. (1990). Longevity, Senescence and the Genome. University of Chicago Press, Chicago.

Forsberg, J.-G. (1979). Developmental mechanism of estrogen-induced irreversible changes in the mouse cervicovaginal epithelium. *Natl. Cancer Inst*.

Monogr. 51: 41–56.

——— (1985). Treatment with different antiestrogens in the neonatal period and effects in the cervicovaginal epithelium and ovaries of adult mice: A comparison to estrogen-induced changes. *Biol. Reprod.* 32: 427–441.

———, Tenenbaum, A., Rydberg, C., and Sernvi, S. (1985). Ovarian structure and function in neonatally estrogen treated mice. In *International Symposium on Estrogen in the Environment.* (J.A. McLachlan, ed.), pp. 327–346. Elsevier, New York.

Fry, D.M. (1995). Reproductive effects in birds exposed to pesticides and industrial chemicals. *Environ. Health Perspect.* 103 (Suppl. 7): 165–171.

Fukazawa, Y., Suzuki, A., Iguchi, T., Takasugi, N., and Bern, H.A. (1992). Sexual dimorphism and strain difference in mouse anococcygeus muscle. *Zool. Sci.* 9: 1273.

———, Nobata, S., Katoh, M., Tanaka, M., Kobayashi, S., Ohta, Y., Hayashi, Y., and Iguchi, T. (1996). Effect of neonatal exposure to diethylstilbestrol and tamoxifen on pelvis and femur in male mice. *Anat. Rec.* 244: 416–422.

Gardner, W.U. (1936). Sexual dimorphism of the pelvis of the mouse, the effect of estrogenic hormones upon the pelvis and upon the development of scrotal hernias. *Am. J. Anat.* 59: 459–483.

Ghanuddi, S.A., Shareha, A.M., El Samannoudy, F.A.H.A., and Elmougy, S.A.Ê. (1978). Adverse effect of phytoestrogens: 2. distribution of alkaline phosphatase in the immature uterus of the rabbit after β-sitosterol treatment. *Libyan J. Sci.* 8: 27–34.

Gibori, G., Kalison, B., Basuray, R., Rao, M.C., and Hunzicker-Dunn, M. (1984). Endocrine role of the decidual tissue: Decidual luteotropin regulation of luteal adenylyl cyclase activity, luteinizing hormone receptors, and steroidogenesis. *Endocrinology* 115: 1157–1163.

Gierthy, J.F., Silkworth, J.B., Tassinari, M., Stein, G.S., and Lian, J.B. (1994). 2,3,7,8-Tetrachlorodibezo-*p*-dioxin inhibits differentiation of normal diploid rat osteoblasts *in vitro. J. Cell. Biochem.* 54: 231–238.

Giusti, R., Iwamoto, K., and Hatch, E.E. (1995). Diethylstilbestrol revisited: A review of the long-term health effects. *Ann. Int. Med.* 122: 778–788.

Goldstein, D.P. (1978). Incompetent cervix in offspring exposed to diethyl-stilbestrol *in utero. Obstet. Gynecol. Suppl.* 52: 73–75.

Gorski, J., and Hou, Q. (1995). Embryonic estrogen receptors: Do they have a physiological function? *Environ. Health Perspect.* 103 (Suppl. 7), 69–72.

Gorki, R.A. (1971). Gonadal hormones and the perinatal development of neuroendocrine function. In *Frontiers in Neuroendocrinology.* (L. Martini, and W.F. Ganong, eds.), pp. 237–290. Oxford University Press, New York.

———, Harlan, R.E., and Christensen, L.W. (1977). Perinatal hormonal exposure aand the development of neuroendocrine regulatory processes. *J.*

Toxicol. Environ. Health 3: 97–121.

Graham, C.E., and Bradley, C.F. (1971). Polyovular follicles in squirrel monkeys after prolonged diethylstilbestrol treatment. *J. Reprod. Fertil.* 27: 181–185.

Greco, T.L., Furlow, J.D., Duello, T.M., and Gorski, J. (1991). Immunodetection of estrogen receptors in fetal and neonatal female mouse reproductive tracts. *Endocrinology* 129: 1326–1332.

Gronemeyer, H., Benhamou, B., Berry, M., Bocquel, M.T., Gofflo, D., Garcia, T., Lerouge, T., Metzger, D., Meyer, M.E., Tora, L., Vergezac, A., and Chambon, P. (1992). Mechanisms of antihormone action. *J. Steroid Biochem. Molec. Biol.* 41: 217–221.

Guillette, L.J., Jr., Gross, T.S., Masson, G.R., Matter, J.M., Percival, H.F., and Woodward, A.R. (1994). Developmental abnormalities of the gonad and abnormal sex hormone concentrations in juvenile alligators from contaminated and control lakes in Florida. *Environ. Health Perspect.* 102: 680–688.

—————— (1995). Endocrine disrupting environmental contaminants and developmental abnormalities in embryos. *Human Ecol. Risk Assess.* 1: 12–36.

—————— and D. A. Crain (1995). Endocrine-disrupting contaminants and reproductive abnormalities in reptiles. *Comments on Toxicology* 5: 381–399.

——————, Gross, T.S., Gross, D., Rooney, A.A., and Percival, H.F. (1995). Gonadal steroidogenesis *in vitro* from juvenile alligators obtained from contaminated or control lakes. *Environ. Health Perspect.* 103 (Suppl. 4): 31–36.

Halling, A., and Forsberg, J.-G. (1989). Plasma testosterone levels and ovarian testosterone content in adult mice treated with diethylstilbestrol neonatally. *J. Steroid Biochem.* 32: 439–443.

—————— and —————— (1990). Ovarian reproductive function after exposure to diethylstilbestrol in neoantal life. *Biol. Reprod.* 43: 472–477.

—————— and —————— (1991). Effects of neonatal exposure to diethylstilbestrol on early mouse embryo development *in vivo* and *in vitro*. *Biol. Reprod.* 45: 157–162.

—————— (1992a). Steroid synthesis in ovarian homogenates from hypophysectomized adult female mice treated with diethylstilbestrol in neonatal life. *J. Toxicol. Environ. Health* 36. 341–353.

—————— and Forsberg, J.-G. (1992b). The functional importance of the oviduct in neonatally estrogenized mouse females for early embryo survival. *Teratology* 45: 75–82.

——————, von Mecklenburg, C., and Forsberg, J.-G. (1993). Factors of importance for decreased early embryo survival in female mice treated neonatally with diethylstilbestrol. *J. Reprod. Fert.* 99: 291–297.

Haney, A.F., Newbold, R.R., and McLachlan, J.A. (1984). Prenatal diethylstilbestrol exposure in the mouse: Effects on ovarian histology and steroidogenesis *in vitro*. *Biol. Reprod.* 30: 471–478.

Herbst, A.L., and Scully, R.E. (1970). Adenocarcinoma of the vagina: A report of 7 cases including 6 clear cell carcinomas (so called mesonephromas). *Cancer* 25: 745–757.

———, Ulfelder, H., and Poskanzer, D.C. (1971). Adenocarcinoma of the vagina: An association of maternal stilbestrol therapy with tumour appearing in young women. *N. Engl. J. Med.* 284: 878–881.

———, Kurman, R.J., and Scully, R.E. (1972). Clear-cell adenocarcinomas of the genital tract in young females. Registry report. *N. Engl. J. Med.* 287: 1259–1264.

———, Poskanzer, D.C., Robboy, S.J., Friedlander, L., and Scully, R.E. (1975). Prenatal exposure to stilbestrol: A prospective comparison of exposed female offspring with unexposed controls. *N. Engl. J. Med.* 292: 334–339.

———, Hubby, M.M., Blough, R.R., and Azizi, F. (1980). A comparison of pregnancy experience in DES-exposed and DES-unexposed daughters. *J. Reprod. Med.* 24: 62–69.

——— and Bern, H.A. eds. (1981). *Developmental Effects of Diethylstilbestrol (DES) in Pregnancy.* p. 203.Thieme-Stratton, New York.

———, Hubby, M.M., Azizi, F., and Makii, M.M. (1981). Reproductive and gynecologic surgical experience in diethylstilbestrol-exposed daughters. *Am. J. Obstet. Gynecol.* 141: 1019–1028.

Holt, J.T., Venkat Gopal, T., Moulton, A.D., and Nienhuis, A.W. (1986). Inducible production of *c-fos* antisense RNA inhibits 3T3 cell proliferation. *Proc. Natl. Acad. Sci. U.S.A.* 83. 4794–4798.

Hou, Q., and Gorski, J. (1993). Estrogen receptor and progesterone receptor genes are expressed differentially in mouse embryos during preimplantation development. *Proc. Natl. Acad. Sci. U.S.A.* 90: 9460–9464.

Hornsby, P.P., Wilcox, A.J., and Herbst, A.L. (1995). Onset of menopause in women exposed to diethylstilbestrol *in utero. Am. J. Obstet. Gynecol.* 172: 92–95.

Ignar-Trowbridge, D.M., Nelson, K.G., Ross, K.A., Washburn, T.F., Korach, K.S., and McLachlan, J.A. (1991). Localization of the estrogen receptor in uterine cells by affinity labeling with [3H] tamoxifen aziridine. *J. Steroid Biochem. Molec. Biol.* 39: 131–132.

———, ———, K.G., Bidwell, M.C., Curtis, S.W., Washburn, T.F., McLachlan, J.A., and Korach, K.S. (1992). Coupling of dual signaling pathways: Epidermal growth factor action involves the estrogen receptor. *Proc. Natl. Acad. Sci. U.S.A.* 89: 4658–4662.

Iguchi, T., Ohta, Y., and Takasugi, N. (1976). Mitotic activity of vaginal epithelial cells following neonatal injections of different doses of estrogen in mice. *Devel. Growth Differ.* 18: 69–78.

——— (1985). Occurrence of polyovular follicles in ovaries of mice treated neonatally with diethylstilbestrol. *Proc. Japan Acad.* 61B, 288–291.

———, Iwase, Y., Kato, H., and Takasugi, N. (1985a). Prevention by vitamin A of the occurrence of permanent vaginal and uterine changes in ovariectomized adult mice treated neonatally with diethylstilbestrol and its nullification in the presence of ovaries. *Exp. Clin. Endocrinol.* 85: 129–137.

———, Hirokawa, M., and Takasugi, N. (1986a). Occurrence of genital tract abnormalities and bladder hernia in female mice exposed neonatally to tamoxifen. *Teratology* 42: 1–11.

——— and Takasugi, N. (1986). Polyovular follicles in the ovary of immature mice exposed prenatally to diethylstilbestrol. *Anat. Embryol.* 175: 53–55.

———, ———, Bern, H.A., and Mills, K.T. (1986b). Frequent occurrence of polyovular follicles in ovaries of mice exposed neonatally to diethylstilbestrol. *Teratology* 34: 29–35.

——— and ———(1987). Postnatal development of uterine abnormalities in mice exposed to DES *in utero*. *Biol. Neonate* 52: 97–103.

———, Ohta, Y., Fukazawa, Y., and Takasugi, N. (1987). Strain differences in the induction of polyovular follicles by neonatal treatment with diethylstilbestrol in mice. *Med. Sci. Res.* 15: 1407–1408.

———, Todoroki, R., Takasugi, N., and Petrow, V. (1988). The effect of an aromatase-and a 5a-reductase-inhibitor upon the occurrence of polyovular follicles, persistent anovulation, and permanent vaginal stratification in mice treated neonatally with testosterone. *Biol. Reprod.* 39: 689–697.

———, Irisawa, S., Fukazawa, Y., Uesugi, Y., and Takasugi, N. (1989). Morphometric analysis of the development of sexual dimorphism of the mouse pelvis. *Anat. Rec.* 224: 490–494.

———, Fukazawa, Y., Uesugi, Y., and Takasugi, N. (1990). Polyovular follicles in mouse ovaries exposed neonatally to diethylstilbestrol *in vivo* and *in vitro*. *Biol. Reprod.* 43: 478–484.

———, Kamiya, K., Uesugi, Y., Sayama, K., and Takasugi, N. (1991). In vitro fertilization of oocytes from polyovular follicles in mouse ovaries exposed neonatally to diethylstilbestrol. *In Vivo* 5: 359–364.

——— (1992). Cellular effects of early exposure to sex hormones and antihormones. *Int. Rev. Cytol.* 139: 1–57.

———, Edery, M., Tsai, P.-S., Ozawa, S., Sato, T., and Bern, H.A. (1993). Epidermal growth factor receptor levels in reproductive organs of female mice exposed neonatally to diethylstilbestrol. *Proc. Soc. Exp. Biol. Med.* 204: 110–116.

———, Fukazawa, Y., and Bern, H.A. (1995). Effects of sex hormones on oncogene expression in the vagina and on the development of sexual dimorphism of the pelvis and anococcygeus muscle. *Environ. Health Perspec.* 103 (Suppl. 7): 79–82.

—————— and Ohta, Y. (1996). Cellular effects of early exposure to tamoxifen. In *Tamoxifen Beyond The Antiestrogen*. (J.A. Kellen, ed.), pp. 179–199. Birkhauser, Boston.

—————— and Bern, H.A. (1996). Transgenerational effects: Intrauterine exposure to diethylstilbestrol (DES) in humans and the neonatal mouse model. *Comments on Toxicology*. 5: 367–360.

Irisawa, S., and Iguchi, T. (1990). Critical period of induction by tamoxifen of genital organ abnormalities in female mice. *In Vivo* 4: 175–180.

Jani, J.P., Patel, J.S., Shah, M.P., Gupta, S.K., and Kashyap, S.K. (1988). Levels of organochlorine pesticides in human milk in Ahmedabad, India. *Int. Arch. Occupat. Environ. Health* 60: 111–114.

Jones, L.A., and Bern, H.A. (1979). Cervicovaginal and mammary gland abnormalities in BALB/cCrgl mice treated neonatally with progesterone and estrogen, alone or in combination. *Cancer Res.* 39: 2560–2567.

Kaldas, R.S., and Hughes, C.L, Jr. (1989). Reproductive and general metabolic effects of phtoestrogens in mammals. *Reprod. Toxicol.* 3: 81–89.

Kamiya, K., Sato, T., Nishimura, N., Goto, Y., Kano, K., and Iguchi, T. (1996). Expression of estrogen receptor and proto-oncogene messenger ribonucleic acids in reproductive tracts of neonatally diethylstilbestrol-exposed female mice with or without postpuberal estrogen administration. *Exp. Clin. Endocrinol*. Diabetes 104: 111–122.

Katzenellenbogen, J.A., Carlson, K.E., Heiman, D.F., Robertson, D.W., Wei, L.L., and Katzenellenbogen, B.S. (1983). Efficient and highly selective covalent labeling of the estrogen receptor with [3H] tamoxifen aziridine. *J. Biol. Chem.* 258: 3487–3495.

Kaufman, R.H., Adam, E., Binder, G.L., and Gerthoffer, E. (1980). Upper genital tract changes and pregnancy outcome in offspring exposed *in utero* to diethylstilbestrol. *Am. J. Obstet. Gynecol.* 137: 299–308.

——————, ——————, Noller, K., Irwin, J.F., and Gray, M. (1986). Upper genital tract changes and infertility in diethylstilbestrol-exposed women. *Am. J. Obstet. Gynecol.* 154: 1312–1318.

Kelce, W.R., Monosson, E., and Gray, L.E., Jr. (1995a). An environmental antiandrogen. *Rec. Prog. Hormone Res.* 50: 449–453.

——————, Stone, C.R., Laws, S.C., Gray, L.E., Kemppainen, J.A., and Wilson, E.M. (1995b). Persistent DDT metabolite p,p′-DDE is a potent androgen receptor antagonist. *Nature* 375: 581–585.

Khan, S.A., and Stancel, G.M., eds. (1994). *Protooncogenes and Growth Factors in Steroid Hormone Induced Growth and Differentiation*, p. 277. CRC, Boca Raton.

Kincle, F.A. (1990). *Hormones and Toxicity in the Neonate*, pp. 334. Springer-Verlag, Berlin.

Kirkland, J.L., Murphy, L., and Stancel, G.M. (1993). Tamoxifen stimulates expression of the *c-fos* proto-oncogene in rodent uterus. *Mol. Pharmacol.* 43: 709–714.

Korach, K.S., Metzler, M., and McLachlan, J.A. (1978). Estrogenic activity *in vivo* and *in vitro* of some diethylstilbestrol metabolites and analogs. *Proc. Natl. Acad. Sci. U.S.A.* 75: 468–471.

Korach, K.S., Horigome, T., Tomooka, Y., Yamashita, S., Newbold, R.R., and McLachlan, J.A. (1988). Immunodetection of estrogen receptor in epithelial and stromal tissues of neonatal mouse uterus. *Proc. Natl. Acad. Sci. U.S.A.* 85: 3334–3337.

Lamartiniere, C.A., Moore, J., Holland, M., and Barnes, S. (1995). Neonatal genistein chemoprevents mammary cancer. *Proc. Soc. Exp. Biol. Med.* 208: 120–123.

Leavitt, W.W., and Meismer, D.M. (1968). Sexual development altered by non-steroidal oestrogen. *Nature* 218: 181–182.

Lee, H.P., Gourley, L., Duffy, S.W., Esteve, J., Lee, J., and Day, N.E. (1991). Dietary effects on breast cancer risk in Singapore. *Lancet* 337: 1197–1200.

Levy, J.R., Faber, K.A., Ayyash, L., and Hughes, C.L., Jr. (1995). The effect of prenatal exposure to the phytoestrogen genistein on sexual differentiation in rats. *Proc. Soc. Exp. Biol. Med.* 208: 60–66.

Mably, T.A., Moore, R.W., and Peterson, R.E. (1992a). In utero lactational exposure of male rats to 2,3,7,8-tetrachlorodibenzo-*p*-dioxin, 1: Effects on androgenic status. *Toxicol. Appl. Pharmacol.* 114: 97–107.

———, ———, Goy, R.W., and Peterson, R.E. (1992b). In utero lactational exposure of male rats to 2,3,7,8-tetrachlorodibenzo-*p*-dioxin: 2. Effects on sexual behavior and the regulation of luteinizing hormone secretion in adulthood. *Toxicol. Appl. Pharmacol.* 114: 108–117.

———, Bjerke, D.L., Moore, R.W., Gendron-Fitzpatrick, A., and Peterson, R.E. (1992c). *In utero* lactational exposure of male rats to 2,3,7,8-tetra-chlorodibenzo-*p*-dioxin: 3. Effects on spermatogenesis and reproductive capability. *Toxicol. Appl. Pharmacol.* 114: 118–126.

MacLatchy, D.L., and van der Kraak, G.J. (1995). The phytoestrogen b-sitos-terol alters the reproductive endocrine status of goldfish. *Toxicol. Appl. Pharmacol.* 134: 305–312.

Mair, D.B., Newbold, R.R., and McLachlan, J.A. (1985). Prenatal diethylstilbe-strol exposure alters murine uterine responses to prepubertal estrogen stimulation. *Endocrinology* 116: 1878–1886.

Malini, T., and Vanithakumari, G. (1993). Effect of b-sitosterol on uterine bio-chemistry: A comparative study with estradiol and progesterone. *Biochem. Mol. Biol. Int.* 31: 659–668.

Mangan, C.E., Borow, L., Burtnett-Rubin, M.M., Egan, V., Giuntoli, R.L., and

Mikuta, J.J. (1982). Pregnancy outcome of women exposed to diethylstilbestrol *in utero*, their mothers, and unexposed siblings. *Obstet. Gynecol.* 59: 315–319.

McLachlan, J.A., Newbold, R.R., Shah, H.C., Hogan, M.D., and Dixon, R.L. (1982). Reduced fertility in female mice exposed transplacentally to diethylstilbestrol (DES). *Fertil. Steril.* 38: 364–371.

—— and ——(1996). Cellular and molecular mechanisms of cancers of the uterus in animals. In *Cellular and Molecular Mechanisms of Hormonal Carcinogenesis: Environmental Influences.* (J. Huff, J. Boyd, and J.C. Barrett, eds.), pp. 175–182. Wiley-Liss, New York.

Medlock, K.L., Lyttle, C.R., Kelepouris, N., Newman, E.D., and Sheehan, D.M. (1991). Estradiol down-regulation of rat uterine estrogen receptor. *Poc. Soc. Exp. Biol. Med.* 196: 293–300.

——, Branham, W.S., and Sheehan, D.M. (1995). Effects of coumestrol and equol on the developing reproductive tract of the rat. *Proc. Soc. Exp. Biol. Med.* 208: 67–71.

Menczer, J., Dulitzky, M., Ben-Baruch, G., and Modan, M. (1986). Primary infertility in women exposed to diethylstilbestrol *in utero. Br. J. Obstet. Gynecol.* 93: 503–507.

Messina, M.J., Persky, V., Setchell, K.D.R., and Barnes, S. (1994). Soy intake and cancer risk: A review of the *in vitro* and *in vivo* data. *Nutr. Cancer* 21: 113–131.

Migliacchio, S., Newbold, R.R., Bullock, B.C., McLachlan, J.A., and Korach, K.S. (1992). Developmental exposure to estrogens induces persistent chages in skeletal tissue. *Endocrinology* 130: 1756–1758.

——, ——, McLachlan, J.A., and Korach, K.S. (1995). Alterations in estrogen levels during development affects the skeleton: Use of an animal model. *Environ. Health Perspec.* 103 (Suppl. 7): 95–97.

Miller, H.J., Cucos, S., Wassermann, D., and Wassermann, M. (1979). Organochlorine insecticides and polychlorinated biphenyls in human milk. *Dev. Toxicol. Environ. Sci.* 4: 379–386.

Miller, M.A., Sheen, Y.Y., Mullick, A., and Katzenellenbogen, B.S. (1986). Antiestrogen binding to estrogen receptors and additional antiestrogen binding sites in human breast cancer cells. In *Estrogen/Antiestrogen Action and Breast Cancer Therapy.* (V.C. Jordan, ed.), pp. 127–148. University of Wisconsin Press, Madison.

Mittendorf, R. (1995). Teratogen update: Carcinogenesis and teratogenesis associated with exposure to diethylstilbestrol (DES) *in utero. Teratology* 51: 435–445.

Mori, T., Nagasawa, H., and Bern, H.A. (1980). Long-term effects of perinatal exposure to hormones on normal and neoplastic mammary growth in rodents: A review. *J. Environ. Pathol. Toxicol.* 3: 191–205.

—— (1986). Abnormalities in the reproductive system of aged mice after neonatal estradiol exposure. *J. Endocrinol. Invest.* 9: 397–402.

—— and Nagasawa, H., eds. (1988). *Toxicity of Hormones in Perinatal Life*, pp. 184. CRC, Boca Raton.

Murphy, L.C., and Dotzlaw, H. (1989). Variant estrogen receptor mRNA species detected in human breast cancer biopsy samples. *Mol. Endocrinol.* 3: 687–693.

Nagasawa, H., Mori, T., and Nakajima, Y. (1980). Long-term effects of progesterone or diethylstilbestrol with or without estrogen after maturity on mammary tumorigenesis. *Eur. J. Cancer* 16: 1583-1589.

—— and ——(1988). Long-term effects of perinatal exposure to hormones and related substances on normal and neoplastic growth of murine mammary glands. In *Toxicity of Hormones in Perinatal Life.* (T. Mori, and H. Nagasawa, eds.), pp. 81–87. CRC, Boca Raton.

Nelson, K.G., Takahashi, T., Lee, D.C., Luetteke, N.C., Bossert, N.L., Ross, K., Eitzman, B.E., and McLachlan, J.A. (1992). Transforming growth factor-a is a potential mediator of estrogen action in the mouse uterus. *Endocrinology* 131: 1657–1664.

——, Sakai, Y., Eitzman, B., Steed, T., and McLachlan, J. (1994). Exposure to diethylstilbestrol during a critical developmental period of the mouse reproductive tract leads to persistent induction of two estrogen-regulated genes. *Cell Growth Differ.* 5: 595–606.

Nephew, K.P., Polek, T.C., Akcali, K.C., and Khan, S.A. (1993). The antiestrogen tamoxifen induces *c-fos* and jun-B, but not *c-jun* or jun-D, protooncogenes in the rat uterus. *Endocrinology* 133: 419–422.

Newbold, R.R., Tyrey, S., Haney, A.F., and McLachlan, J.A. (1983). Developmentally arrested oviduct: A structural and functional defect in mice following prenatal exposure to diethylstilbestrol. *Teratology* 27: 417–426.

—— and McLachlan, J.A. (1996). Transplacental hormonal carcinogenesis: Diethylstilbestrol as an example. In *Cellular and Molecular Mechanisms of Hormonal Carcinogenesis: Environmental Influences.* (J. Huff, J. Boyd, and J.C. Barrett, eds.), pp. 131–147. Wiley-Liss, New York.

Nishimura, N., Goto, Y., and Iguchi, T. (1993). Tamoxifen induces expressions of oncogenes and estrogen receptor in genital tracts of female mice. *Zool. Sci.* (Suppl. 10): 128.

——, Fukazawa, Y., Uchiyama, H., and Iguchi, T. (1997). Effects of estrogenic hormones on early development of *Xenopus laevis. J. Exp. Zool.* 278: 221–233.

Ohta, Y. (1982). Deciduoma formation in rats ovariectomized at different ages. *Biol. Reprod.* 27: 303–311.

—— (1985). Deciduomal response in prepubertal rats adrealectomizedovariectomized at different ages of early postnatal life. *Zool. Sci.* 2: 89–93.

———, Iguchi, T., and Takasugi, N. (1989). Deciduoma formation in rats treated neonatally with the anti-estrogens, tamoxifen and MER-25. *Reprod. Toxicol.* 3: 207–212.

——— (1995). Sterility in neonatally androgenized female rats and the decidual cell reaction. *Int. Rev. Cytol.* 160: 1–52.

———, Uesugi, Y., and Iguchi, T. (1995). Steroid synthesis in testicular and ovarian homogenates from adult mice treated neonatally with diethylstilbestrol. *Med. Sci. Res.* 23: 763-764.

———, Fukazawa, Y., Sato, T., Suzuki, A., Nishimura, N., and Iguchi, T. (1996). Effect of estrogen on ontogenic expression of progesterone receptor in rat uterus. *Zool. Sci.* 13: 143–149.

Pardee, A.B. (1989). G1 events and regulation of cell proliferation. *Science* 246: 603–608.

Persico, E., Scalona, M., Cicatiello, L., Sica, V., Bresciani, F., and Weisz, A. (1990). Activation of "immediate-early" genes by estrogen is not sufficient to achieve stimulation of DNA synthesis in rat uterus. *Biochem. Biophys. Res. Commun.* 171: 287–292.

Peterson, R.E., Moore, R.W., Mably, T.A., Bjerke, D.L., and Goy, R.W. (1992). Male reproductive system ontogeny: Effects of perinatal exposure to 2,3,7,8-tetrachlorodibenzo-*p*-dioxin. In *Chemically-Induced Alterations in Sexual and Functional Development: The Wildlife/Human Connection.* (T. Colborn, and C. Clement, eds.), pp 175–193. Princeton Scientific Publications, Princeton.

Raynaud, A. (1961). Morphogenesis of the mammary gland. In *Milk: The Mammary Gland and Its Secretion, Vol. 1.* (S.K. Kon and A.T. Cowie, eds.), pp. 3–46. Academic, New York.

Riabowol, K.T., Vosatka, R.J., Ziff, E.B., Lamb, N.J., and Feramisco, J.R. (1988). Microinjection of Fos-specific antibodies blocks DNA synthesis in fibroblast cells. *Mol. Cell Biol.* 8: 1670–1676.

Roman, B.L., Sommer, R.J., Shinomiya, K., and Peterson, R.E. (1995). In utero and lactational exposure of the male rat to 2,3,7,8-tetrachlorobenzo-*p*-dioxin: Impaired prostate growth and development without inhibited androgen production. *Toxicol. Appl. Pharmacol.* 134: 241–250.

Rosenblum, E.R., Stauber, R.E., Van Thiel, D.H., Campbell, I.M., and Gavaler, J.S. (1993). Assessment of the estrogenic activity of phytoestrogens isolated from bourbon and beer. *Alcohol Clin. Exp. Res.* 17: 1207–1209.

Rothschild, T.C., Boylan, E.S., Calhoon, R.E., and Vonderhaar, B.K. (1987). Transplacental effects of diethylstilbestrol on mammary development and tumorigenesis in female ACI rats. *Cancer Res.* 47: 4508–4516.

Rotmensch, J., Frey, K., and Herbst, A.L. (1988). Effects on female offspring and mothers after exposure to diethylstilbestrol. In *Toxicity of Hormones in Perinatal Life*. (T. Mori, and H. Nagasawa, eds.), pp. 143–159. CRC, Boca Raton.

Rustia, M., and Shubik, P. (1979). Effects of transplacental exposure to diethylstilbestrol on carcinogenic susceptibility during postnatal life in hamster progeny. *Cancer Res.* 39: 4636–4647.

Ryan, K.J., Naftolin, F., Reddy, V., Flores, F., and Petro, Z. (1972). Estrogen formation in the brain. *Am. J. Obstet. Gynecol.* 114: 454–460.

Safe, S.H. (1995). Modulation of gene expression and endocrine response pathways by 2,3,7,8-tetrachlorodibenzo-*p*-dioxin and related compounds. *Pharmac. Ther.* 67: 247–281.

Sato, T., Okamura, H., Ohta, Y., Hayashi, S., Takamatsu, Y., Takasugi, N., and Iguchi, T. (1992). Estrogen receptor expression in the genital tract of female mice treated neonatally with diethylstilbestrol. *In Vivo* 6: 151–156.

———, Ohta, Y., Okamura, H., Hayashi, S., and Iguchi, T. (1996a). Estrogen receptor (ER) and its messenger ribonucleic acid expression in the genital tract of female mice exposed neonatally to tamoxifen and diethylstilbestrol. *Anat. Rec.* 244: 374–385.

———, Fukazawa, Y., Kojima, H., Enari, M., Iguchi, T., and Ohta, Y. (1997). Apoptotic cell death during the estrous cycle in the rat uterus and vagina. *Anat. Rec.* 248: 76–83.

Schuermann, M., Neuberg, M., Hunter, J.B., Jenuwein, T., Ryseck, R., Bravo, R., and Myller, R. (1989). The leucine repeat motif in Fos protein mediates complex formation with Jun/AP-1 and is required or transformation. *Cell* 56: 507–516.

Scrocchi, L.A., and Jones, L.A. (1991). Alteration of proto-oncogene *c-fos* expression in neonatal estrogenized BALB/c female mice and murine cervicovaginal tumor LJ6195. *Endocrinology* 129: 2251–2253.

Selye, H., Collip, J.B., and Thomson, D.L. (1933). On the effect of the anterior pituitary-like hormone on the ovary of the hypophysectomized rat. *Endocrinology* 17: 494–500.

Setchell, K.D.R., Welsh, M., Lim, C.K. (1987). High-performance liquid chromatographic analysis of phytoestrogens in soy protein preparations with ultraviolet, electrochemical and thermospray mass spectrometric detection. *J. Chromatogr.* 386: 315–323.

——— and Adlercreutz, H. (1988). Mammalian lignans and phytoestrogens: Recent studies on their formation, metabolism and biological role in health and disease. In *The Role of Gut Microflora in Toxicity and Cancer*. (I.A. Rowland, ed.), pp. 315–345. Academic, New York.

Sheehan, D.M., Branham, W.S., Medlock, K.L., and Shanmugasundaram, E.R.B. (1984). Estrogenic activity of zearalenone and zearalanol in the neonatal rat uterus. *Teratology* 29: 383–392.

———— (1995). Introduction: The case for expanded phytoestrogen research. *Proc. Soc. Exp. Biol. Med.* 208: 3–5.

————, ed. (1995). Presentations from the 2d International Congress on Phytoestrogens, Little Rock, AR, Oct. 17–20, 1993. *Proc. Soc. Exp. Biol. Med.* 208: 1–138.

Shellabarger, C.J., and Soo, V.A. (1973). Effects of neonatally administered sex steroids on 7,12-dimethylbenz(a)anthracene-induced mammary neoplasia in rats. *Cancer Res.* 33: 1567–1569.

Shupnik, M.A., Gordon, M.S., and Chin, W.W. (1989). Tissue-specific regulation of rat estrogen receptor mRNAs. *Mol. Endocrinol.* 3: 660–665.

Soto, A.M., Lin, T.-M., Justicia, H., Silvia, R.M., and Sonnenschein, C. (1992). An "in culture" bioassay to assess the estrogenicity of xenobiotics (E-screen). In *Chemically-induced Alterations in Sexual and Functional Development: The Wildlife/Human Connenction.* (T. Colborn, C. Clement, eds.), pp. 295–309. Princeton Scientific Publications, Princeton.

Sumpter, J.P., and Jobling, S. (1995). Vitellogenesis as a biomarker for estrogenic contamination of the aquatic environment. *Environ. Health Perspect.* 103 (Suppl. 7): 173–178.

Taguchi, O., and Nishizuka, Y. (1985). Reproductive tract abnormalities in female mice treated neonatally with tamoxifen. *Am. J. Obstet. Gynecol.* 151: 675–678.

Takasugi, N. (1959). Endocrinological studies of the persistent-estrous animals. *Jap. J. Exp. Morphol.* 13: 20–48.

————, Bern, H.A., and DeOme, K.B. (1962). Persistent vaginal cornification in mice. *Science* 138: 438–439.

———— and ————(1964). Tissue changes in mice with persistent vaginal cornification induced by early postnatal treatment with estrogen. *J. Natl. Cancer Inst.* 33: 855–865.

———— (1976). Cytological basis for permanent vaginal changes in mice treated neonatally with steroid hormones. *Int. Rev. Cytol.* 44: 193–224.

———— (1979). Development of permanently proliferated and cornified vaginal epithelium in mice treated neonatally with steroid hormones and the implication in tumorigenesis. *Natl. Cancer Inst. Monogr.* 51: 57–66.

———— and Bern, H.A. (1988). Introduction: Abnormal genital tract development in mammals following early exposure to sex hormones. In *Toxicology of Hormones in Perinatal Life.* (T. Mori, and S. Nagasawa, eds.), pp. 1–7. CRC, Boca Raton.

Takewaki, K. (1962). Some aspects of hormonal mechanism involved in persistent estrus in the rat. *Experientia* 18: 1–6.

Tenenbaum, A., and Forsberg, J.-G. (1985). Structural and functional changes in ovaries from adult mice treated with diethylstilbestrol in the neonatal period. *J. Reprod. Fert.* 73: 465–477.

Teng, C., Pentecost, B., Chen, Y., Newbold, R.R., Eddy, E., and McLachlan, J.A. (1989). Lactoferrin gene expression in the mouse uterus and mammary gland. *Endocrinology* 124: 992–999.

Tominaga, S. (1985). Cancer incidence in Japanese in Japan, Hawaii, and western United States. *Natl. Cancer Inst. Monogr.* 69: 83–92.

Tomooka, Y., and Bern, H.A. (1982). Growth of mouse mammary glands after neonatal sex hormone treatment. *J. Natl. Cancer Inst.* 69: 1347–1352.

———, DiAugustine, R.P., and McLachlan, J.A. (1986). Proliferation of mouse uterine epithelial cells in vitro. *Endocrinology* 118: 1011–1018.

Tuinstra, L.M.G. (1971). Organochlorine insecticide residues in human milk in the Leiden region. *Netherlands Milk Dairy J.* 25: 24–32.

Uchima, F.-D.A., Edery, M., Iguchi, T., and Bern, H.A. (1991). Growth of mouse endometrial epithelial cells in vitro: Functional integrity of the oestrogen receptor system and failure of oestrogen to induce proliferation. *J. Endocrinol.* 128: 115–120.

Uesugi, Y., Taguchi, O., Noumura, T., and Iguchi, T. (1992). Effects of sex steroids on the development of sexual dimorphism in mouse innominate bone. *Anat. Rec.* 234: 541–548.

Vessey, M.P., Fairweather, D.V., Norman–Smith, B., and Buckley, J. (1983). A randomized double-blind controlled trial of the value of stilboestrol therapy in pregnancy: Long-term follow-up of mothers and their offspring. *Br. J. Obstet. Gynecol.* 90: 1007–1017.

vom Saal, F.S., Nagel, S.C., Palanza, P., Boechler, M., Parmigiani, S., and Welshons, W.V. (1995). Estrogenic pesticides: Binding relative to estradiol in MCF-7 cells and effects of exposure during fetal life on subsequent territorial behaviour in male mice. *Toxicol. Lett.* 77: 343–350.

Walker, B.E. (1983). Complications of pregnancy in mice exposed prenatally to DES. *Teratology* 27: 73–80.

——— (1984). Transplacental exposure to diethylstilbestrol. In *Issues and Review in Teratology, Vol 2.* (H. Kalter, ed.), pp. 157–187. Plenum, New York.

——— and Kurth, L.A. (1995). Multi-generational carcinogenesis from diethylstilbestrol investigated by blastocyst transfers in mice. *Int. J. Cancer* 61: 249–252.

Walters, L.M., Rourke, A.W., and Eroschenko, V.P. (1993). Purified methoxychlor stimulates the reproductive tract in immature female mice. *Reprod. Toxicol.* 7: 599–606.

Webb, D.K., Moulton, B.C., and Khan, S.A. (1993). Estrogen induces expression of *c-jun* and jun-B protooncogenes in specific rat uterine cells. *Endocrinology* 133: 20–28.

Weisz, A., and Bresciani, F. (1988). Estrogen induces expression of *c-fos* and *c-myc* protooncogenes in rat uterus. *Mol. Endocrinol.* 2: 816–824.

————, Cicatiello, L., Persico, M., and Bresciani, F. (1990). Estrogen stimulates transcription of *c-jun* proto-oncogene. *Mol. Endocrinol.* 4: 1041–1050.

Williams, B.A., Mills, K.T., Burroughs, C.D., and Bern, H.A. (1989). Reproductive alterations in female C57BL/Crgl mice exposed neonatally to zearalenone, an estrogenic mycotoxin. *Cancer Lett.* 46: 225–230.

Wordinger, R.J., and Derrenbacker, J. (1989). *In utero* exposure of mice to diethylstilbestrol alters neonatal ovarian follicle growth and development. *Acta Anat.* 134: 312–318.

————, Brown, D., Atkins, E., and Jackson, F.L. (1989). Superovulation and early embryo development in the adult mouse after prenatal exposure to diethylstilbestrol. *J. Reprod. Fert.* 85: 383–388.

Xing, H., and Shapiro, D.J. (1993). An estrogen receptor mutant exhibiting hormone-independent transactivation and enhanced affinity for the estrogen response element. *J. Biol. Chem.* 268: 23227–23233.

Yamashita, S., Newbold, R.R., McLachlan, J.A., and Korach, K.S. (1989). Developmental pattern of estrogen receptor expression in female mouse genital tracts. *Endocrinology* 125: 2888–2896.

————, ————, ————, and ————(1990). The role of estrogen receptor in uterine epithelial proliferation and cytodifferentiation in neonatal mice. *Endocrinology* 127: 2456–2463.

Zimmer, H.J., and Maddrey, W.C. (1987). Toxic and drug-induced hepatitis. In *Diseases of the Liver.* (L. Schiff, and E.R. Schiff, eds.), pp. 635–636. Lippincott, Philadelphia.

Chapter 10

ENDOCRINE DISRUPTION IN MALE HUMAN REPRODUCTION

Jorma Toppari[1,2] and Niels E. Skakkebæk[1]

[1] Department of Growth
and Reproduction
The National University Hospital
Copenhagen, Denmark

[2] Departments of Pediatrics
and Physiology
University of Turku
Turku, Finland

Introduction

Male reproduction involves complex and sensitive processes. Normal function in the adult depends on normal development and organization of the organs in fetal life, as well as normal postnatal growth and pubertal development. Thus, although failure of reproduction is often not manifest until a man is 30 to 40 years old, the cause of infertility occurred perhaps unnoticed in fetal life or early childhood. Important abnormalities associated with reproductive failure include impairment of spermatogenesis (leading to poor semen quality), undescended testes, and other genital abnormalities. Testicular cancer, which most often occurs in young men, should also be considered as a reproductive disorder.

Many lines of research indicate that testicular cancer, in a yet unknown way, is biologically associated with other types of abnormalities of the reproductive organs. First, it has recently been shown that men with testicular cancer often have disruption of spermatogenesis (including Sertoli cell–only pattern and spermatogenic arrest), even in the contralateral testis. Secondly, men with various types of abnormalities of the reproductive organs are at particular risk of developing testicular cancer. The best-documented association is between undescended testes and testicular cancer. A recent study also showed a link between hypospadias and cancer of the testis. Similarly, it has been recognized for decades that abnormally developed sexual organs (i.e., undescended testis) often resulted in poor reproductive function in adulthood.

Although it is well known that the adult germ cells are sensitive to toxicants and other factors, the fetal period seems to be particularly vulnerable. Thus, recent research seems to indicate that not only so-called developmental disorders like maldescent of the testis and hypospadias but also testicular cancer and decreased semen quality may originate in fetal life. In our search for a better understanding of etiology and pathogenesis, we should therefore not only look at the disorders individually but also consider the possibility that they sometimes may have a common cause. There is, in fact, both epidemiological and experimental evidence that administration of DES and other estrogens causes different types of abnormalities within an individual. In the following, we shall, with regard to the human male, review the literature on trends in these disorders, of which some—perhaps all—are increasing in the western world.

Semen Quality

Good semen quality is essential for reproductive success. All mammals produce a huge excess of sperm compared to what is needed for procreation. However, there are large differences between species in the quality of sperm. In most mammals more than 60% of sperm is morphologically and functionally normal, whereas in humans, at least in the industrialized world, more than 50% of sperm is either morphologically or functionally abnormal. Furthermore, the quantitative marginal between normal and subnormal sperm counts is much narrower in humans compared to other mammals. Thus, humans may be more vulnerable than other mammals to environmental insults that affect spermatogenesis.

Declining semen quality has been a topic of vivid discussion in the 1990s. Since the 1970s there have been reports suggesting a decline in semen quality. (Bostofte et al., 1983; Leto and Frensilli, 1981; Nelson and Bunge, 1974). At first the decline was presumed to reflect changes in the treatment of infertility patients or selection biases rather than a true biological trend, because the reports were based on data from men attending infertility clinics or on very selected groups of fertile men. Attitudes changed when a comprehensive meta-analysis of sperm studies that included only healthy normal men indicated a significant decrease in sperm concentration (113 million/ml vs. 66 million/ml) and semen volume (3.40 ml vs. 2.75 ml) over the period 1938–1990 (Carlsen et al., 1992). This report stimulated a series of new studies. The meta-analysis was even repeated and republished by another group (Olsen et al., 1995) that agreed with the general conclusion that there has been a significant decline in sperm concentrations over the last half century. However, the latter study emphasized that the adverse trend did not appear to be continuing, which was not claimed in the first paper either (Skakkebæk and Keiding, 1994). A recent reanalysis and new analysis indicates that sperm count has continued to decline in North America and Western Europe, although local variation is present (Swan et al., 1997; Swan and Elkin, 1999; Swan, 2000).

Table 10-1. Geographic variation in mean sperm concentration in the 1990s

Location	Mean (millions/ml)	Reference
Finland, Kuopio	133.9	Vierula et al., 1996
U.K., Edinburgh	104.5	Irvine et al., 1996
France, Toulouse	83.1	Bujan et al., 1996
France, Paris	60.0	Auger et al., 1995
Denmark, Zealand	69.2	Jensen et al., 1996
Belgium, Ghent	58.6	Van Waeleghem et al., 1996
U.S., New York	131.5	Fisch et al., 1996
U.S., Minnesota	100.8	Fisch et al., 1996
U.S., California	72.7	Fisch et al., 1996
U.S., Seattle	52.0	Paulsen et al., 1996

Temporal trends in semen quality may have been hidden behind the large geographical variations during the last 3 decades (Table 10-1). Several studies demonstrate that there are areas with a clear decline in semen quality, whereas some countries show no apparent change. For example, in Finland from the 1950s to the 1990s, sperm counts have stayed well above 100 million/ml (Suominen and Vierula, 1993; Vierula et al., 1996), whereas studies from Paris (Auger et al., 1995), Belgium (Van Waeleghem et al., 1996), and Greece (Adamopoulos et al., 1996) showed significant declines. In a Parisian study of 1,351 healthy sperm donors, a 2.1% yearly decrease in sperm concentration from 89 million/ml in 1973 to 60 million/ml in 1992 ($p < 0.001$) was found (Auger et al., 1995). Concentration is not the single most important factor influencing the fertilizing capacity of sperm; other parameters, such as motility and morphology, play a significant role. In the French study, the percentages of motile and normal spermatozoa also decreased significantly, whereas semen volume did not change. Interestingly, the year of birth of the men affected the results significantly. Multiple–regression analysis, which allows for separate effects of age and year of birth, revealed yearly decreases of 2.6% in sperm concentration, 0.3% in motility percentage, and 0.7% in the percentage of normal spermatozoa according to the year of birth of the men (all changes, $p < 0.001$) (Auger et al., 1995). Similar results were obtained in a Scottish study (Irvine et al., 1996) of semen donors, where a correlation was found between the median sperm concentration and the year of birth during 1940–1969. Sperm concentration decreased from 120 million/ml to 75 million/ml. When the birth cohort effect was ignored, no change was found. The association between declining semen quality and a more recent year of birth lends support to the concept that adverse effects during prenatal and childhood period may influence sperm production capacity in adult life. Deterioration of sperm counts as well as motility among semen donor candidates during the past 2 decades was also observed in a Belgian study (Van Waeleghem et al., 1996). Ginsburg and Hardiman (1992) found a decrease in sperm concentrations (105 million/ml in 1978–1983 vs. 76 million/ml in 1984–1989) of the partners of women treated for infertility and living in the Thames water supply area of London, whereas no decrease was found among those who lived in other water supply areas of London. At the same time, the mean percentage of abnormal spermatozoa increased in all water supply areas (18% to 19% vs. 30% to 32%) (Ginsburg and Hardiman, 1992). Notably, the data in the studies cited above originated in laboratories that used consistently the same methods for semen analysis throughout the period. Thus, longitudinal data from the same laboratory can reliably reflect local temporal trends, whereas geographic differences may partly be dependent on methodological variation in laboratory techniques. Bearing that in mind, four European sperm laboratories from Denmark, Finland, France, and Scotland recently compared results of their semen analyses using the same samples. The results of sperm concentrations and semen volume were very consistent among the laboratories, whereas motility and morphology assessment varied considerably (Niels Jørgensen, personal communication).

In addition to Finnish studies, stable sperm concentrations have been reported from other countries. No decline was observed in Toulouse, France, where mean sperm concentration was 83 million/ml (Bujan et al., 1996), or in American centers from California, Minnesota, New York City (Fisch et al., 1996), and Seattle (Paulsen et al., 1996). Some of these areas (Minnesota, New York, Seattle) showed even slight increases in the mean sperm concentrations. However, large geographical differences were found in the United States. (Table 10-1). The highest concentrations were found in New York, 131.5 million/ml; the mean value from Minnesota was 100.8 million/ml; in California, 72.7 million/ml; and in Seattle, the mean sperm concentrations ranged from 46.5 to 89 million/ml. Thus far, the study of Fisch and coworkers is the only one indicating high sperm concentrations among New Yorkers in the 1990s, whereas several studies showed high sperm counts in Finland (for references, see Suominen and Vierula, 1993; Vierula et al., 1996). For adequate analysis, it will be necessary to conduct future studies with a more rigorous experimental design. Currently, direct comparisons between the European and American studies cannot be made, because the study populations differed markedly. Most of the European studies included only healthy sperm donors, whereas the men in the United States were either prevasectomy patients (Fisch et al., 1996) or volunteers in drug studies (Paulsen et al., 1996). Although different, all these groups were selected, and therefore may not reflect the general population that may have much different semen quality.

Currently there are only hypothetical explanations for the large geographic differences in semen quality. Environmental endocrine disruption is certainly a biologically plausible cause for deteriorating semen quality, but that remains to be proved. Only in a few cases has it been possible to measure exposures to endocrine disruptors reliably to correalate exposure to any outcome. However,

DES exposure has been well documented in studies following the so-called Dieckmann cohort (Dieckmann et al., 1953) which also included a control group. Semen quality of men exposed to DES *in utero* was significantly worse than that of nonexposed controls (Gill et al., 1977, 1979). Similar results have been obtained in several other studies. However, despite lower semen quality, fertility of the DES-exposed men did not differ significantly from controls (Wilcox et al., 1995). This suggests that there was a wide enough safety margin for these men to avoid infertility. If the exposure occurred today, such a safety margin might not exist in several areas of the world. Interestingly, there is a significant difference between England and Finland in fertility, as measured by the the time to pregnancy (Joffe, 1996). This difference may partly result from the large difference in sperm concentrations in these two countries; Finnish men have higher sperm counts than English men.

Many environmental contaminants elicit testicular toxicity by disturbing spermatogenesis. The best-known example of this is DBCP, which was found to cause azoospermia in men who were occupationally exposed to the chemical (Whorton et al., 1977, 1979; Potashnik et al., 1978). A clue to this was discovered through unusually close contacts between the wives of the workers, and numerous experimental studies verified the cause-effect relationship. More subtle changes caused by disturbed endocrine homeostasis may never become as obvious.

Testicular Cancer

Testicular cancer is an endocrine-related neoplasm that has a peak incidence during young adulthood (25–35 years). There is accumulating evidence that the disease originates early during development and becomes manifest after pubertal hormonal stimulation at a relatively young age (Skakkebæk et al., 1987). Gonadal dysgenesis, which occurs in the absence of normal sequence of genetic and endocrine regulation of sex differentiation, leads to high susceptibility to testicular cancer (Müller et al., 1985). Another well-known risk factor is cryptorchidism (maldescent of testis) that may share a causal factor with cancer rather than influence tumorigenesis itself (Giwercman et al., 1988).

The incidence of testicular cancer has increased rapidly worldwide (Fig. 10-1). It is now the most common malignancy of young men in many countries; for example, in Denmark the lifetime risk of developing testicular cancer approaches 1%. The incidence has steadily increased from the beginning of this century, i.e., as long as there have been cancer statistics (Forman and Møller, 1994). Data from cancer registries indicate significant increases in incidence in Poland, Germany, the Nordic and Baltic countries (Adami et al., 1994; Hakulinen et al., 1986), England and Wales (Pike et al., 1987; Nethersell et al., 1984), Scotland (Boyle et al., 1987), Australia (Stone et al., 1991), New Zealand (Wilkinson et al., 1992; Pearce et al., 1987), and the white population of the United States (Spitz et al.,

Incidence of Testicular Cancer

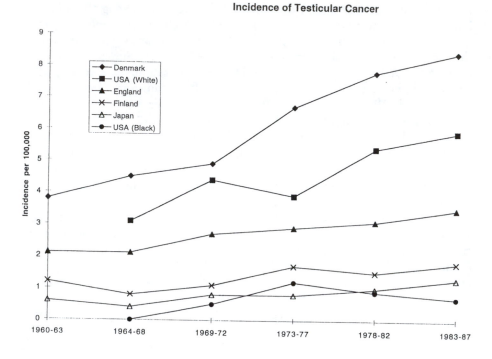

Figure 10-1.
Incidence of testicular cancer between 1960–1987 on the basis of data from "Cancer Incidence in Five Continents," UICC, Geneva & IARC, Lyon, Vols. I–VI. The data for the United States originate from Alameda (very similar to other U.S. regions); the English figures come from the Birmingham region; and Japanese data from Miyagi. Finnish and Danish data are based on the statistics of national cancer registries.

1986). The yearly increase has been approximately 2% to 4% in men under 50 years of age, and has occurred in the same age group in which testicular cancer incidence peaks (i.e., young adults). Incidence rates also vary between rural and urban areas. For example, in Denmark the incidence in Copenhagen and suburbs has been higher than in rural areas (Clemmesen, 1991).

A detailed analysis of testicular cancer incidences in Denmark, Finland, Norway, Sweden, Poland, and former East Germany revealed that the increasing trend for cancer risk follows a birth cohort pattern (Bergström et al., 1996). For example, in East Germany, the incidence at ages 35 to 39 almost doubled for each birth cohort from 1925–1934 through 1935–1944 to 1945–1954 (Bergström et al., 1996). Furthermore, the birth cohort from 1955–1964 experienced a dramatic increase in incidence at ages 20 to 29 when compared to the previous cohort. Similar development was evident in all six countries (Bergström et al., 1996). Notably, deterioration of semen quality was similarly associated to birth cohorts in two recent studies; i.e., the later year of birth correlated to declined sperm concentrations (Auger et al., 1995; Irvine et al., 1996). Existence of this kind of birth cohort pattern strongly suggests that there is an increasing adverse environmental influence on testicular development; whether this environmental influence affects the endocrine system remains to be elucidated.

There are marked racial and geographic differences in the incidence of testicular cancer. For example, incidence in Denmark is at least 4-fold higher than in nearby Finland, and Caucasians are 3-fold more susceptible to this disease than are African Americans. Nevertheless, there is a trend toward an increased incidence in all these populations. The incidence of both seminomas and non-seminomas has increased (Forman and Møller, 1994). Mortality due to testicular cancer increased from the beginning of this century to the early 1970s when, because of developments in cancer chemotherapy, mortality started to decline (Forman and Møller, 1994).

The exact etiology of testicular cancer is still unknown. Patients with gonadal dysgenesis have given some clues that point to the importance of normal endocrine homeostasis in the suppression of tumorigenesis. The role of prenatal hormonal exposures as risk factors has been analyzed in a few case control studies (Gershman and Stolley, 1988; Moss et al., 1986; Brown et al., 1986; Schottenfeld et al., 1980; Depue et al., 1983; Henderson et al., 1979), but the results are not unequivocal. The relative risk of testicular cancer (or odds ratios) in hormone-exposed vs. nonexposed men varied between 0.9 and 8.0 in these studies. The exposures were highly variable in the study populations, which were often too small to show statistically significant differences between groups. When data from the six studies (Gershman and Stolley, 1988; Moss et al., 1986; Brown et al., 1986; Schottenfeld et al., 1980; Depue et al., 1983; Henderson et al., 1979) were combined in a meta-analysis, a marginally significant increase in testicular cancer incidence for the hormone-exposed (including all hormones) patients was found; Mantel-Haenszel estimates of the common odds ratio was 2.1, with 95% confidence intervals of 1.3 to 3.3 (Toppari et al., 1995, 1996). Again, DES-exposed men might serve as a good study population to analyze the role of estrogen action, although DES might also function through other mechanisms. Thus far there is no conclusive evidence to indicate an increased risk of testicular cancer in DES-exposed men, although the incidence of cryptorchidism, which is a risk factor for testicular cancer, is abnormally high in this group (Stillman, 1982). Patient cases with seminoma in DES-exposed men have been reported (Conley et al., 1983), but epidemiological studies have failed to show a statistically significant association between DES-exposure and testicular cancer. However, DES-exposure was a significant risk factor for testicular cancer on the basis of the meta-analysis of the above-mentioned six studies (Gershman and Stolley, 1988; Moss et al., 1986; Brown et al., 1986; Schottenfeld et al., 1980; Depue et al., 1983; Henderson et al., 1979): Odds ratio was 2.6, with 95% confidence limits of 1.1 to 6.1. There are no recent follow-up studies that would have covered the age period when the exposed men would have had the highest risk of developing testicular cancer. Therefore, it is important to obtain new information on the incidence of testicular cancer in men born to mothers who participated in the double-blind, placebo-controlled DES-trial in the 1950s (Dieckmann et al., 1953).

Cryptorchidism

Cryptorchidism is the most common male congenital malformation. Its etiology is not known, but it is clear that normal androgen action is needed for testicular descent. Thus, disruption of androgen action, by either androgen resistance or antiandrogenic agents, can result in cryptorchidism. Estrogens can also act in an antiandrogenic direction by decreasing the relative ratio of androgens to estrogens.

Data on newborns have indicated a substantial increase in the incidence of cryptorchidism. However, reliability of different studies is difficult to assess, because it is often not clear how a cryptorchid testis was defined. Inclusion of different proportions of boys with retractile testes could account for the reported differences. The sources of data used in the studies also differ, and ascertainment of cryptorchidism often varies.

From birth to 1 year of age, the prevalence rates of cryptorchidism have varied between 0.03% and 13.4% on the basis of data from hospital or central registers (often including preterm babies) (Berkowitz et al., 1993; Ansell et al., 1992; Correy et al., 1991; Benson et al., 1991; Choi et al., 1989; Swerdlow and Melzer, 1988; Morley and Lucas, 1987; Campbell et al., 1987; Seddon et al., 1985; Daughaday, 1981; Hsieh and Huang, 1985; Hirasing et al., 1982; Czeizel et al., 1981; Mau and Schnakenburg, 1977; Heinonen et al., 1977; Mital and Garg, 1972; Halevi, 1967; Scorer, 1964; Pitt, 1962; McDonald, 1958; McIntosh et al., 1954; Harris and Steinberg, 1954). School surveys have shown the incidence of cryptorchidism to be 0.16% to 13.3% (Yücesan et al., 1993; Onuora and Evbuomwan, 1989; Blom, 1984; Cour-Palais, 1966; Panayotou, 1965; Ward and Hunter, 1960; Baumrucker, 1946; Johnson, 1939; Williams, 1936; Hsieh and Huang, 1985; Czeizel et al., 1981), and cohort studies based on discharge diagnosis show the incidence to be 2% to 4.7% (Thorup and Cortes, 1990; Chilvers et al., 1984; Campbell et al., 1987). Most studies have included only Caucasian populations. In India (Mital and Garg, 1972), Taiwan (Hsieh and Huang, 1985), and Korea (Choi et al., 1989), incidence of cryptorchidism among newborns was 1.6%, 1.4%, and 0.7%, respectively. A school survey in Nigeria (Onuora and Evbuomwan, 1989) indicated a prevalence of 0.5%. The incidence of cryptorchidism among African Americans was reported to be only one-third of that among whites (Heinonen et al., 1977), although another study (Berkowitz et al., 1993) did not find a significant difference in New York. Racial and ethnic data are pooled in most studies. Evaluation of temporal trends in the incidence of cryptorchidism is difficult, because there are no prospective studies examining longitudinal changes, and only a few studies confined to the same population and geographic area and using consistently the same definition of the condition. Thus, temporal trends have to be analyzed by assessing data sets that are as close to prospective design as possible.

There is a lot of indirect evidence suggesting an increasing trend in the incidence of cryptorchidism. According to the discharge data from the Hospital Inpatient Enquiry from England and Wales, the proportion of boys undergoing orchidopexy (operation to bring the testis into the scrotum) before the age of 15 increased from 1.4% for a 1952 birth cohort to 2.9% for a 1977 birth cohort (Chilvers et al., 1984). However, it is not known whether the criteria for selection of patients to the operation remained the same during the observation period (e.g., number of boys with retractile testes undergoing surgery). In Scotland, the annual number of discharges of boys aged 0 to 14 years with the diagnosis of cryptorchidism also increased significantly during 1961–1985 (Campbell et al., 1987). In Denmark, the incidence of cryptorchidism among male newborns weighing > 2,500 g varied between 1% and 1.8% in three study groups in the late 1950s (Buemann et al., 1961). However, school surveys suggested prevalence rates up to 7% during 1940–1966 (Blom, 1984; Møller et al., 1996). The difference may result from possible inclusion of many retractile testes in the latter study. During the period 1982–1985 the incidence of cryptorchidism in Denmark was approximately 2% according to a cohort analysis of data from the Danish National Register of Hospital In- and Outpatients (Thorup and Cortes, 1990).

The best series of comparable studies on cryptorchidism over a long time period is from England (Ansell et al., 1992; Scorer, 1964; Fig. 10-2). In the late 1950s, Scorer, who used very accurate definitions of the positions of testes, studied more than 3,500 male infants for cryptorchidism in London and followed them up to 1 year of age (Scorer, 1964). The incidences of cryptorchidism at 3 months of age in boys with birthweights < 2,500 g and >2,500 g were 1.74% and 0.91%, respectively. The very same examination technique and definitions of cryptorchidism were used in a similar study comprising 7,441 male infants from Oxford in the late 1980s (Ansell et al., 1992). The prevalence rates of cryptorchidism at the age of 3 months in boys with birthweights < 2,500 g and > 2,500 g were 5.2% and 1.61%, respectively, indicating a significant increase from the 1950s. At the same time (the 1980s), a similar study was also performed in New York (Berkowitz et al., 1993), comprising 6,935 male infants, and the corresponding prevalence rates were 1.94% and 0.91%. However, the data from New York and English cities are not directly comparable, because the American study population was racially and ethnically more heterogeneous than the English populations, and it is known that cryptorchism is more common among Caucasians than among African Americans (Heinonen et al., 1977), although the difference was not apparent in the study from New York. Thus, there has been a significant increase in the incidence of cryptorchidism in England, but one cannot tell anything about temporal trends in New York, since there are no adequate historic data.

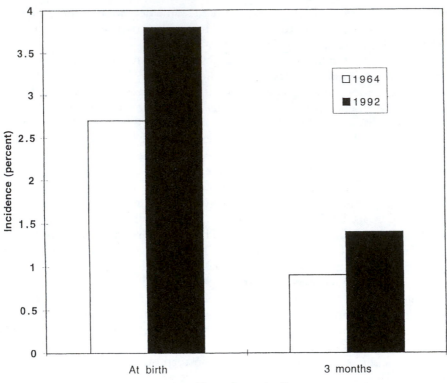

Figure 10-2.
Incidence of cryptorchidism in England. The data are from the study of Scorer (1964), reporting incidence of cryptorchidism in the late 1950s in London, and from the John Radcliffe Hospital Cryptorchidism Study (Ansell et al., 1992), reporting data from the late 1980s in Oxford.

If endocrine disruption (e.g., by xenoestrogens) contributed to the increasing incidence of cryptorchidism, one would assume that men exposed to extra estrogens *in utero* would be prone to have this condition. Looking back to DES-exposed males shows that this is exactly the case (Gill et al., 1977, 1979; Henderson et al., 1976). The data of Gill and coworkers (Gill et al., 1977, 1979), indicating an increased incidence of cryptorchidism among DES-exposed boys, are based on the follow-up studies of the offspring of mothers who took part in the double-blind study of DES effects on pregnancy in 1953 (Dieckmann et al., 1953), and therefore the studies can be considered prospective.

In another large case control study of cryptorchidism patients, no association between the disorder and exposure to estrogens during the pregnancy could be found (Beard et al., 1984). However, this type of study has all the difficulties of retrospective epidemiology, such as missclassification of cases due to difficulties in exposure reporting. The aforementioned prospective study showing unequivocally adverse effects adds more to the weight of evidence than does a retrospective study showing no effect. In the recent follow-up study of DES-exposed males, it was found that the men exposed to DES before the 11th gestation week had twice as high a frequency of genital anomalies than did those exposed only later (Wilcox et al., 1995), indicating the importance of the timing of the exposure (time of organogenesis).

Hypospadias

Hypospadias is a urogenital malformation where the urethral opening is located underneath the penis instead of at the tip. Hypospadias is typical for patients whose androgen metabolism (production or conversion from testosterone to dihydotestosterone) or action (receptor defects) is abnormal. Thus, all antiandrogenic compounds can disturb normal urethral development. Estrogenic compounds also act similarly. Hypospadias is registered in national birth defect registries, but ascertainment of the milder forms of these malformations is very variable. Nevertheless, any adverse trends in the incidence of hypospadias could give clues of the changing occurrence of underlying etiologic factors.

Incidence of hypospadias has increased according to several reports (W.H.O., 1991; Bjerkedal and Bakketeig, 1975; Källén et al., 1986; Källén and Winberg, 1982; Czeizel et al., 1986; Czeizel, 1985; Matlai and Beral, 1985). Birth prevalence of hypospadias varies between 0.37 and 41 per 10,000 infants in the world (Sweet et al., 1974; Hohlbein, 1959). The figures are difficult to compare because of varying levels of ascertainment, different inclusion of minor forms of hypospadias, and differences in ethnic origin of the population. Similar to cryptorchidism, very few longitudinal studies confined to the same population and geographic area exist. England and Wales (Matlai and Beral, 1985), Hungary (Czeizel et al., 1986; Czeizel, 1985), Sweden (W.H.O., 1991; Källén et al., 1986; Källén and Winberg, 1982), Norway (W.H.O., 1991; Bjerkedal and Bakketeig, 1975) and Denmark (W.H.O., 1991; Källén et al., 1986) have shown an increasing incidence of hypospadias, whereas no trend was found in Finland, Spain, New Zealand, Australia, and Czechoslovakia (W.H.O., 1991).

In England and Wales, there was an increase in the prevalence of hypospadias from 7.3 per 10,000 births in 1964 to approximately 16 per 10,000 births in the early 1980s, when the number of cases stabilized. In 1990 the prevalence of hypospadias decreased to 11.7 per 10,000. In Hungary, there was a rapid increase in the prevalence of hypospadias from 5.5 to 23.9 per 10,000 births during 1964–1978; thereafter the prevalence has remained at approximately the same level.

In Denmark, the incidence of hypospadias increased from 7.5 to 12 per 10,000 births during 1970–1981 (Källén et al., 1986), and a further similar increase was noticed during the period 1982–1988 (W.H.O., 1991). However, before the latter increase, a new registration system was introduced. The Swedish data also indicated a marked increase in the early 1970s: Prevalence of hypospadias at birth was 40% higher between 1974–1982 compared to the period 1965–1968 (Källén et al., 1986; Avellán, 1975). Furthermore, the data obtained in the earlier period could be more complete, because data included both hospital records and registry data. In Norway the prevalence of hypospadias at birth increased from 7 to 8 per 10,000 births between 1967–1971 to 13 per 10,000 in 1973 (Bjerkedal and Bakketeig, 1975), and further to 20.7 per 10,000 births in 1988 (W.H.O., 1991).

There are apparent ethnic differences in the incidence of hypospadias that may be due to varying genetic predisposition. Caucasians in the United States have a higher incidence than African Americans: The ratio is 1.3 to 3.9:1 (Finley et al., 1994; Shapiro et al., 1958; Chung et al., 1968; Wallace et al., 1953). In British Columbia, Canada, Native Americans were reported to have a lower prevalence of hypospadias than the general Caucasian population, with a ratio of 1:6.7 (Lowry et al., 1986; Leung et al., 1985). Thus, Caucasians may be more exposed to harmful factors (for example in their diet), or they may be more susceptible to endocrine disruption than are some other ethnic populations.

Interestingly, the same populations that have a low incidence of testicular cancer (e.g., Finland; Adami et al., 1994) also have a very low prevalence of hypospadias. Furthermore, there is also variation in the prevalence of hypospadias within different countries (Källén et al., 1986; Hautau, 1960).

Earlier, we referred to the studies of Gill and coworkers to illustrate how endocrine disruptors might influence urogenital development in males. Not surprisingly, the incidence of hypospadias was dramatically increased in DES-exposed male subjects compared to controls (4.4% vs. 0%) (Gill et al., 1977, 1979; Henderson et al., 1976). This lends credence to the hypothesis that exposure to endocrine-disrupting compounds can increase the incidence of hypospadius.

Prostatic Diseases

Two prostatic diseases, benign prostatic hyperplasia and prostatic cancer, are major health problems among aging males. Prostate cancer is the most common malignancy of men in many countries, and it is the second leading cause of cancer death among men in the United States (Hsing and Devesa, 1994). Both incidence and mortality show an increasing trend worldwide, e.g., in England, Italy, Japan, Sweden, and the United States (Whittemore, 1994). The highest incidence of prostate cancer is found in the United States among African Americans whose prevalence rate is approximately double that in Caucasians (Muir et al., 1991). The incidence is also high in northern Europe, intermediate in southern Europe, and low in Asia. Epidemiology of benign prostatic hyperplasia is poorly characterized, and its relationship to prostate cancer is uncertain. However, it has been suggested that these diseases share etiologic factors (Santti et al., 1994).

Environmental and dietary factors appear to be important determinants of the geographic, secular, and racial differences in the incidence of prostate cancer (for review, see Santti et al., 1994). Because both androgens and estrogens influence the growth and regulation of the prostate, endocrine disruption has been hypothesized to be responsible for adverse effects resulting in hyperplasia or cancer (Santti et al., 1994). Neonatal estrogen treatment of rodents induces dysplastic changes in prostatic epithelium that may later result in malignancy (Pylkkänen et al., 1993). Similarly, squamous metaplasia was found in prostatic ducts of infants exposed to exogenous estrogens *in utero* (Driscoll and Taylor, 1980). The significance of this finding is uncertain, because the relationship between metaplasia and malignancy is not clear. The male population that was exposed to DES *in utero* has not yet reached the age when prostatic cancer commonly occurs, and therefore epidemiologic studies of these men do not yet exist to prove or disprove the causal relationship between early estrogenization and prostatic tumorigenesis.

Male Breast Cancer

Male breast cancer is so rare that it can almost be considered a curiosity. However, because xenoestrogens are suggested to contribute to the pathogenesis of breast cancer in women (Davis et al., 1993), the trends in the incidence of this disease in males are also of interest. There are only a few studies on geographical and temporal trends in the incidence. In Denmark, the incidence of male breast cancer slightly increased during 1943–1982, whereas no changes were apparent in Sweden (1958–1992), Norway, or Finland (1953–1982) (Ewertz et al., 1989). Notably, the incidence in Denmark was about twice that of Finland.

Conclusions

It is evident that there are adverse trends in male reproductive health. Incidence of testicular cancer has increased dramatically in several countries during this century, and this increase has continued until today. Similarly, incidence of prostate cancer has increased. Testicular cancer occurs at a young age, and therefore ascertainment is high, whereas prostate cancer is typical for older men, and therefore the increase of its incidence might partly be due to the aging of the population. Several studies indicate a decline in semen quality in different European countries. However, stable sperm concentrations have been reported also, and great variability exists between different geographic areas in the steady-state levels. Both the geographic differences and temporal trends suggest that environmental factors could contribute to human semen quality. The incidence of hypospadias and cyrptorchidism appears also to have increased, although these data have been more controversial than those on testicular cancer because of uncertainties in ascertainment and/or reporting. Nevertheless, the best available evidence indicates a substantial increase in the incidence of these disorders. All of these trends are compatible with adverse action of environmental endocrine disruptors influencing androgen and estrogen actions during human growth and differentiation. Causal relationship between adverse outcome and specific exposure in humans has been demonstrated only for diethylstilbestrol, and it may be difficult to pinpoint individual exposures that could affect human reproduction. We are constantly exposed to a complex mixture of factors influencing our endocrine system, and we may have to rely on animal and *in vitro* experiments when identifying potential endocrine disruptors. Fortunately, although numerous differences exist between species, animal experiments have been very informative and predictive concerning adverse human effects. Thus, all data on endocrine disruption are also relevant for male reproductive health. Risk assessment for human reproductive health will therefore be based on weight of evidence, considering not only data about humans but also all relevant information from wildlife and laboratory animals, *in vitro* experimentation, and such new mathematical models as quantitative structure activity relationships.

References

Adami, H., Bergström, R., Möhner, M., Zatonski, W., Storm, H., Ekbom, A., Tretli, S., Teppo, L., Ziegler, H., Rahu, M., Gurevicius, R., and Stengrevics, A. (1994). Testicular cancer in nine Northern European countries. *Int. J. Cancer* 59: 33–38.

Adamopoulos, D., Nicopoulou, S., Pappa, A., Andreou, E., Michopoulos, I., Deligianni, V., and Karamertzanis, M. (1996). Sperm number and seminal volume trends in a selected male population from the greater Athens area during the period 1977–1993. In *Miniposter Book of the 9th European Workshop on Molecular and Cellular Endocrinology of the Testis*, (R. Solborg and V. Hansson, eds.). Miniposter G1, Geilo, Norway.

Ansell, P.E., Bennett, V., Bull, D., Jackson, M.B., Pike, L.A., Pike, M.C., Chilvers, C.E.D., Dudley, N.E., Gough, M.H., Griffiths, D.M., Redman, C., Wilkinson, A.R., Macfarlane, A., and Coupland, C.A.C. (1992). Cryptorchidism: A prospective study of 7500 consecutive male births, 1984–88. *Arch. Dis. Child.* 67: 892–899.

Auger, J., Kunstmann, J.M., Czyglik, F., and Jouannet, P. (1995). Decline in semen quality among fertile men in Paris during the past 20 years. *N. Engl. J. Med.* 332: 281–285.

Avellán, L. (1975). The incidence of hypospadias in Sweden. *Scand. J. Plast. Reconstr. Surg.* 9: 129–139.

Baumrucker, G.O. (1946). Incidence of testicular pathology. *Bull. U.S. Army Med. Dept.* 4: 312–314.

Beard, M., Melton, L.J., O'Fallon, W.M., Noller, K.L., and Benson, R.C. (1984). Cryptorchism and maternal estrogen exposure. *Am. J. Epidemiol.* 120: 707–716.

Benson, R.C.J., Beard, C.M., Kelalis, P.P., and Kurland, L.T. (1991). Malignant potential of the cryptorchid testis. *Mayo. Clin. Proc.* 66: 372–378.

Bergström, R., Adami, H.-O., Möhner, M., Zatonski, W., Storm, H., Ekbom, A., Tretli, S., Teppo, L., Akre, O., and Hakulinen, T. (1996). Increase in testicular cancer incidence in six European countries: A birth cohort phenomenon. *J. Natl. Cancer Inst.* 88: 727–733.

Berkowitz, G.S., Lapinski, R.H., Dolgin, S.E., Gazella, J.G., Bodian, C.A., and Holzman, I.R. (1993). Prevalence and natural history of cryptorchidism. *Pediatrics* 92: 44–49.

Bjerkedal, T., and Bakketeig, L.S. (1975). Surveillance of congenital malformations and other conditions of the newborn. *Int. J. Epidemiol.* 4: 31–36.

Blom, K. (1984). Undescended testis and the time of spontanous descent in 2516 schoolboys. *Ugeskr. Laeger* 146: 616–617.

Bostofte, E., Serup, J., and Rebbe, H. (1983). Has the fertility of Danish men declined through the years in terms of semen quality? A comparison of semen qualities between 1952 and 1972. *Int. J. Fertil.* 28: 91–95.

Boyle, P., Kaye, S.B., and Robertson, A.G. (1987). Changes in testicular cancer in Scotland. *Eur. J. Cancer Clin. Oncol.* 23: 827–830.

Brown, L.M., Pottern, L.M., and Hoover, R.N. (1986). Prenatal and perinatal risk factors for testicular cancer. *Cancer Res.* 46: 4812–4816.

Buemann, B., Henriksen, H., Villumsen, Å.L., Westh, Å., and Zachau-Christiansen, B. (1961). Incidence of undescended testis in the newborn. *Acta Chir. Scand.* (Suppl. 283): 289–293.

Bujan, L., Mansar, A., Pontonnier, F., and Mieusset, R. (1996). Time series analysis of sperm concentration in fertile men in Toulouse, France between 1977 and 1992. *Br. Med. J.* 312: 471–473.

Campbell, D.M., Webb, J.A., and Hargreave, T.B. (1987). Cryptorchidism in Scotland. *Br. Med. J.* 294: 1237–1238.

Carlsen, E., Giwercman, A., Keiding, N., and Skakkebæk, N.E. (1992). Evidence for decreasing quality of semen during the past 50 years. *Br. Med. J.* 304: 609–613.

Chilvers, C., Pike, M.C., Forman, D., Fogelman, K., and Wadsworth, M.E.J. (1984). Apparent doubling of frequency of undescended testis in England and Wales in 1962–81. *Lancet* ii (no. 8398): 330–332.

Choi, H., Kim, K.M., Koh, S.K., Kim, K.S., Woo, Y.N., Yoon, J.B., Choi, S.K., and Kim, S.W. (1989). A survey of externally recognizable genitourinary anomalies in Korean newborns. *J. Kor. Med. Sci.* 4: 13–21.

Chung, C.S., Myrianthopoulos, N.C., and Yoshizaki, H. (1968). Racial and prenatal factors in major congenital malformations. *Am. J. Hum. Genet.* 20: 44–60.

Clemmesen, J. (1991). Testis cancer incidence—suggestion of a world pattern. *Int. J. Androl. Suppl.* 4: 111–120.

Conley, G.R., Sant, G.R., Ucci, A.A., and Mitcheson, H.D. (1983). Seminoma and epididymal cysts in a young man with known diethylstilbestrol exposure *in utero*. *JAMA* 249: 1325–1326.

Correy, J.F., Newman, N.M., Collins, J.A., Burrows, E.A., Burrows, R.F., and Curran, J.T. (1991). Use of prescription drugs in the first trimester and congenital malformations. *Aust. N.Z. J. Obstet. Gynaec.* 31: 340–344.

Cour-Palais, I.J. (1966). Spontaneous descent of the testicle. *Lancet* i: 1403–1405.

Czeizel, A. (1985). Increasing trends in congenital malformations of male external genitalia. *Lancet* i: 462–463.

———, Erödi, E., and Tóth, J. (1981). An epidemiological study on undescended testis. *J. Urol.* 126: 524–527.

———, Tóth, J., and Czvenits, E. (1986). Increased birth prevalence of isolated hypospadias in Hungary. *Acta Paediatr. Hung.* 27: 329–337.

Daughaday, W.H. (1981). Growth hormone and the somatomedins. In *Endocrine Control of Growth.* (W.H. Daughaday, ed.), pp. 1-24. Elsevier, New York.

Davis, D.L., Bradlov, H.L., Wolff, M., Woodruff, T., Hoel, D.G., and Anton-Culver, H. (1993). Medical hypothesis: Xenoestrogens as preventable causes of breast cancer. *Environ. Health Perspec.* 101: 372–377.

Depue, R.H., Pike, M.C., and Henderson, B.E. (1983). Estrogen exposure during gestation and risk of testicular cancer. *J. Natl. Cancer Inst.* 71: 1151–1155.

Dieckmann, W.J., Davis, M.E., Rynkiewicz, L.M., and Pottinger, R.E. (1953). Does administration of diethylstilbestrol during pregnancy have therapeutic value? *Am. J. Obstet. Gynecol.* 66: 1062.

Driscoll, S.G., and Taylor, S.H. (1980). Effects of prenatal maternal estrogen on the male urogenital system. *Obstet. Gynecol.* 56: 537–542.

Ewertz, M., Holmberg, L., Karjalainen, S., Tretli, S., and Adami, H.-O. (1989).

Incidence of male breast cancer in Scandinavia, 1943–1982. *Int. J. Cancer* 43: 27–31.

Finley, W.H., Gustavson, K.-H., Hall, T.M., Hurst, D.C., Bargainer, C.M., and Wiedmeyer, J.A. (1994). Birth defects surveillance: Jefferson County, Alabama, and Uppsala County, Sweden. *South. Med. J.* 87: 440–445.

Fisch, H., Goluboff, E.T., Olson, J.H., Faldshuh, J. Broder, S.J., and Barad, D.H. (1996). Semen analyses in 1283 men from the United States over a 25-year period: No decline in quality. *Fertil. Steril.* 64: 1009–10014.

Forman, D., and Møller, H. (1994). Testicular cancer. *Cancer Surv.* 19/20: 323–341.

Gershman, S.T., and Stolley, P.D. (1988). A case-control study of testicular cancer using Connecticut tumour registry data. *Int. J. Epidemiol.* 17: 738–742.

Gill, W.B., Schumacher, G.F.B., and Bibbo, M. (1977). Pathological semen and anatomical abnormalities of the genital tract in human male subjects exposed to diethylstilbestrol *in utero*. *J. Urol.* 117: 477–480.

———, ———, ———, Straus, F.H.I., and Schoenberg, H.W. (1979). Association of diethylstilbestrol exposure *in utero* with cryptorchidism, testicular hypoplasia and semen abnormalities. *J. Urol.* 122: 36–39.

Ginsburg, J., and Hardiman, P. (1992). Decreasing quality of semen. *Br. Med. J.* 304: 1229.

Giwercman, A., Müller, J., and Skakkebæk, N.E. (1988). Carcinoma in situ of the undescended testis. *Semin. Urol.* 6: 110–119.

Hakulinen, T., Andersen, A., Malker, B., Pukkala, E., Schou, G., and Tulinius, H. (1986). Trends in cancer incidence in the Nordic countries. *APMIS* 94 (Suppl. 288): 7–166.

Halevi, H.S. (1967). Congenital malformations in Israel. *Brit. J. Prev. Soc. Med.* 21: 66–77.

Harris, L.E., and Steinberg, A.G. (1954). Abnormalities observed during the first six days of life in 8,716 live-born infants. *Pediatrics* 14: 314–326.

Hautau, E.R. (1960). Congenital malformations in infants born to Michigan residents in 1958. *J. Mich. State Med. Soc.* 1833–1836.

Heinonen, O.P., Slone, D., and Shapiro, S. (1977). Birth Defects and Drugs in Pregnancy. Littleton, Publishing Sciences.

Henderson, B.E., Benton, B., Cosgrove, M., Baptista, J., Aldrich, J., Townsend, D., Hart, W., and Mack, T.M. (1976). Urogenital tract abnormalities in sons of women treated with diethylstilbestrol. *Pediatrics* 58: 505–507.

———, ———, Jing, J., Yu, M.C., and Pike, M.C. (1979). Risk factors for cancer of the testis in young men. *Int. J. Cancer* 23: 598–602.

Hirasing, R.A., Grimberg, R., and Hirasing, H.D. (1982). De frequentie van niet normaal ingedaalde testes bij jonge kinderen. *Ned. T. Geneesk.* 126: 2294–2296.

Hohlbein, R. (1959). Missbildungsfrequenz in Dresden. *Zentralblatt für Gynäkologie* 18: 719–731.

Hsieh, J.-T., and Huang, T.-S. (1985). A study on cryptorchidism. *J. Formos. Med. Assoc.* 84: 953–959.

Hsing, A.W., and Devesa, S.S. (1994). Prostate cancer mortality in the United States by cohort year of birth, 1865–1940. *Cancer Epidemiol. Biomarkers Prev.* 3: 527–530.

IARC. (1976). *Cancer Incidence in Five Continents, Vol.3.* (J. Waterhouse, C. Muir, P. Correa, and J. Powell, eds.). IARC Scientific Pub. No. 15. International Agency for Research on Cancer, Lyon, France.

———— (1982). *Cancer Incidence in Five Continents, Vol. 4.* (J. Waterhouse, C. Muir, K. Shanmugaratnam, and J. Powell, eds.). IARC Scientific Pub. No. 42. International Agency for Research on Cancer, Lyon, France.

———— (1987). *Cancer Incidence in Five Continents, Vol. 5.* (C. Muir, J. Waterhouse, T. Mack, J. Powell, and S. Whelan, eds.). IARC Scientific Pub. No. 88. International Agency for Research on Cancer, Lyon, France.

———— (1992). *Cancer Incidence in Five Continents, Vol. 6.* (D.M. Parkin, C.S. Muir, S.L. Whelan, Y.T. Gao, J. Ferlay, and J. Powell, eds.). IARC Scientific Pub. No. 120. International Agency for Research on Cancer, Lyon, France.

Irvine, S., Cawood, E., Richardson, D., MacDonald, E., and Aitken, J. (1996). Evidence of deteriorating semen quality in the United Kingdom: Birth cohort study in 577 men in Scotland over 11 years. *Br. Med. J.* 312: 467–471.

Jensen, T.K., Giwercman, A., Carlsen, E., Scheike, T., and Skakkebæk, N.E. (1996). Semen quality among members of organic food associations in Zealand, Denmark. *Lancet* 347: 1844.

Joffe, M. (1996). Decreased fertility in Britain compared with Finland. *Lancet* 347: 1519–1522.

Johnson, W.W. (1939). Cryptorchidism. *JAMA* 113: 25–27.

Källén, B., Bertollini, R., Castilla, E., Czeizel, A., Knudsen, L.B., Martinez-Frias, M.L., Mastroiacovo, P., and Mutchinick, O. (1986). A joint international study on the epidemiology of hypospadias. *Acta Paediatr. Scan.* (Suppl.) 324: 5–52.

———— and Winberg, J. (1982). An epidemiological study of hypospadias in Sweden. *Acta Paediatr. Scand.*(Suppl.) 293: 1–21.

Leto, S., and Frensilli, F.J. (1981). Changing parameters of donor semen. *Fertil. Steril.* 36: 766–770.

Leung, T.J., Baird, P.A., and McGillivray, B. (1985). Hypospadias in British Columbia. *Am. J. Med. Genet.* 21: 39–48.

Lowry, R.B., Thunem, N.Y., and Silver, M. (1986). Congenital anomalies in American Indians of British Columbia. *Gen. Epidemiol.* 3: 455–467.

Matlai, P., and Beral, V. (1985). Trends in congenital malformations of external genitalia. *Lancet* i: 108.

Mau, G., and Schnakenburg, K.v. (1977). Maldescent of the testes—an epidemiological study. *Eur. J. Pediatr.* 126: 77–84.

McDonald, A.D. (1958). Maternal health and congenital defect. *N. Engl. J. Med.* 258: 767–773.

McIntosh, R., Merritt, K.K., Richards, M.R., Samuels, M.H., and Bellows, M.T. (1954). The incidence of congenital malformations: A study of 5,964 pregnancies. *Pediatrics* 14: 505–521.

Mital, V.K., and Garg, B.K. (1972). Undescended testicle. *Indian J. Pediatr.* 39: 171–174.

Møller, H., Prener, A., and Skakkebæk, N.E. (1996). Testicular cancer, cryptorchidism, inguinal hernia, testicular atrophy, and genital malformations: Case-control studies in Denmark. *Cancer Causes and Control* 7: 264–274.

Morley, R., and Lucas, A. (1987). Undescended testes in low birthweight infants. *Br. Med. J.* 295: 753.

Moss, A.R., Osmond, D., Bacchetti, P., Torti, F.M., and Gurgin, V. (1986). Hormonal risk factors in testicular cancer: A case-control study. *Am. J. Epidemiol.* 124: 39–52.

Muir, C.S., Nectoux, J., and Staszewski, J. (1991). The epidemiology of prostatic cancer: Geographic distribution and time-trends. *Acta Oncol.* 30: 133–140.

Müller, J., Skakkebæk, N.E., Ritzén, M., Plöen, L., and Petersen, K.E. (1985). Carcinoma in situ of the testis in children with 45,X/46,XY gonadal dysgenesis. *J. Ped.* 106: 431–436.

Nelson, C.M.K., and Bunge, R.G. (1974). Semen analysis: Evidence for changing parameters of male fertility potential. *Fertil. Steril.* 24: 503–507.

Nethersell, A.B.W., Drake, L.K., and Sikora, K. (1984). The increasing incidence of testicular cancer in East Anglia. *Br. J. Cancer* 50: 377–380.

Olsen, G.W., Bodner, K.M., Ramlow, J.M., Ross, C.E., and Lipshultz, L.I. (1995). Have sperm counts been reduced 50% in 50 years? A statistical model revisited. *Fertil. Steril.* 63: 887–893.

Onuora, V.C., and Evbuomwan, I. (1989). Abnormal findings associated with undescended testis in Nigerian children. *Indian J. Pediatr.* 56: 509–511.

Panayotou, P.C. (1965). The incidence of undescended testis in boys attending elementary schools in Greece. *Br. J. Clin. Pract.* 19: 501–507.

Paulsen, C.A., Berman, N.G., and Wang, C. (1996). Data from men in greater Seattle area reveals no downward trend in semen quality: Further evidence that deterioration of semen quality is not geographically uniform. *Fertil. Steril.* 64: 1015–1020.

Pearce, N., Sheppard, R.A., Howard, J.K., Fraser, J., and Lilley, B.M. (1987). Time trends and occupational differences in cancer of the testis in New Zealand. *Cancer* 59: 1677–1682.

Pike, M.C., Chilvers, C.E.D., and Bobrow, L.G. (1987). Classification of testicular cancer in incidence and mortality statistics. *Br. J. Cancer* 56: 83–85.

Pitt, D.B. (1962). A study of congenital malformations. *Aust. N.Z. J. Obstet. Gynaec.* 2: 23–30.

Potashnik, G., Ben-Aderet, N., Israeli, R. Yanai-Inbar, I., and Sober, I. (1978). Suppressive effect of 1,2-Dibromo-3-chloropropane on human spermatogenesis. *Fertil. Steril.* 30: 444–447.

Pylkkänen, L., Mäkelä, S., Valve, E., Härkönen, P., Toikkanen, S., and Santti, R. (1993). Prostatic dysplasia associated with increased expression of c-myc in neonatally estrogenized mice. *J. Urol.* 149: 1593–1601.

Santti, R., Newbold, R.R., Mäkelä, S., Pylkkänen, L., and McLachlan, J.A. (1994). Developmental estrogenization and prostatic neoplasia. *Prostate* 24: 67–78.

Schottenfeld, D., Warshauer, M.E., Sherlock, S., Zauber, A.G., Leder, M., and Payne, R. (1980). The epidemiology of testicular cancer in young adults. *Am. J. Epidemiol.* 112: 232–246.

Scorer, C.G. (1964). The descent of the testis. *Arch. Dis. Child.* 39: 605–609.

Shapiro, R.N., Eddy, W., Fitzgibbon, J., and O'Brien, G. (1958). The incidence of congenital anomalies discovered in the neonatal period. *Am. J. Surg.* 96: 396–400.

Skakkebæk, N.E., Berthelsen, J.G., Giwercman, A., and Müller, J. (1987). Carcinoma-in-situ of the testis: Possible origin from gonocytes and precursor of all types of germ cell tumours except spermatocytoma. *Int. J. Androl.* 10: 19–28.

———— and Keiding, N. (1994). Changes in semen and the testis. *Br. Med. J.* 309: 1316–1317.

Spitz, M.R., Sider, J.G., Pollack, E.S., Lynch, H.K., and Newell, G.R. (1986). Incidence and descriptive features of testicular cancer among United States whites, blacks, and hispanics, 1973–1982. *Cancer* 58: 1785–1790.

Stillman, R.J. (1982). In utero exposure to diethylstilbestrol: Adverse effects on the reproductive tract and reproductive performance in male and female offspring. *Am. J. Obstet. Gynecol.* 142: 905–921.

Stone, J.M., Cruickshank, D.G., Sandeman, T.F., and Matthews, J.P. (1991). Trebling of the incidence of testicular cancer in Victoria, Australia (1950-1985). *Cancer* 68: 211–219.

Suominen, J., and Vierula, M. (1993). Semen quality of Finnish men. *Br. Med. J.* 306: 1579.

Swan, S.H., Elkin, E.P., and Fenster, L. (1997). Have sperm densities declined?: A reanalysis of global trend data. *Environ Health Perspec.* 105: 1228–1232.

———— and ———— (1999). Declining sperm quality: Can the past inform the present? *BioEssays* 21: 614–621.

————. Geographical variability in semen quality. *OrGyn* 2000. In press.

Sweet, R.A., Schrott, H.G., Kurland, R., and Culp, O.S. (1974). Study of the incidence of hypospadias in Rochester, Minnesota, 1940–1970, and a case-control comparison of possible etiologic factors. *Mayo. Clin. Proc.* 49: 52–58.

Swerdlow, A.J., and Melzer, D. (1988). The value of England and Wales congenital malformation notification scheme data for epidemiology: Male genital tract malformations. *J. Epidemiol. Community Health* 42: 8–13.

Toppari, J., Larsen, J.C., Christiansen, P., Giwercman, A., Grandjean, P., Guillette, L.J., Jr., Jégou, B., Jensen, T.K., Jouannet, P., Keiding, N., Leffers, H., McLachlan, J.A., Meyer, O., Müller, J., Rajpert-de Meyts, E., Scheike, T., Sharpe, R., Sumpter J., and Skakkebæk, N.E. (1995). Male reproductive health and environmental chemicals with estrogenic effects. Ministry of Environment and Energy, Danish Environmental Protection Agency, Copenhagen, Denmark. *Miljøprojekt* nr. 290: 1–166.

————, ————, ————, ————, ————, ————, ————, ————,
————, ————, ————, ————, ————, ————, ————, ————,
and ———— (1996). Male reproductive health and environmental xenoestrogens. *Environ Health Perspec.* 104 (Suppl. 4): 741–801.

Thorup, J., and Cortes, D. (1990). The incidence of maldescended testes in Denmark. *Pediatr. Surg. Int.* 4: 2–5.

UICC. (1966). *Cancer Incidence in Five Continents, Vol. 1. A Technical Report.* (R. Doll, P. Payne, and J. Waterhouse, eds.). International Union Against Cancer, Geneva. (Distributed by Springer Verlag, Berlin.).

———— (1970). *Cancer Incidence in Five Continents, Vol. 2.* (R. Doll, C.S. Muir, and J. Waterhouse, eds.). International Union Against Cancer, Geneva.

Van Waeleghem, K., De Clercq, N., Vermeulen, L., Schoonjans, F., and Comhaire, F. (1996). Deterioration of sperm quality in young Belgian men. *Hum. Reprod.* 11: 325–329.

Vierula, M., Niemi, M., Keiski, A., Saaranen, M., Saarikoski, S., and Suominen, J. (1996). High and unchanged sperm counts of Finnish men. *Int. J. Androl.* 19: 11–17.

Wallace, H.M., Baumgartner, L., and Rich, H. (1953). Congenital malformations and birth injuries in New York City. *Pediatrics* 12: 525–535.

Ward, B., and Hunter, W.M. (1960). The absent testicle. *Br. Med. J.* Apr. 9: 1110–1111.

Whittemore, A.S. (1994). Prostate cancer. *Cancer Surv.* 19–20: 1117–1126.

W.H.O. (1991). *Congenital Malformations Worldwide: A Report from The International Clearinghouse for Birth Defects Monitoring Systems,* p. 113–118. Elsevier, Oxford.

Whorton, D., Krauss, R.M., Marshall, S., and Milby, T.H. (1977). Infertility in male pesticide workers. *Lancet* ii: 1259–1261.

Whorton, M.D., Milby, T.H., Krauss, R.M., and Stubbs, H.A. (1979). Testicular function of DBCP-exposed pesticide workers. *J. Occup. Med.* 21: 161–166.

Wilcox, A.J., Baird, D.D., Weinberg, C.R., Hornsby, P.P., and Herbst, A.L. (1995). Fertility in men exposed prenatally to diethylstilbestrol. *N. Engl. J. Med.* 332: 1411–1416.

Wilkinson, T.J., Colls, B.M., and Schluter, P.J. (1992). Increased incidence of germ cell testicular cancer in New Zealand Maoris. *Br. J. Cancer* 64: 769–771.

Williams, P. (1936). The imperfectly migrated testis. *Lancet* i: 426–427.

Yücesan, S., Dindar, H., Olcay, I., Okur, H., Kilicaslan, S., Ergören, Y., Tüysüz, C., Koca, M., Civilo, B., and En, I.S. (1993). Prevalence of congenital abnormalities in Turkish school children. *Eur. J. Epidemiol.* 9: 373–380.

Chapter 11

XENOESTROGENS IN THE CONTEXT OF CARCINOGENESIS

Ana M. Soto and Carlos Sonnenschein

Tufts University School of Medicine
Department of Anatomy and Cellular Proliferation
136 Harrison Avenue, Boston, MA 02111

Introduction

Control of cell proliferation is a topic of central relevance to the understanding of carcinogenesis. We have recently reviewed the premises adopted by experimentalists to study the control of cell proliferation, as well as the application of these premises to the study of the role of sex steroids in the proliferation of their target cells (Sonnenschein and Soto, 1999). In this chapter, we will review the role of estrogens in the development of neoplasias within their target organs. Researchers have postulated that environmental estrogens may be the underlying cause for the increase of testicular cancer, undescended testis, and malformations of the male genital tract during the last half of this century (Sharpe et al., 1993; see Chap. 10). The role of other sex steroids, such as that of androgens in relation to carcinogenesis, will not be discussed. Although androgens are a main factor in the development of prostate cancer at the present time, there is no evidence that environmental contaminants act as androgen mimics. Some environmental contaminants, however, do possess antiandrogenic properties (Kelce et al., 1994, 1995). Although these antiandrogenic compounds may disrupt the development of the male genital tract, it is not yet known whether they play a role in carcinogenesis.

Xenoestrogens

Xenobiotics of widely diverse chemical structure have estrogenic properties (Hammond et al., 1979; Meyers et al., 1977). This diversity makes it difficult to predict the estrogenicity of chemicals solely on structural bases. Hence, their identification as estrogens has relied on bioassays, using diverse endpoints on which estrogens play a direct or indirect role on cell proliferation, uterine growth, or the induction of specific genes. Hertz (1985) argued convincingly that the *proliferative* effect of natural estrogens on the female genital tract is the hallmark of estrogen action. Thus, this property was adopted as the one that determines whether or not a chemical is an estrogen in animals *(in vivo)* or in cell culture models *(in vitro*; Hertz, 1985). This requires measuring the increase of proliferative activity in tissues of the female genital tract after estrogen administration. An equally reliable, simple, and rapidly performed method was developed using estrogen-target, serum-sensitive breast cells, a method that measures cell proliferation as a specific marker for estrogenicity (estrogens-SCREEN assay) (Soto et al., 1991, 1992). Other *in vitro* assays rely on the estrogen induction of endogenous genes, such as PS2 and progesterone receptor (PgR) (Soto et al., 1995b), or transfected reporter genes (Jobling et al., 1995; Arnold et al., 1996).

Cumulative Effect of Xenoestrogens

Humans and wildlife are exposed to a variety of chemicals *simultaneously* (Fox, 1992; Thomas et al., 1992). Residues of diverse estrogenic xenobiotics coexist in the fat and body fluids of exposed individuals (Thomas et al., 1992). Hence, it is likely that they may become bioavailable. At such time, they may act cumulatively; that is, when present at individual levels lower than those needed to express overt estrogenicity, their activity may add up to a level sufficient to trigger a full estrogenic response. We explored this concept and found that xenoestrogens indeed act cumulatively in the E-SCREEN assay (Soto et al., 1994, 1997); the effect of some mixtures is more than additive by 3- to 5-fold. Thus, measuring the total estrogenic burden due to environmental contaminants present in plasma/tissue samples may be more meaningful than measuring the levels of each of the known xenoestrogens. This has significant implications for the study of human conditions suspected to be caused by xenoestrogens, such as undescended testis (cryptorchidism), testicular and breast cancer, and the decrease of sperm counts and quality during the last 50 years (Giwercman et al., 1993; Sharpe et al., 1993; Wolff et al., 1993; see Chap. 10).

Novel Xenoestrogens

Numerous types of compounds have been identified as novel xenoestrogens, including antioxidants (alkylphenols, butylhydroxyanisole), plasticizers (bisphenol-A, dibutylphthalate, butylbenzylphthalate), PCB congeners, disinfectants (o-phenylphenol), and pesticides (toxaphene, dieldrin, endosulfan, and lindane) (see Table 11-1; Soto et al., 1995b, 1996). The newly identified xenoestrogens tested so far not only induced cell proliferation but also increased the expression of pS2 and PgR. This confirms their estrogen-mimicking properties and the specificity of the E-SCREEN assay as a tool to identify estrogens. These xenoestrogens compete with estradiol for binding to its receptor; their relative binding affinities to the estrogen receptor correlated well with their potency to induce cell proliferation, and with the induction of marker gene products such as pS2 and PgR (Soto et al., 1995b).

Most of these xenoestrogens are several orders of magnitude less potent than natural estrogens when assayed *in vitro*. However, BPA, the only one tested so far for endocrine disruption, was found to produce effects on animals that were exposed *in utero* at doses lower than expected from its potency measured in adult animals and *in vitro* (vom Saal et al., 1997; Nagel et al., 1997).

Neoplasia

One of the most vexing biological problems is finding an accurate definition of cancer. Virchow is often quoted as having said, "no man, even under torture, can

Table 11-1. Estrogenic xenobiotics commonly used as insecticides, fungicides, plasticizers, antioxidants, and other uses

Insecticides	Fungicides	Plastizicers	Antioxidants	Other[a]
o,p′DDT	o-phenyl phenol	bisphenol-A	propylphenol	p-phenylpnenol
p,p′DDT	bisphenol-A	bisphenol-A dimethacrylate	butylphenol	2,3,4 TCB
p,p′DDE		dibutylphthalate	penthylphenol	2,2′,4,5 TCB
o,p′DDD		benzylbutyl phthalate	octylphenol	2,3,4,5 TCB
p,p′DDD			butylhydroxy anisole	2,4,4′,6 TCB
Chlordecone			4 OH-biphenyl	2,2′,3,3′,6,6′HCB
Dieldrin				2 OH-2′,5′DCB
endosulfan				3 OH-2′,5′DCB
1-hydroxy chlordene				4 OH-2′,5′DCB
Lindane				4 OH-2: 2′,5 tCB
Methoxychl or				4 OH-2′,4′,6′tCB
Toxaphene				3 OH-2′,3′,4′,5′TCB
				4 OH-2′,3′,4′,5′TCB

[a]DCB: dichlorobiphenyl, TCB: trichlorobiphenyl, TCB: tetrachlorobiphenyl, HCB: hexachlorobiphenyl.

say exactly what a tumor is" (Uwing, 1916). Cancer is both a biological problem and a medical one. Biologically speaking, "neoplasia" is a better term than cancer or tumor because it denotes an accumulation of cells different from that of "hyperplasia." However, each one of the properties of neoplasias could also be expressed in normal cells (i.e., invasiveness, ability to proliferate, etc.). From an evolutionary perspective, neoplasias appear with the advent of multicellular organisms; its purview spans several hierarchical levels of organization from the cellular to the organismal.

Neoplasias are viewed in two contexts: (a) as an aberration of development and (b) as an aberration of the control of cell proliferation. Definitions are contradicted by either the behavior of a particular neoplasm or by that of a normal cell type. For example, that proposed by Willis—"a tumor is an abnormal mass of tissue, the growth of which exceeds and is uncoordinated with that of the normal tissues, and persists in the same excessive manner after cessation of the stimuli which evoked the change" (Willis, 1967)—does not take into consideration the phenomenon of regression. Regression often occurs in hormonal carcinogenesis after hormone withdrawal; a comparable phenomenon is that of true and complete spontaneous regression of early-stage melanomas (Clark, 1991) and some neuroblastomas (Turkel et al., 1974). Another drawback of Willis's definition is that it uses the word *growth* as a synonym of "cell proliferation"; this is imprecise and misleading.

Theories on the Mechanisms of Carcinogenesis

Genetic Origin

A major reason to invoke a genetic origin for certain cancers is the existence of familial cancers that are inherited through the germinal line. The premise of this hypothesis is that cancer arises through mutations in putative oncogenes (positive mediators) and in *tumor-suppressor genes* (negative mediators) which regulate cell proliferation. As mentioned above, the lethal (2) giant larvae mutant [l(2)gl] in *Drosophila* is the best-studied model and the only bona fide suppressor gene, since injection of the wild-type gene into early embryos results in the development of normal flies (Mechler et al., 1991).

Research on chemical-induced carcinogenesis is based on the premise that, in addition to their intoxicating effects, these chemicals induce somatic mutations. This led to the two-stage model of carcinogenesis whereby an "initiating" agent causes permanent DNA damage, and a "promoting" agent induces proliferation of the genetically altered cells (Beremblum et al., 1947). Although this view is somewhat consistent with some experiments on skin carcinogenesis, there is a substantial body of experimental evidence that contradicts this simplistic interpretation of the data (Iversen, 1993; Rubin, 1993). Currently, this hypothesis has many followers among researchers studying hormonal carcinogenesis. Despite the many reports favoring this notion, the unequivocal identification of the candidate mutated gene(s) responsible for tumor formation has been elusive. This much has been formally acknowledged by Varmus (1984) and by Bishop (1983), the original proponents of a crucial role for oncogenes in carcinogenesis.

Epigenetic Origin

Reasons to invoke an epigenetic origin are based on two experimental observations that rule out DNA mutation as the ultimate cause. First, blastocysts implanted in the abdominal cavity of mice cause the development of embryonal carcinoma. When a few of these tumor cells were injected into normal blastocysts they contributed to normal progeny in different tissues of the "mosaic" mice, including oocytes and spermatozoa that, in turn, generated normal individuals (Mintz et al., 1975). Second, normal tadpoles develop when nuclei of tetraploid frog renal carcinoma cells are transplanted into enucleated diploid eggs (McKinnel et al., 1969). In addition, certain tumors, arising during development from fetal tissue, appear in association with developmental anomalies (nephroblastoma with horseshoe kidney, hypospadias, and cryptorchidism), suggesting that neoplasias may develop when normal development is affected (Foulds, 1969). The frequent regression of neuroblastoma in infants suggests that cancer cells may revert to normalcy when placed in a permissive environment.

The major obstacle for the acceptance of a developmental mishap as causation is the absence of a working hypothesis that would point to underlying mechanisms. Meanwhile, the mutational hypothesis points to research at the genetic level, an endeavor in which powerful technologies may be used to study mutational phenomena. The problem with this approach is that, in the long run, if the identified gene is not directly responsible for the cancer, as in the case of the *lethal (2) giant larvae* mutant, one is again at a loss for finding out where and when the ultimate cause acts. Neither the mutational/genetic theory nor the epigenetic theory has produced a compelling mechanistic explanation of the events that lead to carcinogenesis. This difficulty probably arises from the fact that phenomena at the whole-organism level cannot be reconstructed from the properties of its components; emergent properties arise, or disappear, when hierarchical levels of complexity are crossed (Soto et al., 1993; Rubin, 1992).

A significant shortcoming in establishing a successful research program in carcinogenesis is the lack of resolution of the fundamental controversy about which is the default state of cells in metazoa. There are only two choices to be made: The first proposes that the default is quiescence; the other one posits that *proliferation* is, instead, the default state of these cells. The merits of both alternatives are discussed in Sonnenschein and Soto (1999). Briefly, the former requires the search for growth factors and other possible stimulators of cell proliferation, while the latter implicitly centers on the search for intracellular and extracellular inhibitory factors.

Hormonal Carcinogenesis

It is difficult to unambiguously establish the role played by hormones in the development of neoplasias. Within the genetic hypothesis, two roles have been postulated for hormones; namely, that they induce mutations and that they act as promoters. For proponents of the epigenetic perspective, extemporaneous exposure to hormones is considered teratogenic; neoplasia is the result of altered development. Finally, some think that the genetic and epigenetic options are not mutually exclusive; sex hormones would contribute to the development of neoplasia by acting at these three endpoints (mutation, control of cell proliferation, and differentiation).

Hormones as Mutagens

Supporters of the genetic causation hypothesis (two-step model of carcinogenesis) propose that certain estrogens are able to form DNA adducts. This would lead to mutations in yet-to-be-identified genes that in turn, through yet-to-be-defined pathways, would result in neoplasms. These inferential pathways involve entities such as "oncogenes" and suppressor genes.

The main research program in this endeavor has been to elucidate metabolic pathways leading to the formation of estrogen metabolites that could form DNA adducts in estrogen-target tissues. It was expected that these DNA adducts, when misrepaired, would originate mutations. A prediction of the mutagenic hypothesis is that not all estrogenic compounds are carcinogenic; only those that are mutagenic are expected to induce tumor formation. For example, 2-fluoroestradiol, a compound with an estrogenic potency similar to estradiol, does not induce tumorigenesis in the Syrian hamster model, while estradiol does. This is explained by the ability of estradiol to be metabolized to 2-hydroxy metabolites, while 2-fluoroestradiol is not metabolized (Liehr et al., 1986). From a similar perspective, DES is metabolized to an unstable semiquinone that can react with DNA (Bhat et al., 1994; Roy et al., 1992); others have postulated that DES may interact with spindle formation, causing aneuploidy (Barrett et al., 1981). Bradlow et al. suggested that estradiol is metabolized through two mutually exclusive pathways resulting in a 2-OH estrone and 16α-estrone; they propose that their genotoxic activity is entirely due to 16α-estrone (Bradlow et al., 1995; Davis et al., 1993). From this perspective, carcinogenesis may be induced by chemicals that affect the metabolism of natural estrogens enhancing the formation of 16α-estrone. Estrogens have also been implicated in the development of prostate cancer in rats (Noble, 1977) where mutational mechanisms were invoked (Han et al., 1995). These hypotheses are based on circumstantial evidence; a firm demonstration of causality is still missing.

Xenoestrogens and the Mutational Hypothesis

If xenoestrogen exposure is a risk factor for neoplasms of the female genital tract, breast, and prostate, and if it is postulated that mutation is the first step in carcinogenesis, how are xenoestrogens expected to act? Accumulation in a target cell would be proportional to the binding affinity for the xenoestrogens by estrogen receptors. Therefore, a linear dose-response curve may be assumed when associating exposure to effects. However, once xenoestrogens are accumulated in the target tissue, the rate of conversion to metabolites able to produce DNA adducts must be dependent on their affinity for the enzymes involved in this pathway. If xenoestrogens act by altering the metabolism of endogenous estrogens, their mutagenic activity would be disassociated from the estrogenic activity and linked, instead, with the ability of these xenoestrogens to induce or activate enzymes that regulate the metabolism of endogenous estrogens. In this instance, a linear dose-response curve with ER binding would not be expected. However, potency as a mutagen may not be directly related to estrogenic potency. In conclusion, while the estrogenic potency of xenoestrogens may be important, it does not seem to be the main determinant for their potential mutagenicity.

Hormones as Promoters

Animal models and observational data in humans indicate that tumors in estrogen and androgen target organs do not arise in individuals that had been gonadectomized before or during early adulthood. It is postulated that the role of sex steroids in this context is to sustain cell proliferation in genetically susceptible individuals (i.e., only certain strains develop tumors on sustained hormone exposure).

Other supporters of the mutational theory propose that hormones act as "promoters," inducing cell proliferation of target cells that were "mutated" or "initiated" by other carcinogens. In normal sex hormone–sensitive tissues, a tight regulation of cell number operates, whereby sex steroids both induce (step 1) and inhibit (step 2) cell proliferation. The "initiated" cells must overcome these restraining mechanisms of step 2 to proliferate selectively and become a hyper-plasia first and, later on, a tumor. Once the tumor develops, it may or may not require hormones to propagate further (hormone-sensitive or insensitive, respectively). In this view, a cell "mutated" in its ability to proliferate would acquire a selective advantage to multiply over those impervious to the carcino-gen mutagen. This is a rarely analyzed paradox. The paradox could be reconciled if those mutations are shown to be only in suppressor genes or colyogenes (see Sonnenschein and Soto, 1999). Even the proponents of oncogenes state that cancer cells have lost the ability to respond to organismal signals that inhibit cell proliferation (Alberts et al., 1994).

Hormones not only are necessary during the process of carcinogenesis but may also play a role in the propagation of hormone-sensitive tumors. Thus, breast and prostate cancers in humans regress after estrogens or androgens, respectively, are withdrawn or suppressed. In animal models, regression may "cure" the tumor. In humans, clinical regressions are temporary due to the selec-tion of "hormone-insensitive" phenotypes. This process is called tumor progres-sion; this recurrence has been attributed to genetic (additional mutations) and/or epigenetic (adaptive) mechanisms due to short-lived therapeutic regimes (Szelei et al., 1997; Soto et al., 1995a; Sonnenschein et al., 1991, 1994).

Xenoestrogens and the Promotional Hypothesis

As explained above, the issue regarding how estrogen levels may affect prolif-eration and carcinogenesis remains unsolved. To assess whether or not xenoestrogens significantly increase exposure to estrogens in normal adult women, one first has to ask how ovarian estrogen levels affect proliferation in their target organs. For example, ductal cell proliferation in the breast is maximal from the late follicular phase throughout the luteal phase, i.e., when endogenous estrogen levels are high (Meyer, 1977). Further increases in estrogen levels may not affect cell proliferation, since the endogenous levels of estrogens at this point are already triggering a full proliferative response followed by a proliferative shutoff. The ubiquitous presence of xenoestrogens in foods, their persistence in the environment, lack of binding to the plasma carrier protein SHBG, and cumulative action may increase the "basal" levels of estrogens during the early follicular phase of the menstrual cycle (Bradlow et al., 1995; Davis et al., 1993; Brotons et al., 1994; Olea et al., 1996; Soto et al., 1994, 1995b, 1997). This may result in an early onset of proliferative activity of the organs of the female genital tract and breast, prolonging the period of proliferative activity during each cycle, and leading to a higher inci-dence of breast tumors in later years. Hence, the assumption of a linear dose-response curve is not appropriate when evaluating the role of xenoe-strogens as promoters.

Hormones as Teratogens

Developmental biologists consider that the role of extemporaneous hormonal activity is essentially teratogenic. That is, hormones favor the persistence of cell populations past the point in which they should disappear during normal development (Takasugi, 1976; Ozawa et al., 1991; Bern, 1992a, b.).

The DES model

Genital tract organogenesis occurs during the first trimester in humans and at gestational days 9 to 16 in mice. The role of estrogenic hormones in the normal development of the mammalian reproductive tract is not completely understood, although it is clear that estrogen receptors must be present for estradiol to mediate biological activity (Lubahn et al., 1993). Exposure to exogenous estrogens during early development results in several anomalies of the genital tract, including neoplasia. Some of these effects entail the persistence of tissues that regress or express different cellular markers during development. For example, Mullerian ducts are the structures that give rise to organs of the female genital tract. They regress during normal development in males, but persist in males exposed to estrogens during development.

A small number of women exposed to DES *in utero* manifested a series of anomalies of the genital tract (adenosis, ectropion, and anomalies of the cervix) and an increased incidence of clear-cell adenocarcinoma of the vagina (risk from birth to 34 years of age is 1:1,000) (Mittendorf, 1995; Newbold et al., 1985). Exposure to DES occurred before the 13th week of pregnancy in women that developed clear-cell adenocarcinoma. The fact that 90% of the cases were diagnosed between ages 15 and 27 suggests that, in addition to *in utero* exposure, the hormonal milieu present at puberty is also required for the development of this lesion. It is unknown whether the anomalies of the genital tract mentioned above predispose to neoplasia at a later point. Interestingly, in the mouse model, a main effect of DES exposure was that animals developed adenocarcinomas of the uterus after 4 months of age. Before this time in development, exposed mice had exhibited hypoplastic uteri with few or no glands (Newbold, 1995). This suggests that the primary carcinogenic effect may neither be a proliferative one nor a mutational one. Instead, the primary effect may be developmental (epigenetic?) in the sense that cell populations that should have disappeared do not.

DES-exposed women are just now entering the 5th decade of life. Therefore, it is not yet known whether *in utero* exposure to DES also increases the risk of hormone-related neoplasms (endometrium, breast) at the age in which these neoplasms appear in the unexposed population. Bern (1992c; Bern et al., 1976) and Newbold and McLachlan (Newbold et al., 1985) studied the effect of prenatal and early postnatal exposure to DES on the genital tract of mice and found that the most important feature of this syndrome is that some of the morphological alterations are not readily recognizable at birth, but they manifest themselves during puberty and adult life. In the uterus, cystic endometrial hyperplasia, leiomyomas, adenocarcinomas, and stromal cell sarcomas were observed. In the vagina, the proliferative lesions reported were basal cell hyperplasia combined with hyperkeratinization, epidermoid tumors, and adenocarcinomas. It should be noted that vaginal adenocarcinomas appeared when mice were exposed to relatively low doses (2.5 µg/kg/day), whereas uterine adenocarcinomas appeared at higher exposure levels (100 µg/kg/day) on days 9 to 16 of gestation (Newbold et al., 1990). Ovariectomy before puberty inhibited the development of these neoplasias. Prolactinomas were observed in mice exposed on days 16 to 17 of prenatal development (Walker et al., 1993). An interesting consequence of neonatal exposure to DES in mice infected with mouse mammary tumor viruses (MMTV) is shortened latency period and increased incidence of mammary tumors. In the CD-1 strain of mice, which has a high incidence of spontaneous mammary tumors, offspring of females exposed *in utero* had a significantly higher incidence of ovarian and mammary tumors than offspring of females exposed to vehicle (Walker, 1990). DES-induced vaginal adenocarcinomas and squamous cell carcinomas in Wistar rats treated on days 18 to 20 of gestation (Baggs et al., 1991). More recently, it was reported that male mice exposed *in utero* to DES can transmit a carcinogenic effect to their offspring (Turusov et al., 1992). Walker and Kurth showed that female mice exposed *in utero* can transmit a carcinogenic effect to their offspring (Walker et al., 1995). With the use of blastocyst transfers, it was shown that offspring of normal blastocysts that had been transferred to mice exposed prenatally to DES developed uterine adenocarcinomas (7%); offspring from blastocysts from female mice exposed to DES *in utero* transferred to mice exposed to vehicle also developed endometrial adenocarcinomas (16%). Hence, the neoplastic effects of intrauterine exposure may be due to "germ cell modification" (mutation or gene imprinting), as well as to alteration of the maternal environment.

Xenoestrogens and the Developmental Hypothesis

Time of exposure appears to be crucial for eliciting developmental mishaps. In addition, some of the developmental alterations mediated by estrogens occur at significantly lower doses than those necessary for causing estrogenic effects in adults. For example, Burroughs et al. have found that hypoplasia of uterine glands occurs after neonatal exposure to extremely low doses of coumestrol, a phytoestrogen (Burroughs et al. 1985, 1990). In addition, vom Saal observed significant increases in the size of the prostate in adult animals exposed *in utero* to higher levels of estrogen due to a positional effect (a male between two females vs. a male between two males) (vom Saal et al., 1992; Nonneman et al., 1992). Moreover, *in utero* exposure to low doses of BPA also resulted in increased prostate size in the adult (Nagel et al., 1997). It should be noted that in Vom Saal's experiments, the dose-response curve looks like an inverted U; i.e., the highest doses were less effective in inducing these effects than the lower ones (Nagel et al., 1997; vom Saal et al., 1997). In summary, there are stages of particular vulnerability during development, and the developing organism seems to be far more sensitive to minute variations of hormone levels than the adult organism.

Animal Models for Hormonal Carcinogenesis

"Spontaneous" Neoplasia of Estrogen-Target Organs in Animal Models

Neoplasia of endocrine and reproductive organs seldom occur in wildlife and laboratory animals subjected to a restricted diet. Long-term studies in laboratory animals revealed that mammary tumors occur spontaneously in some rat and mouse strains. For example, Sprague-Dawley and ACI aging virgin females develop mammary tumors spontaneously. Ovariectomy and multiple pregnancies during early adulthood significantly decreased the incidence of these tumors. Endometrial tumors also develop in Han:Wistar, BDII/Han, and Donryu strains (Nagaoka et al., 1990). Ovariectomy inhibits the development of these neoplasias (Deerberg et al., 1987). Spontaneous prolactinomas develop in certain rat (Sprague-Dawley) and mouse strains (C57BL/6) (Greenman et al., 1990). Adenoma and adenocarcinoma of the magnum of the oviduct and leiomyoma of the ventral ligament of the oviduct are the most frequent spontaneous neoplasias in the reproductive tract of hens (Anjum et al., 1988b); they correlate with high plasma estrogen levels (Anjum et al., 1988a).

Experimental Neoplasia as a Result of Hormonal Manipulation in Animal Models

"Hormonal carcinogenesis" developed as a discipline within endocrinology as a result of studies on the function of endocrine organs involving organ ablation-hormone replacement experiments (Foulds, 1969). Ovarian hormones were found to play a role in tumor development of the mammary gland. Pituitary tumors could be induced by estrogen treatment in rats, as well as by thyroidectomy in mice. Thyroid tumors in mice were induced by goitrogens, through an increase in TSH plasma levels. Ovarian tumors were induced by transplanting ovaries into the spleen of ovariectomized rats and mice (presumably gonadotropin-induced). And gonadectomy induced adrenocortical tumors in some strains of guinea pigs, rats, mice, and hamsters.

Ovarian Hormones and Neoplasia

The search for a role of ovarian hormones in breast neoplasia can be traced to the end of the 19th century when Beatson reported that ovariectomy resulted in clinical regression of advanced breast cancer (Beatson, 1896). This result may be interpreted today as evidence for the trophic role of estrogens in tumor growth and cell survival; ovariectomy drastically reduced the incidence of breast cancer. Endometrial cancer in humans is also related to estrogen exposure. Vaginal clear-cell carcinoma in young women appears to be a consequence of *in utero* exposure to DES. Understandably, experimentation in humans is severely restricted by ethical concerns. On the other hand, animal models provide valuable insight on the problem but are not always directly extrapolable to humans. Estrogens were found to induce pituitary neoplasia (rat, mouse, European hamster), mammary cancer (rat and mouse), and kidney tumors in male Syrian and European hamsters. These kidney tumors are estrogen-sensitive. Although they do not seem to have an equivalent in human pathology, they are currently used to explore the role of estrogens as mutagens.

Endometrial Tumors

Endometrial tumors occur at a relatively high incidence in certain strains of rats. DES exposure throughout adult life results in a 1.7% incidence of uterine adenocarcinoma in mice, while neonatal (days 1–5) administration results in 90% incidence. Tumors did not develop in animals ovariectomized before puberty (Newbold et al., 1990). These tumors required estrogens for continuous growth when transplanted. Adenocarcinomas of the uterus may also be developed by administration of the carcinogen n-nitroso-n-methylurea (NMU) to intact adult mice (Niwa et al., 1993); progestagens inhibited the development of tumors in estrogen-treated animals (Niwa et al., 1995). Estrogens induce cell proliferation in the endometrium, and progesterone inhibits the proliferative effect of estrogens in rodents and humans.

In humans, endometrial adenocarcinoma rates increased in women taking estrogen-replacement therapy (Feldhoff et al., 1983; Bos et al., 1988). Simultaneous administration of estrogens and progestagens (hormone-replacement therapy) results in a much lower incidence of this type of cancer (almost similar to those in untreated women) (Brinton et al., 1993; Hulka et al., 1980).

Recently, it has been suggested that the xenoestrogen hypothesis should be tested by focusing on endometrial rather than breast cancer (Adami et al., 1995). This is predicated on the rapid increase of the incidence of endometrial cancer in postmenopausal women treated with unopposed estrogens and on the low incidence of this malignancy when compared with that of breast cancer. However, it is likely that xenoestrogen exposure does not increase the risk of endometrial cancer in mature, cycling women because their ovaries produce progesterone. A large percentage of the population of postmenopausal women are taking hormone-replacement therapy to avoid osteoporosis and heart disease. Although this group will not be affected, those postmenopausal women not taking progesterone may be at risk.

Mammary tumors

In 1928, Murray demonstrated that mammary cancer could be induced to appear in male mice transplanted with ovaries belonging to a strain in which almost 100% of the females developed mammary cancer (Murray, 1928). Lacassagne reproduced these results by treating male mice with ovarian extracts ("folliculin") (Lacasagne, 1932). While it is possible to obtain a high tumor yield by prolonged treatment with estrogens in susceptible rats, in mice this only happens in the presence of MMTV.

Mammary tumors can be induced in the Sprague-Dawley and other rat strains by prolonged treatment with estrogens. They shorten the latency period and increase the incidence of tumors that otherwise would appear if the animals were observed for their entire lifespan. For example, the spontaneous incidence of mammary adenocarcinomas in female ACI rats was reported to be 7%, whereas treatment with 5-mg pellets of DES at 80 days of age increased the incidence to 52% after 200 days of observation (Shellabarger et al., 1978). Regardless of whether or not the resulting neoplasia behaves as a hormone-sensitive tumor, it develops only in intact, nonovariectomized animals (Nandi et al., 1995).

Experimental estrogen-induced mammary tumors required prolonged treatment with hormones; the latency periods were extremely long, and the incidence was usually low. The discovery that chemical carcinogens such as methylcholanthrene and 7,12-dimethylbenzanthracene (DMBA)-induced mammary carcinomas in some rat strains greatly facilitated the study of these tumors, since the latency period was shortened and the percentage of incidence was higher than that of estrogen-induced tumors. Interestingly, the tumors obtained with DMBA or nitroso-methylurea (NMU) were histologically similar to human breast tumors. Several factors play a role in the induction of mammary carcinomas by these agents: (a) Genetic background is a factor, since tumors only develop in certain strains. Over 80% of these carcinogen-treated rats present tumors after 90 days of observation when they belong to the Fisher, Wistar/Furth, Sprague/Dawley, and other inbred and outbred strains; this incidence is significantly lowered, or nonexistent, when other strains, like Copenhagen, are used. (b) Exposure to estrogens also influences carcinomas. In nulliparous Sprague/Dawley rats, mammary cancer develops spontaneously. Similarly, estrogens are necessary for the development of DMBA- and NMU-induced mammary cancer in rats. Ovariectomy prior to carcinogen treatment inhibits tumor formation; estrogen treatment of ovariectomized animals results in comparable tumor incidence rates and latency periods like those observed in intact animals. Paradoxically, high doses of estrogens increase the latency period, decrease the size of tumors, and result in a lower tumor yield per animal., Hence, estrogens also have a biphasic effect on the induction of mammary carcinoma. (c) Pituitary hormones influence carcinomas, as hypophysectomy prevents the development of DMBA-induced tumors. Prolactin appears to stimulate the growth of DMBA-induced mammary carcinomas in ovariectomized, adrenalectomized, and hypophysectomized rats (Talwalker et al., 1961). However, estrogens seem to be essential for the prolonged growth of these tumors (Leung et al., 1975; Welsch, 1985), which contradicts the hypothesis that estrogens act by inducing the secretion of prolactin. (d) Developmental factors influence the induction of mammary carcinomas. In DMBA-induced mammary cancer in rats, a "window of vulnerability" was identified between the 45th and 55th day of life (Russo et al., 1978); the carcinogen administration during this period significantly increases the incidence of carcinomas and decreases the latency period. Multiparity further decreases the incidence of carcinomas. Tumor growth is stimulated by pregnancy. These effects are explained tentatively by the intense proliferative activity of structures called terminal end buds from which new gland ducts are originated during this "window of vulnerability" (Russo et al., 1987). Further development of the gland produces structures that become "carcinogen-resistant" (Russo et al., 1978).

In humans, only a small percentage (5%–10%) of breast cancer is attributed to genetic inheritance. Otherwise, risk factors are mostly related to cumulative lifetime exposure to endogenous ovarian hormones (Hulka et al., 1995) such as early menarche and late menopause. Pregnancy also plays a role; nulliparous women who have a higher risk than those that have undergone full-term pregnancies in their early 20s, whereas first pregnancy in the late 30s and 40s increases the risk of breast cancer (Sonnenschein et al., 1996). There is also some evidence that the level of estrogen exposure during development *in utero* may influence the risk of breast cancer (Ekbom et al., 1992). Exposure to radiation during adolescence and early adulthood is also a risk factor. Epidemiological studies have, for the most part, studied exposure to estrogens from the viewpoint that they act as promoters. For example, a study by Toniolo et al. showed a significant positive correlation between free estrogen levels (estrogens not bound to sex steroid binding globulin) in postmenopausal women and their incidence of breast cancer a few years later (Toniolo et al., 1995).

Pituitary tumors

Chronic estrogen treatment in Fischer, ACI, or Wistar-Furth rats results in the development of pituitary adenomas and transplantable neoplasms that grow as estrogen-sensitive tumors. The proliferative response to estrogens ceased after a few days, in spite of the continuous presence of estrogens in rats from strains that did not develop adenomas (proliferative shutoff) (Wiklund et al., 1981). Rats from strains in which the proliferative response was maintained as long as estrogens were administered developed neoplasms. Hence, tumors developed in animals that have apparently lost the ability to express the estrogen-induced proliferative shutoff.

Testicular neoplasias

There are no animal species that adequately model human testicular neoplasias. Hence, we will discuss current thoughts about the genesis of this disease in humans. Germ cell tumors develop from carcinoma *in situ* (Skakkebaek et al., 1987). The age-specific pattern of tumor incidence in males shows a small peak from birth to 4 years of age, and a second increase after puberty reaches a peak between 20 to 30 years of age for malignant teratoma, and at 30 to 40 years for seminoma (Horwich et al., 1995). While tumor incidence has increased recently in young men, there is no clear evidence of an increase in boys (Moller et al., 1995). Although histological examination of testicular parenchyma adjacent to tumors in adult men revealed the presence of carcinoma *in situ*, this association was not found in tissue from boys (Jorgensen et al., 1995); this suggests separate etiologies. The age distribution for incidence of testicular cancer suggests that exposure to risk factors occurs early in life, and progression from carcinoma *in situ* to clinical cancer is influenced by androgens and/or pituitary hormones. The risk of testicular cancer is increased in men with a history of testicular maldescent, gonadal dysgenesis, androgen insensitivity, intersex states, and infertility. Maldescent is associated with a 5- to 10-fold relative risk increase; when undescent is unilateral, testicular cancer may arise in the contralateral testicle or in both (Foley et al., 1995). In addition, orchidopexy may not prevent testicular cancer (Senturia, 1987). This suggests that an inherent germinal defect present in the germinal epithelium may be responsible for the two pathologies (Sohval, 1956). Testicular dysgenesis is an etiologic factor in cryptorchidism (Sohval, 1953). Also, dysgeneic tissue is frequently found in undescended testis (Paulson et al., 1982). In addition, approximately 20% of the cases of testicular cancer have a history of maldescent. Tumors arise in the contralateral (normally descended) testis as well (Johnson et al., 1968). A 2-fold increase in incidence of undescended testis has been reported from 1950 to 1970 (Chilivers et al., 1984); similar increases in hypospadias have been reported (Kallen, 1988; Kallen et al., 1992). (See the discussion of these pathologies in Chap. 10.) The hypothesis that high estrogen exposure *in utero* may be a risk factor for testicular cancer is supported by the increased incidence of this pathology in dizygotic twins (a condition that results in increased estrogen exposure) (Braun et al., 1995).

Mouse strains where testicular cancer arises spontaneously were described (Horwich et al., 1995). Mice exposed *in utero* to DES offer a model for testicular maldescent (Newbold, 1995). This pathology is strongly correlated with testicular cancer of germinal cell origin in humans. However, the testicular cancer associated with DES exposure in mice originates in the rete testis (nongerminal origin). For the most part, hormone-induced testicular cancers in laboratory rodents are Leydig cell adenomas; DES treatment induces these tumors in European hamsters (Reznik-Schuller, 1979).

Prostate cancer

The etiology of human prostate cancer is unknown. However, like other cancers of the genital tract, its incidence is practically nonexistent in men who were castrated before they reached 40 years of age (Paulson et al., 1982). In addition, in most cases, clinical prostate cancer regresses after castration. Models for prostate cancer have been developed in animal systems. Most pathologists believe that rat ventral prostate tumors are not representative of the human disease, while those of the dorsolateral prostate are good models for human prosatate carcinoma (Bosland, 1992).

Spontaneous cancer: Cribriform carcinoma of the ventral prostate develops frequently in the aging ACI rat (Bosland, 1992) and the AXC rat (Shain et al., 1977). Adenocarcinoma of the dorsolateral prostate develops in the Lobund-Wistar rat (Bosland, 1992). One interesting feature of the ACI tumors is that their incidence rate increases in animals exposed to high fat diet (Iizumi et al., 1987); this is consistent with correlations derived from human studies.

Hormone-induced cancer: Prostate cancer may be induced in rats by treatment with chemical carcinogens, androgens, carcinogens plus androgens, and androgens plus estrogens. Treatment with chemical carcinogens, in otherwise normal males, resulted in ventral prostate tumors in the F344 and MRC rat (Bosland, 1992). Lobund-Wistar rats developed spontaneous prostate cancer. Prolonged treatment with testosterone increased the tumor yield, and the latency period decreased (Pollard et al., 1985). Interestingly, increasing the fat content in the diet resulted in further shortening of the latency period. Combinations of chemical carcinogens and testosterone in various protocols increased the tumor incidence over that obtained with carcinogen alone. Moreover, carcinomas also appeared in the seminal vesicles and coagulating glands. However, the most striking results are those obtained with a combination of estradiol or ethynyl-estradiol and testosterone. Noble originally found that estradiol plus testosterone was more effective than testosterone alone (Noble, 1982). These hormones induce epithelial dysplasia and, subsequently, adenocarcinoma in the dorsolateral prostate of NBL rats (Noble, 1982) and F344 rats treated with the carcinogen 3,2′-dimethyl-4 aminobiphenyl (DMAB) (Shirai et al., 1994). An interesting feature of the DMAB model is that testosterone alone, as well as testosterone plus ethynyl-estradiol, significantly decreases the incidence of ventral prostate carcinoma below that obtained with DMAB alone, while increasing carcinoma incidence of the lateral, dorsal, and anteriorprostates. Testosterone and estrogen levels increased 2- to 3-fold during this treatment. The role of estrogens in this process is unknown. However, several studies found a cooperative effect of estrogens given together with androgens in normal prostate growth (Walsh et al., 1976; Ofner et al., 1992). Others have suggested that the role of estrogens is to produce DNA damage (Han et al., 1995). Most interestingly, the androgen 5α-dihydrotestosterone, which is not metabolized into estrogens, fails to induce prostate cancer in Lobund-Wistar rats (Pollard et al., 1987) and in the DMAB model (Shirai et al., 1994).

Endocrine Disruptors and Neoplasia in Animal Models

DDT and Estrogen-Sensitive Tumor Growth

Estrogen-sensitive mammary MT2 cells grow as a tumor when inoculated into ovariectomized syngenic hosts treated with estradiol. The full estrogen agonist o,p′DDT sustained tumor growth at the same rate achieved with estradiol pellets. The congener p,p′DDD, which is a partial agonist and less potent than p,p′DDT, did not increase the tumor size over that found in ovariectomized controls (Robison et al., 1985).

Neoplasias in Animals Treated with Estrogenic Pesticides

As reviewed above, natural estrogens induce neoplasias in reproductive and endocrine organs. There is no consensus on whether they do so through nonhormonal mechanisms (i.e., as mutagens) or due to hormonal activity ("promotional" effect). In addition, carcinogens devoid of hormonal activity, such as NMU, induce mammary neoplasias that behave as estrogen-sensitive tumors. Therefore, carcinogenicity studies done by long-term exposure to maximal tolerable doses of a chemical are unsuitable for addressing the question of whether or not their carcinogenicity is mediated by their hormonal activity. Several carcinogenicity studies have been conducted using both mice and rats. Most of the reports claim that there are no statistically significant differences between controls and experimentals. It should be noted that, for the most part, interpretation of these long-term studies is obfuscated by high mortality due to general toxicity, sample loss, high incidence of tumors in the control animals, or insufficient sampling (Anonymous, 1977, 1978a, b, c, 1979; Reuber, 1980, 1981). Results of the few studies reporting pesticide-induced increased tumor incidence in reproductive or endocrine organs are summarized in Table 11-2.

Pesticides and Breast Cancer

Among the estrogenic xenobiotics, PCBs and DDT have been suspected as inducers of breast cancer because they were released massively into the environment beginning approximately 50 years ago, and they are persistent (i.e., their presence in serum may represent cumulative exposure during a lifetime). Early studies that showed no correlation between breast cancer incidence and

Table 11-2. Pesticide-induced neoplasias in reproductive or endocrine organs of laboratory species

Neoplasia	Pesticide	Species/Strain	Reference
Mammary gland	Methoxychlor	Osborne-Mendel rats	Reuber, 1980
Mammary gland	Endosulfan	Osborne-Mendel rats	Reuber, 1981
Endometrium	Endosulfan	Osborne-Mendel rats	Reuber, 1981
Pituitary	Methoxychlor	Osborne-Mendel rats	Reuber, 1980
Pituitary	Lindane	Osborne-Mendel rats	Reuber, 1979a
Ovary	DDT	Osborne-Mendel rats	Reuber, 1978
Ovary	Methoxychlor	Osborne-Mendel rats	Reuber, 1980
Ovary	Lindane	Osborne-Mendel rats	Reuber, 1979a
Testis	Methoxychlor	Balb/c mice	Reuber, 1980
Adrenal	Methoxychlor	Osborne-Mendel rats	Reuber, 1980
Adrenal	Lindane	Osborne-Mendel rats	Reuber, 1979a
Thyroid	Lindane	Osborne-Mendel rats	Reuber, 1979a
Thyroid	DDE	Osborne-Mendel rats	Anonymous, 1978c
Thyroid	Toxaphene	Osborne-Mendel rats	Anonymous, 1979

xenoestrogen levels were comprised of a small number of cases and controls that were not matched for other risk factors. However, three recent studies do show a correlation between the occurrence of breast cancer and the levels of xenoestrogens. Wolff et al. found that serum DDE levels correlated with breast cancer incidence in a study of 58 breast cancer patients and 171 controls that were well matched for risk factors and age (Wolff et al., 1993). Another study documented that estrogen receptor-positive breast cancer correlated with higher concentrations of DDE in their tissues (Dewailly et al., 1994). Krieger et al. studied 150 women with breast cancer and 150 controls. Each set was made up of 50 African-American women, 50 Caucasian women, and 50 Asian-American women. When the data from all ethnic groups were pooled, no significant correlation was observed between plasma levels of DDE and breast cancer (Krieger et al., 1994). However, when the cases and matching controls were evaluated separately, according to their ethnic group, high-serum DDE levels were correlated with breast cancer incidence in Caucasian and African-American women, but there was no significant correlation in Asian-American women. Evidence of a link between exposure to PCBs and breast cancer incidence is equivocal (Wolff et al., 1995). However, these studies correlated exposure to total PCBs rather than to the levels of specific congeners. To clarify the relevance of these findings, epidemiological studies should be conducted involving larger sample sizes.

A recent study of 717 women, 240 of whom had breast cancer, revealed that elevated serum levels of dieldrin were associated with a significantly increased dose-related risk of breast cancer, whereas no relationship was observed for DDT, its metabolites, or for a number of PCBs (Hoyer et al., 1998). These data suggest that a broader array of contaminants needs to be examined. It would be impractical to test only for the presence of DDT metabolites and PCBs, since not all of the PCB congeners are estrogenic, and many other environmental estrogens may also play a role. Newly identified xenoestrogens may be less persistent than estrogenic PCBs and DDT metabolites, but they are widely used, and it may be inferred that exposure occurs steadily due to their presence in foods (Soto et al., 1995b). Although measuring the levels of each one of the xenoestrogens in blood is a better approximation to real exposure at the target organ level, the question of whether xenoestrogens are a risk factor for breast cancer has to be addressed by measuring the estrogen burden of the sum of these chemicals.

Conclusions

The development of neoplasias has been a subject of intense research during the 20th century. However, little is known about mechanisms underlying this phenomenon. In the last 4 decades, due to the technological advances that allow the study of DNA, research focused primarily on the search for genes involved in cancer (oncogenes/suppressor genes), in spite of the lack of hypotheses regarding how mutations are supposed to cause neoplasia. Epigenetic causation has not generated a comparable research program, because a working hypothesis that would single out discrete underlying mechanisms has not yet been formulated. This is so in spite of the strong evidence gathered by B. Mintz, H. Rubin, and others regarding the reversibility of the malignant phenotype (Mintz et al., 1975; Rubin, 1993; Rubin et al., 1992). In addition, as cell proliferation is central to the development of tumors, research in cancer is heavily dependent on the premises adopted to study control of cell proliferation. We have suggested that a reappraisal of the premises adopted when studying control of cell proliferation is needed, namely, the idea of adopting *proliferation* in lieu of *quiescence* as the default state of cells (see Sonnenschein and Soto, 1999). This may generate new and more successful paradigms for studying carcinogenesis.

Endogenous sex steroids are a major causal agent in the development of neoplasias in their target organs. The underlying mechanisms of resistance and susceptibility are presently unknown. Experimental carcinogenesis studies in animal models and observational studies in humans have produced data consistent with the notion that sex steroids are causal agents, as they have been shown to control the development of their target organs and the proliferation of their target cells. In addition, estrogens induce the formation of DNA adducts that may result in DNA mutations.

Models for estrogen-induced carcinogenesis were developed to obtain high tumor yields with short latency periods, rather than to mimic the patterns of hormone exposure that may induce neoplasias in humans and in susceptible strains of lab animals. Thus, researchers have resorted to "unnatural" means to provoke carcinogenesis, such as administering pharmacological doses of estrogens, exposing normal females to carcinogens such as DMBA (Wiklund et al., 1982; Russo et al., 1987), or a combination of both. To the best of our knowledge, no detailed studies aimed at determining the dose-response curve of estrogen for tumor induction have been conducted. Moreover, the conditions used to generate these tumors do not try to approximate the hormonal pattern acting in susceptible strains, including the fluctuations due to the menstrual and estrous cycles, and the pulsatility of hormone release.

In addition to hormone exposure, the incidence of hormone-induced tumors may be enhanced in animals fed high-fat diets without any apparent changes in the plasma levels of hormones. Human studies also suggest that diet plays an important role in the development of these neoplasias.

It has been proposed that extemporaneous administration of sex hormones during development leads to permanent lesions in the genital tract. In turn, this teratogenic effect also predisposes tissues to develop neoplasia. Remarkably, developmental effects occur at doses lower than those needed to trigger responses in adult animals. Recent experiments indicate that females exposed to DES *in utero* transmit a neoplastic phenotype to blastocysts from normal animals (Walker et al., 1995). These data indicate that alterations in the maternal environment may lead to neoplasia.

Whether or not hormonally active agents in the environment have a causal role in the development of neoplasias of estrogen and androgen-target organs in humans has not been explored thoroughly. Consequently, it is premature to draw definitive conclusions regarding their contribution to the increase of breast, testicular, and prostate cancer incidence. The plausibility of this hypothesis is based on evidence that (a) exposure to natural estrogens is a main risk factor for endometrial and breast cancers; (b) exposure to androgens is a risk factor for prostate cancer; and (c) estrogenic pesticides induce endocrine and reproductive tumors in some rodent strains. As explained above, our understanding of neoplastic development in general, as well as the role of hormones in this process, is rudimentary.

The main criticism raised against the xenoestrogen hypothesis is that these chemicals are, in general, less potent than natural estrogens and that current exposure levels are supposedly insignificant when compared to the levels of endogenous hormones. These objections are not supported by data. First, the data revealing low potency of xenoestrogens were gathered by receptor-binding assays and studies done in tissue culture. Since these studies do not address metabolic rates and bioavailability, they may underestimate the potency of xenoestrogens in whole organisms. Second, the few epidemiological studies addressing the role of xenoestrogens in carcinogenesis have dealt mostly with exposure to DDT metabolites and organochlorines that are not estrogenic. Moreover, DDT is just one of the many xenoestrogens to which humans are exposed. Evidence of human exposure to the newly identified xenoestrogens (such as phthalates) has yet to be gathered, and thus the actual level of cumulative exposure to xenoestrogens is still unknown. Implicit in the concept that xenoestrogens are "weak" estrogens is the assumption of linear dose responses for these chemicals. In this regard, the dose response to estradiol is not linear but biphasic. It induces cell proliferation at low doses and inhibits it at high doses (The 2-step mechanism has been reviewed in detail elsewhere; see Sonnenschein and Soto, 1999.) From this perspective, the pattern of exposure may be very relevant. Additionally, developmental effects seem to occur at lower doses than those effective in adult animals. The scant dose-response data on developmental effects shows an inverted U shape, with deleterious effects occurring at the lowest doses. In summary, a large research program is needed to find out whether or not current levels of xenoestrogen exposure increase the incidence of hormone-related neoplasias.

Although there are many uncertainties about basic mechanisms in the control of cell proliferation and carcinogenesis at the present time (the shape of the dose-response curve, levels of exposure, and timing of xenoestrogen exposure), we are still faced with the question of what should be done from the perspective of public and environmental health. Exploring these basic questions will take years of intense research. Should we wait for the basic science to be done, or should we adopt a preventive approach by diminishing exposures to endocrine disruptors now? The two are not mutually exclusive, as the enactment of an aggressive preventive approach is not incompatible with the initiation of a meticulous reappraisal of the accepted premises within this field.

Acknowledgments

This work was supported in part by PHS NIH grants CA-55574 and CA-13410 and NSF DCB-8711746. The assistance of the Center for Reproductive Research at Tufts University (P30 HD 28897) is acknowledged. We are grateful to Maria Luizzi, Cheryl Michaelson, and Beau Weill for their editorial assistance.

References

Adami, H.O., Lipworth, L., Titus-Ernstoff, L., Hsieh, C.C., Hanberg, A., Ahlborg, U., Baron, J., and Trichopoulos, D. (1995). Organochlorine compounds and estrogen-related cancers in women. *Cancer Causes & Control* 6: 551–566.

Alberts, B., Bray, D., Lewis, J., Raff, M., Roberts, K., and Watson, J.D. (1994). *Molecular Biology of the Cell, 3d ed.* Garland, New York.

Anjum, A.D., and Payne, L.N. (1988a). Concentration of steroid sex hormones in the plasma of hens in relation to oviduct tumours. *Brit. Poultry Sci.* 29: 729–734.

—— and —— (1988b). Spontaneous occurrence and experimental induction of leiomyoma of the ventral ligament of the oviduct of the hen. *Res. Vet. Sci.* 45: 341–348.

Anonymous (1977). Bioassay of lindane for possible carcinogenicity. *Tech. Rep. Series* 14.

—— (1978a). Bioassay of endosulfan for possible carcinogenicity. *Tech. Rep. Series* 62.

—— (1978b). Bioassay of methoxychlor for possible carcinogenicity. *Tech. Rep. Series* 35.

—— (1978c). Bioassays of DDT, TDE, and p,p'-DDE for possible carcinogenicity. *DHEW pub*. no. 1386.

—— (1979). Bioassay of toxaphene for possible carcinogenicity. *Tech. Rep. Series* 37.

Arnold, S., Robinson, M.K., Notides, A.C., Guillette, L.J., Jr., and McLachlan, J.A. (1996). A yeast estrogen screen for examining the relative exposure of cells to natural and xeno estrogens. *Environ. Health Perspec.* 104: 544–548.

Baggs, R.B., Miller, R.K., and Odoroff, C.L. (1991). Carcinogenicity of diethylstilbestrol in the Wistar rat: Effect of postnatal oral contraceptive steroids. *Cancer Res.* 51: 3311–3315.

Barrett, J.C., Wong, A., and McLachlan, J.A. (1981). Diethylstilbestrol induces neoplastic transformation of cells in culture without measurable somatic mutation at two loci. *Science* 212: 1402–1404.

Beatson, G.T. (1896). On the treatment of inoperable cases of carcinoma of the mamma: Suggestions for a new method of treatment with illustrative cases. *Lancet* 2: 104–106.

Beremblum, I., and Shubik, P. (1947). A new quantative approach to the study of the stages of chemical carcinogenesis in the mouse's skin. *Brit. J. Cancer* 1346: 383–391.

Bern, H.A., Jones, L.A., and Mills, K.T. (1976). Use of the neonatal mouse in studying long-term effects of early exposure to hormones and other agents. *J. Toxicol. & Environ. Health* (Suppl. 1): 103–116.

—— (1992a). The development of the role of hormones in development: A double remembrance. *Cancer Res. Lab.* 131(5): 2037–2038.

—— (1992b). *Diethylstilbestrol Syndrome: Present Status of Animal and Human Studies in Hormonal Carcinogenesis*. Springer Verlag, New York.

—— (1992c). The fragile fetus. In *Chemically-Induced Alterations in Sexual and Functional Development: The Wildlife/Human Connection*. (T. Colburn and C. Clement, ed.), pp. 9–15. Princeton: Princeton Scientific Publishing, Princeton.

Bhat, H.K., Han, X., Gladek, A., and Liehr, J.G. (1994). Regulation of the formation of the major diethylstilbestrol-DNA adduct and some evidence of its structure. *Carcinogenesis* 15: 2137–2142.

Bishop, J.M. (1983). Cellular oncogenes and retroviruses. *Ann. Rev. Biochem.* 52: 301–354.

Bos, O.J.M., Fischer, M.J.E., Wilting, J., and Janssen, L.H.M. (1988). Drug-binding and other physiochemical properties of a large tryptic and a large peptic fragment of human serum albumin. *Biochemica et Biophysica Acta* 953: 37–47.

Bosland, M.C. (1992). Animal models for the study of prostate carcinogenesis. *J. Cellular Biochem.* (Suppl. 16H): 89–98.

Bradlow, H.L., Davis, D.L., Lin, G., Sepkovic, D., and Tiwari, R. (1995). Effects of pesticides on the ratio of 16 alpha/2-hydroxyestrone: A biologic marker of breast cancer risk. *Environ. Health Perspec.* 103 (Suppl. 7): 147–150.

Braun, M.M., Ahlbom, A., Floderus, B., Brinton, L.A., and Hoover, R.N. (1995). Effect of twinship on incidence of cancer of the testis, breast, and other sites (Sweden). *Cancer Causes & Control* 6: 519–524.

Brinton, L.A., and Schairer, C. (1993). Estrogen replacement therapy and breast cancer risk. *Epidemiol. Rev.* 15: 66–79.

Brotons, J.A., Olea-Serrano, M.F., Villalobos, M., and Olea, N. (1994). Xenoestrogens released from lacquer coating in food cans. *Environ. Health Perspec.* 103(6), 608–612.

Burroughs, C.D., Bern, H.A., and Stokstad, E.L. (1985). Prolonged vaginal cornification and other changes in mice treated neonatally with coumestrol, a plant estrogen. *J. Toxicol. & Environ. Health* 15: 51–61.

———, Mills, K.T., and Bern, H.A. (1990). Long-term genital tract changes in female mice treated neonatally with coumestrol. *Reproductive Toxicol.* 4: 127–135.

Chilvers, C., Pike, M.C., Forman, D., Fogelman, K., and Wadsworth, M.E.J. (1984). Apparent doubling of frequency of undescended testis in England and Wales in 1962–81. *Lancet* ii (no. 8398) 330–332.

Clark, W.H. (1991). Tumour progression and the nature of cancer. *Brit. J. Cancer* 64: 631–644.

Davis, D.L., Bradlow, H.L., Wolff, M., Woodruff, T., Hoel, D.G., and Anton-Culver, H. (1993). Medical hypothesis: Xenoestrogens as preventable causes of breast cancer. *Environ. Health Perspec.* 101: 372–377.

Deerberg, F., and Kaspareit, J. (1987). Endometrical carcinoma in BD II/Han rats: Model of a spontaneous hormone-dependent tumor. *J. Natl. Cancer Inst.* 78: 1245–1251.

Dewailly, E., Dodin, S., Verreault, R., Ayotte, P., Sauve, L., Morin, J., and Brisson, J. (1994). High organochlorine body burden in women with estrogen receptor positive breast cancer. *J. Natl. Cancer Inst.* 86: 232–234.

Ekbom, A., Trichopoulos, D., Adami, H.-O., Hsieh, C.-C., and Lan, S.-J. (1992). Evidence of prenatal influences on breast cancer risk. *Lancet* 340: 1015–1018.

Feldhoff, R., and Ledden, D.J. (1983). Evidence for the spontaneous formation of interspecies hybrid molecules of human, rat and bovine serum albumin. *Biochem. Biophys. Res. Communications* 114: 20–27.

Foley, S., Middleton, S., Stitson, D., and Mahoney, M. (1995). The incidence of testicular cancer in Royal Air Force personnel. *Brit. J. Urol.* 76: 495–496.

Foulds, L. (1969). *Neoplastic Development.* Academic, New York.

Fox, G.A. (1992). Epidemiological and pathobiological evidence of contaminant-induced alterations in sexual development in free-living wildlife. In *Chemically Induced Alterations in Sexual and Functional Development: The Wildlife/Human Connection.* (T. Colborn and C. Clement, eds.), pp. 147–158. Princeton Scientific Publishing, Princeton.

Giwercman, A., Carlsen, E., Keiding, N., and Skakkebaek, N.E. (1993). Evidence for increasing incidence of abnormalities of the human testis: A review. *Environ. Health Perspec.* 101: 65–71.

Greenman, D.L., Highman, B., Chen, J., Sheldon, W., and Gass, G. (1990). Estrogen-induced thyroid follicular cell adenomas in C57BL/6 mice. *J. Toxicol. & Environ. Health* 29: 269–278.

Hammond, B., Katzenellenbogen, B.S., Kranthammer, N., and McConnell, J. (1979). Estrogenic activity of the insecticide chlordecone (kepone) and interaction with uterine estrogen receptors. *Proc. Natl. Acad. Sci. USA* 76: 6641–6645.

Han, X., Liehr, J.G., and Bosland, M.C. (1995). Induction of a DNA adduct detectable by 32P-postlabeling in the dorsolateral prostate of NBL/Cr rats treated with estradiol-17β and testosterone. *Carcinogenesis* 16: 951–954.

Hertz, R. (1985). The estrogen problem—retrospect and prospect. In *Estrogens in the Environment II—Influences on Development.* (J.A. McLachlan, ed.), pp. 1-11. Elsevier, New York.

Horwich, A., Mason, M.D., and Hendry, W.F. (1995). Urological cancer. In *Oxford Textbook of Oncology.* (M. Peckham, H. Pinedo, and U. Veronesi, eds.), pp. 1407–1530. Oxford University Press, New York.

Hoyer, A.P., Grandjean, P., Jorgersen, T., Brock, J.W., and Hartvig, H.B. (1998). Organochlorine exposure and risk of breast cancer. *Lancet* 352: 1816–1820.

Hulka, B.S., Kaufman, D.G., Fowler, W.C., Jr., Grimson, R.C., and Greenberg, B.G. (1980). Predominance of early endometrial cancers after long-term estrogen use. *JAMA* 244: 2419–2422.

——— and Stark, A.T. (1995). Breast cancer: Cause and prevention. *Lancet* 346: 883–887.

Iizumi, T., Yazaki, T., Kanoh, S., Kondo, I., and Koiso, K. (1987). Establishment of a new prostatic carcinoma cell line (TSU-Pr1). *J. Urol.* 137: 1304–1306.

Iversen, O.H. (1993). The reverse experiment in two-stage skin carcinogenesis. *APMIS* 101: 1–96.

Jobling, S., Reynolds, T., White, R., Parker, M.G., and Sumpter, J.P. (1995). A variety of environmentally persistent chemicals, including some phthlate plasticizers, are weakly estrogenic. *Environ. Health Perspec.* 103: 582–587.

Johnson, D.E., Woodhead, D.M., Pohl, D.R., and Robison, J.R. (1968). Cryptorchism and testicular tumorigenesis. *Surgery* 63: 919–922.

Jorgensen, N., Muller, J., Giwercman, A., Visfeldt, J., Moller, H., and Skakkebaek, N.E. (1995). DNA content and expression of tumour markers in germ cells adjacent to germ cell tumours in childhood: Probably a different origin for infantile and adolescent germ cell tumours. *J. Pathol.* 176: 269–278.

Kallen, B. (1988). Case control study of hypospadias, based on registry information. *Teratology* 38: 45–50.

———, Castilla, E.E., Robert, E., Lancaster, P.A., Kringelbach, M., Martinez-Frias, M.L., and Mastroiacovo, P. (1992). An international case-control study on hypospadias: The problem with variability and the beauty of diversity. *European J. Epidemiol.* 8: 256–263.

Kelce, W.R., Monosson, E., Gamcsik, M.P., Laws, S.C., and Gray, L.E., Jr. (1994). Environmental hormone disruptors: Evidence that vinclozolin developmental toxicity is mediated by antiandrogenic metabolites. *Toxicol. & Appl. Pharmacol.* 126: 276–285.

———, Stone, C.R., Laws, S.C., Gray, L.E., Kemppainen, J.A., and Wilson, E.M. (1995). Persistent DDT metabolite p,p'-DDE is a potent androgen receptor antagonist. *Nature* 375: 581–585.

Krieger, N., Wolff, M.S., Hiatt, R.A., Rivera, M., Vogelman, J., and Orentreich, N. (1994). Breast cancer and serum organochlorines: A prospective study among white, black, and Asian women. *J. Natl. Cancer Inst.* 86: 589–599.

Lacasagne, A. (1932). Aparition de cancers de la mammelle chez la souris male, soumise à des injections de folliculine. *C.R. Hebd. Seanc. Acad. Sci.* 195: 630–632.

Leung, B.S., and Sasaki, G.H. (1975). On the mechanism of prolactin and estrogen action in 7,12-dimethylbenzanthracene-induced mammary carcinoma in the rat. II: In vivo tumor responses and estrogen receptor. *Endocrinology* 97: 564–572.

Liehr, J.G., Stancel, G.M., Chorich, L.P., Bousfield, G.R., and Ulubelen, A.A. (1986). Hormonal carcinogenesis: Separation of estrogenicity from carcinogenicity. *Chem. Biol. Interactions* 59: 173–184.

Lubahn, D.B., Moyer, J.S., Golding, T.S., Couse, J.F., Korach, K.S., and Smithies, O. (1993). Alteration of reproductive function but not prenatal sexual development after insertional disruption of the mouse estrogen receptor gene. *Proc. Natl. Acad. Sci. USA* 90: 11162–11166.

McKinnel, R.G., Deggins, B.A., and Labbat, D.D. (1969). Transplantation of pluripotent nuclei from triploid frog tumors. *Science* 165: 394–396.

Mechler, B.M., Strand, D., Kalmes, A., Merz, R., Schmidt, M., and Torok, I. (1991). Drosophila as a model system for molecular analysis of tumorogenesis. *Environ. Health Perspec.* 93: 63–71.

Meyer, J.S. (1977). Cell proliferation in normal human breast ducts, fibroadenomas, and other ductal hyperplasias measured by nuclear labeling with tritiated thymidine. *Hum. Pathol.* 8: 67–81.

Meyers, C.Y., Matthews, W.S., Ho, L.L., Kolb, V.M., and Parady, T.E. (1977). Carboxylic acid formation from kepone. In *Catalysis in Organic Synthesis.* (G.W. Smith, ed.), pp. 213–255. Academic, New York.

Mintz, B., and Ilmensee, K. (1975). Normal genetically mosaic mice produce from malignant teratocarcinoma cells. *Proc. Natl. Acad. Sci. USA* 72: 3585–3589.

Mittendorf, R. (1995). Teratogen update: Carcinogenesis and teratogenesis associated with exposure to diethylstilbestrol (DES) in utero. *Teratology* 51: 435–445.

Moller, H., Jorgensen, N., and Forman, D. (1995). Trends in incidence of testicular cancer in boys and adolescent men. *Int. J. Cancer* 61: 761–764.

Murray, W.S. (1928). Ovarian secretion and tumor incidence. *J. Cancer Res.* 12: 18–25.

Nagaoka, T., Onodera, H., Matsushima, Y., Todate, A., Shibutani, M., Ogasawara, H., and Maekawa, A. (1990). Spontaneous uterine adenocarcinomas in aged rats and their relation to endocrine imbalance. *J. Cancer Res. Clin. Oncol.* 116: 623–628.

Nagel, S.C., vom Saal, F.S., Thayer, K.A., Dhar, M.G., Boechler, M., and Welshons, W.V. (1997). Relative binding affinity-serum modified access (RBA-SMA) assay predicts the relative in vivo bioactivity of the xenoestrogens bisphenol A and octylphenol. *Environ. Health Perspec.* 105: 70–76.

Nandi, S., Guzman, R., and Yang, J. (1995). Hormones and mammary carcinogenesis in mice, rats, and humans: A unifying hypothesis. *Proc. Natl. Acad. Sci. USA* 92: 3650–3657.

Newbold, R.R. (1995). Cellular and molecular effects of developmental exposure to diethylstilbestrol: Implications for other environmental estrogens. *Environ. Health Perspec.* 103 (Suppl. 7): 83–87.

——— and McLachlan, J.A. (1985).Diethylstilbestrol associated defects in murine genital tract development. In *Estrogens in the Environment II: Influences on Development.* (McLachlan, J.A., ed.), pp. 288–318. Elsevier, New York.

———, Bullock, B.C., and McLachlan, J.A. (1990). Uterine adenocarcinoma in mice following developmental treatment with estrogens: A model for hormonal carcinogenesis. *Cancer Res.* 50: 7677–7681.

Niwa, K., Murase, T., Furui, T., Morishita, S., Mori, H., and Tanaka, T. (1993). Enhancing effects of estrogens on endometrial carcinogenesis initiated by N-methyl-N-nitrosourea in ICR mice. *Japanese J. Cancer Res.* 84: 951–955.

———, Morishita, S., Murase, T., Itoh, N., Tanaka, T., Mori, H., and Tamaya, T. (1995). Inhibitory effects of medroxyprogesterone acetate on mouse endometrial carcinogenesis. *Japanese J. Cancer Res.* 86: 724–729.

Noble, R.L. (1977). The development of prostatic adenocarcinoma in Nb rats following prolonged sex hormone administration. *Cancer Res.* 37: 1929–1933.

——— (1982). Prostate carcinoma of the Nb rat in relation to hormones. *Internat. Rev. Exper. Pathol.* 23: 113–159.

Nonneman, D.J., Ganjam, V.K., Welshons, W.V., and vom Saal, F.S. (1992). Intrauterine position effects on steroid metabolism and steroid receptors of reproductive organs in male mice. *Biol. Reprod.* 47: 723–729.

Ofner, P., Bosland, M.C., and Vena, R.L. (1992). Differential effects of diethyl-stilbestrol and estradiol-17 beta in combination with testosterone on rat prostate lobes. *Toxicol. & Appl. Pharmacol.* 112: 300–309.

Olea, N., Pulgar, R., Perez, P., Olea-Serrano, F., Rivas, A., Novillo-Fertrell, A., Pedraza, V., Soto, A.M., and Sonnenschein, C. (1996). Estrogenicity of resin-based composites and sealants used in dentistry. *Environ. Health Perspec.* 104(3), 298–305.

Ozawa, S., Iguchi, T., Sawada, K., Ohta, Y., Takasugi, N., and Bern, H.A. (1991). Postnatal vaginal nodules induced by prenatal diethylstilbestrol treatment correlate with later development of ovary-independenttt vaginal and uterine changes in mice. *Cancer Letters* 58: 167–175.

Paulson, D.F., Einhorn, L.H., Peckham, M.J., and Williams, S.D. (1982) Cancer of the testis. In *Cancer Principles & Practice of Oncology.* (V.T. DeVita, S. Hellman, and S.A. Rosenberg, eds.), pp. 786–822. Lippincott, Philadelphia.

Pollard, M., and Luckert, P.H. (1985). Tumorigenic effects of direct- and indi-rect-acting chemical carcinogens in rats on a restricted diet. *J. Natl. Cancer Inst.* 74: 1347–1349.

———, Snyder, D.L., and Luckert, P.H. (1987). Dihydrotestosterone does not induce prostate adenocarcinoma in L-W rats. *Prostate* 10: 325–331.

Reuber, M.D. (1978). Carcinomas of the liver in Osborne-Mendel rats ingest-ing DDT. *Tumori* 64(6), 571–577.

——— (1979a). Carcinogenicity of lindane. *Environ. Res.* 19: 460–481.

——— (1979b). Carcinomas of the liver in Osborne-Mendel rats ingesting methoxychlor. *Life Sci.* 24: 1367–1371.

——— (1980). Carcinogenicity and toxicity of methoxychlor. *Environ. Health Perspec.* 36: 205–219.

——— (1981). The role of toxicity in the carcinogenicity of endosulfan. *Sci.*

Total Environ. 20: 23–47.

Reznik-Schuller, H. (1979). Carcinogenic effects of diethylstilbestrol in male Syrian golden hamsters and European hamsters. *J. Natl. Cancer Inst.* 62: 1083–1088.

Robison, A.K., Sirbasku, D.A., and Stancel, G.M. (1985). DDT supports the growth of an estrogen-responsive tumor. *Toxicol. Lett.* 27: 109–113.

Roy, D., Bernhardt, A., Strobel, H.W., and Liehr, J.G. (1992). Catalysis of the oxidation of steroid and stilbene estrogens to estrogen quinone metabolites by the beta-naphthoflavone-inducible cytochrome P450 IA family. *Arch. Biochem. Biophys.* 296: 450–456.

Rubin, A.L., Sneade-Koenig, A., and Rubin, H. (1992). High rate of diversification and reversal among subclones of neoplasitically transformed NIH 3T3 clones. *Proc. Natl. Acad. Sci. USA* 89: 4183–4186.

Rubin, H. (1992). Cancer development: The rise of epigenetics. *Eur. J. Cancer* 28: 1–2.

――― (1993). Epigenetic nature of neoplastic transformation. In *Developmental Biology and Cancer.* (G.M. Hodges and C. Rowlatt, eds.), pp. 61–84. CRC, Boca Raton.

Russo, J., and Russo, I.H. (1978). DNA labeling index and structure of the rat mammary gland as determinants of its susceptibility to carcinogenesis. *J. Natl. Cancer Inst.* 61: 1451–1459.

――― and ――― (1987). Biological and molecular bases of mammary carcinogenesis. *Lab. Invest.* 57: 112–137.

Senturia, Y.D. (1987). The epidemiology of testicular cancer. *Brit. J. Urol.* 60: 285–291.

Shain, S.A., McCullough, B., Nitchuk, M., and Boesel, R.W. (1977). Prostate carcinogenesis in the AXC rat. *Oncology* 34: 114–122.

Sharpe, R.M., and Skakkebaek, N.E. (1993). Are oestrogens involved in falling sperm count and disorders of the male reproductive tract? *Lancet* 341: 1392–1395.

Shellabarger, C.J., Stone, J.P., and Holtzman, S. (1978). Rat differences in mammary tumor induction with estrogen and neutron radiation. *J. Natl. Cancer Inst.* 61: 1505–1508.

Shirai, T., Imaida, K., Masui, T., Iwasaki, S., Mori, T., Kato, T., and Ito, N. (1994). Effects of testosterone, dihydrotestosterone and estrogen on 3, 2′-dimethyl-4-aminobiphenyl-induced rat prostate carcinogenesis. *Int. J. Cancer* 57: 224–228.

Skakkebaek, N.E., Bethelsen, J.G., Giwercman, A., and Muller, J. (1987). Carcinoma in situ of the testis: possible origin from gonocytes and precursor of all types of germ cell tumors except spermatocytoma. *Int. J. Androl.* 10: 19–28.

Sohval, A.R. (1953). Testicular dysgenesis as an etiologic factor in cryptorchidism. *J. Urol.* 72: 693–702.

—————— (1956). Testicular dysgenesis in relation to neoplasm of the testicle. *J. Urol.* 75: 285–291.

Sonnenschein, C., and Soto, A.M. (1991). Cell proliferation in metazoans: Negative control mechanisms. In *Regulatory Mechanisms in Breast Cancer.* (M.E. Lippman and R.B. Dickson, eds.), pp. 171–194. Kluwer, Boston.

—————— and —————— (1999). *The Society of Cells: Control of Cell Proliferation and Cancer.* New York: Springer.

——————, Szelei, J., Nye, T.L., and Soto, A.M. (1994). Control of cell proliferation of human breast MCF7 cells; serum and estrogen resistant variants. *Oncol. Res.* 6: 373–381.

—————— and Soto, A.M. (1996). Control of estrogen-target cell proliferation, environmental estrogens and breast tumorigenesis. *Comments on Toxicol.* 5: 425–434.

Soto, A.M., Justicia, H., Wray, J.W., and Sonnenschein, C. (1991). p-Nonylphenol: An estrogenic xenobiotic released from "modified" polystyrene. *Environ. Health Perspec.* 92: 167–173.

——————, Lin, T-M., Justicia, H., Silvia, R.M., and Sonnenschein, C. (1992). An "in culture" bioassay to assess the estrogenicity of xenobiotics. In *Chemically Induced Alterations in Sexual Development: The Wildlife/Human Connection.* (T. Colborn and C. Clement, eds.), pp. 295–309. Princeton Scientific Publishing, Princeton.

—————— and Sonnenschein, C. (1993). Regulation of cell proliferation: Is the ultimate control positive or negative? In *New Frontiers in Cancer Causation.* (O.H. Iversen, ed.) pp. 109–123. Taylor & Francis, Washington, D.C.

——————, Chung, K.L., and Sonnenschein, C. (1994). The pesticides endosulfan, toxaphene, and dieldrin have estrogenic effects on human estrogen sensitive cells. *Environ. Health Perspec.* 102: 380–383.

——————, Lin, T.M., Sakabe, K., Olea, N., Damassa, D.A., and Sonnenschein, C. (1995a). Variants of the human prostate LNCaP cell line as a tool to study discrete components of the androgen-mediated proliferative response. *Oncol. Res.* 7: 545–558.

——————, Sonnenschein, C., Chung, K.L., Fernandez, M.F., Olea, N., and Olea-Serrano, M.F. (1995b). The E-SCREEN assay as a tool to identify estrogens: An update on estrogenic environmental pollutants. *Environ. Health Perspec.* 103: 113–122.

—————— and —————— (1996). Environmental sex hormone agonists and antagonists. *Comments on Toxicol.* 5: 329–346.

——————, Fernandez, M.F., Luizzi, M.F., Oles Karasko, A.S., and Sonnenschein, C. (1997). Developing a marker of exposure to xenoestrogen mixtures in human serum. *Environ. Health Perspec.* 105: 647–654.

Szelei, J., Jimenez, J., Soto, A.M., Luizzi, M.F., and Sonnenschein, C. (1997). Androgen-induced inhibition of proliferation in human breast cancer MCF7 cells transfected with androgen receptor. *Endocrinology* 138: 1406–1412.

Takasugi, N. (1976). Cytological basis for permanent vaginal changes on mice treated neonatally with steroid hormones. *Int. Rev. Cytol.* 44: 193–224.

Talwalker, P.K., and Meites, J. (1961). Mammary lobulo-alveolar growth induced by anterior pituitary hormones in adreno-ovariectomized and adreno-ovariectomized-hypophysectomized rats. *Proc. Soc. Exp. Biol. Med.* 107: 880–883.

Thomas, K.B., and Colborn, T. (1992).Organochlorine endocrine dispruptors in human tissue. In *Chemically Induced Alterations in Sexual Development: The Wildlife/Human Connection.* (T. Colborn and C. Clement, eds.), pp. 365–394. Princeton Scientific Publishing, Princeton.

Toniolo, P.G., Levitz, M., Zeleniuch-Jacquotte, A., Banerjee, S., Koenig, K.L., Shore, R.E., Strax, P., and Pasternack, B.S. (1995). A prospective study of endogenous estrogens and breast cancer in postmenopausal women. *J. Natl. Cancer Inst.* 87: 190–197.

Turkel, S.B., and Itabashi, H.H. (1974). The natural history of neuroblastic cells in the fetal adrenal gland. *Am. J. Pathol.* 76(2): 225–236.

Turusov, V.S., Trukhanova, L.S., Parfenov, Y.D., and Tomatis, L. (1992). Occurrence of tumours on the descendants of CBA male mice prenatally treated with diethylstilbestrol. *Int. J. Cancer* 50: 131–135.

Uwing, J. (1916). Pathological aspects of some problems of experimental cancer *Res.. J. Cancer Res.* 1: 71–86.

Varmus, H.E. (1984). The molecular genetics of cellular oncogenes. *Ann. Rev. Genetics* 18: 553–612.

vom Saal, F.S., Montano, M.M., and Wang, M.H. (1992). Sexual differentiation in mammals. In *Chemically-Induced Alterations in Sexual and Functional Development: The Wildlife/Human Connection.* (T. Colborn and C. Clement, eds.), pp. 17–83. Princeton Scientific Publishing, Princeton.

———, Timms, B.G., Montano, M.M., Palanza, P., Thayer, K.A., Nagel, S.C., Ganjam, V.K., Parmigiani, S., and Welshons, W.V. (1997). Prostate enlargement in mice due to fetal exposure to low doses of estradiol or diethylstilbestrol and opposite effects at high doses. *Proc. Natl. Acad. Sci. USA* 94: 2056–2061.

Walker, B.E. (1990). Tumors in female offspring of control and diethylstilbestrol-exposed mice fed high-fat diets. *J. Natl. Cancer Inst.* 82: 50–54.

——— and Kurth, L.A. (1993). Pituitary tumors in mice exposed prenatally to diethylstilbestrol. *Cancer Res.* 53: 1546–1549.

——— and ——— (1995). Multi-generational carcinogenesis from diethylstilbestrol investigated by blastocyst transfers in mice. *Int. J. Cancer* 61: 249–252.

Walsh, P.C., and Wilson, J.D. (1976). The induction of prostatic hypertrophy in the dog with androstanediol. *J. Clin. Invest.* 57: 1093–1097.

Welsch, C.W. (1985). Host factors affecting the growth of of carcinogen-induced rat mammary carcinomas: A review and tribute to Charles Brenton Huggins. *Cancer Res.* 45: 3415–3443.

Wiklund, J.A., Rutledge, J., and Gorski, J. (1981). A genetic model for the inheritance of pituitary tumor susceptibility in F344 rats. *Endocrinology* 109: 1708–1714.

———, Wertz, N., and Gorski, J. (1981). A comparison of estrogen effects on uterine and pituitary growth and prolactin synthesis in F344 and Holtzman rats. *Endocrinology* 109: 1700–1707.

——— and Gorski, J. (1982). Genetic differences in estrogen-induced deoxyribonucleic acid synthesis in the rat pituitary: Correlations with pituitary tumor susceptibility. *Endocrinology* 111: 1140–1149.

Willis, R.A. (1967). *Pathology of Tumors, 4th ed.* Butterworths, London.

Wolff, M.S., Toniolo, P.G., Lee, E.W., Rivera, M., and Dubin, N. (1993). Blood levels of organochlorine residues and risk of breast cancer. *J. Natl. Cancer Inst.* 85: 648–652.

——— and ——— (1995). Environmental organochlorine exposure as a potential etiologic factor in breast cancer. *Environ. Health Perspec.* 103: 141–145.

Chapter 12

AN ANTHROPOLOGICAL INTERPRETATION OF ENDOCRINE DISRUPTION IN CHILDREN

Elizabeth A. Guillette

Bureau of Applied Research in Anthropology
University of Arizona
Tucson, AZ 85721

The first words uttered by parents after the birth of a child reflect their concerns about normalcy. "Is my child all right?" The reply is based on the gross anatomy of having five fingers and five toes, or other normal external features. The hidden internal anatomy and physiological function are unknown. As scientists, we recognize that harmony in external features does not guarantee conformity in internal functioning. This fact grows in importance as environmental contamination becomes increasingly widespread. The possibility of covert effects of endocrine-disrupting contaminants (EDCs), which may have an immediate or delayed internal influence on the child's overall health, have only recently emerged, although gross teratogenic defects have been associated with such EDCs as dioxin and certain herbicides (Sherman, 1995). The purpose of this chapter is to present what is suspected and known about EDCs as obstructing normal childhood physiology and functioning, and to place this knowledge within a framework applicable to all types of EDC research.

Introduction

Other sections of this book reflect on the interactions of evolutionary responses to the environment and how EDC contamination has not allowed sufficient time for a protective evolutionary response to develop for most vertebrates. Temporally, evolutionary responses occur very slowly in humans, reflecting a long reproductive cycle between generations. On the other hand, cultural evolution has occurred at a more rapid pace. Marked technological change has occurred in the western culture over the last hundred years and is increasing rapidly. Developing countries, taking benefit of industrial and agricultural advances, have experienced marked technological change in a matter of decades. The children of today are a product of this cultural evolution as much as they are of biological evolution. As with biological evolution, cultural evolution serves as provocation for continuing action and reaction in future generations. We act and react according to the preceding changes that have occurred, both on an individual level and on a global level.

More ancestral vertebrates are not excluded from this process of "modernization." Specific aspects of both biological evolution and social organization are tied to various aspects of human cultural evolution and social change. Foremost are the pressures from human-induced ecological change and habitat compression. Other diverse factors affecting both animal and human welfare include pressures from population growth, social, economic, and political influences, plus access to the basic necessities of life. Correspondingly, we must remember that an event occurring locally may eventually have a global impact (i.e., the destruction of rain forests). Evolutionary factors are also a two-way street, reflecting the evolutionary interdependence of animal and plant life. Changes in biodiversity are known to lead to previously innocuous insects becoming devastating pests. Zoological and botanical change and/or extinction can easily feed back into the quality and quantity of human life (Epstein et al., 1997). Thus, the assessment of the impact of EDCs must be placed in a holistic, global context, with recognition of the magnitude of events that are capable of shaping the future for both animal offspring and our own children.

Reproductive Rights

The early unsettling hints that EDCs may be disrupting the many loci in the endocrine system are increasingly being accepted as reality. In light of the extensive scope of findings, both in animals and humans, the time has come to place endocrine disruption in a broad-based framework in which to evaluate the future of our children. The foundation of the framework lies in the reformulation of basic rights to reflect the need for sustainable existence, including ongoing reproduction and productivity. Three basic prerogatives, based in terms of reproductive rights to ensure the health and productivity of future children, are necessary: (1) the right to a healthy body for pregnancy and parenting, (2) the right to impregnate or become pregnant when a child is desired, and (3) the right to have the expectation that one's children will be able to express these same reproductive rights without physical or mental liabilities leading to restrictions (Guillette, 1997). Such rights, as stated, decrease the emphasis on the traditional sociobiological paradigm regarding the passage of genes and increase emphasis on a continuation of normal physiological function and intellectual prowess for all generations. Other chapters present what is known about EDCs in relation to reproductive processes. I will discuss reproductive rights as they apply to the children of today, integrated with thoughts on what is needed to ensure that today's generation can expect that future generations will have the same reproductive rights.

The course of the future will reflect the mental status, as well as the physical status, of today's children beginning with their conception and continuing throughout life. The healthy child is defined as born free of contamination-induced defects and who has no greater risk of exhibiting pathology later in life, either in terms of disease or dysfunction, than if never exposed to EDCs, and who has the same, or greater, ability to reproduce in adulthood as his or her fore-fathers. Implied in this statement is the concept that the child will be mentally, as well as physically, fit. With pressures to limit family size because of world population growth and limited resources to care for an excessive number of children with preventable pathology, it is of paramount importance that all children fall within this definition of "healthy."

The Right to a Healthy Pregnancy

Worldwide fertility rates, reflecting the number of births per woman, dropped for the first time in 1996 (Popline, 1997). Population control advocates assert that the decrease reflects an increase in the use of contraception, particularly in developing countries. Other factors are not generally considered. Unfortunately, there is no systematized record of global infertility, but a few statistics are available. In parts of sub-Saharan Africa—including Kenya, Uganda, Cameroon, Zaire, and Babon—infertility rates range between 30% and 40% (Leke et al., 1993). The underlying cause of most of the infertility is unknown. Sexually transmitted disease accounts for only one-third of the cases. Pathology, such as low sperm counts and endometriosis, has been identified in another third, conditions that have already been correlated with toxic exposures. The cause of infertility in the remaining third is unknown, which may be reflective of pathology difficult to diagnosis (Leke et al., 1993). Abnormal ovarian morphology, including polyovular follicles and polynuclear oocytes, is associated with alligators and mice exposed to a number of EDC contaminants (Iguchi, 1992; Guillette, 1994; Guillette and Guillette, 1996). An accelerated onset of reproductive senescence following prenatal exposure to EDCs occurs in rodents, although there is no comparable menopausal data for humans (Gray, 1991). The relatively recent increase in infertility for the sub-Saharan African women described above suggests that it is due to environmental change. Exposure levels in most sub-Saharan human populations, resulting from widespread use of pesticides—particularly DDT in coffee, tea, and cocoa plantations common to these areas—has never been fully determined or documented.

Problems with conception need not result from actual disease. Contamination from exposure to microwaves, industrial chemicals, or pesticides are associated with sexual disturbances. The problems range from decreased libido to erectile and ejaculatory problems in males (Bancroft, 1993). The impact of toxins on female sexual behavior is unknown (Bancroft, 1993). Many studies have shown that when mothers are exposed to high levels of EDCs prior to or with pregnancy, incidence of spontaneous abortion, prematurity, reduced birth weights, and smaller head circumference increase, depending on the type of contaminant exposure (Guo et al., 1993; Karmaus and Wolf, 1995; Guillette et al., 1998). Thus, the EDC-related prenatal health status of the child is frequently assumed to be a reflection of only the maternal exposures and cross-placental transfer. This may not be totally accurate. Men exposed to pesticides through farm work in India produced children with a 300% increase in congenital defects and a 4-fold increase in neonatal death when compared to controls (Rupa et al., 1991). However, neither the mother's exposure nor the history of grandparents was considered in this research. Children of dioxin-exposed mothers continue to have significantly elevated dioxin blood levels 25 years after birth (Schecter and Ryan, 1993). These children, now adult women, are in a position to pass the same EDCs on to the third generation.

Other factors, resulting from cultural evolution but completely unrelated to EDC or other toxic waste contamination, serve to further complicate the right to a healthy pregnancy. The obvious ones of poor diet, alcohol, tobacco, and drug use, poverty, and lack of prenatal medical care are generally considered when evaluating the impact of contamination. We must equally consider psychosocial stressors that impact the outcome of pregnancy. Such stressors may be observable. Loud, ambient noise levels at airports and at some industrial facilities have been correlated with lower birth weights and reduced physical growth during early childhood (Schell, 1997). Many of these same confounders complicating human research apply to wildlife and the stresses of noise, poverty in terms of limited habitat and food supplies, and disrupted social patterns of behavior resulting from human intervention (Epstein et al., 1997). These various confounding variables should not be allowed to become faults in research design. Instead, recognition should be given to their absence or presence within the studied and reference populations, along with the possible role of such factors in pregnancy outcomes and health of the newborn. Comprehensive recognition of all factors involved with pregnancy can provide strength to the correlative evidence relating EDCs to poor postnatal outcomes.

The Right to a Healthy Body for Parenting

It is beyond the scope of this chapter to detail the suspected health changes in adults that result from environmental change. An overview of changes in world health patterns provides basic insight. The increase in various chronic diseases

among younger and younger adults during the last 50 years appears to corre-
spond with the introduction and increased presence of EDCs. Cancer is no
longer a disease of the elderly in modern nations. Half of the world's cancers are
now found in developing nations, all of which have been experiencing modernization and the accompanying increase in EDCs for the last 30 to 50 years
(Polednak, 1989; World Cancer Research Fund, 1997). Since the introduction
of man-made toxic chemicals, cancers of the reproductive track are now occurring early in life, besides having increased 3-fold in incidence (Benedek and
Kiple, 1993).

Research is minimal on the correlation between EDC exposure and adult
infectious disease. We are all aware of the recent outbreaks of both old and
new infectious disease, yet neither pathogen mutations nor increased incidence of disease in adults has been investigated in terms of contaminant
exposure. At the same time, correlation between immune system malfunc-
tioning and EDCs has been documented (National Research Council, 1993;
Colborn et al., 1996).

Environmental change appears to be influencing the gender of the child to be
parented. Slow, mysterious declines in male births have occurred in various parts
of the world. Suspect factors include exposure to dioxin, pesticides, and high
voltage (Knave et al., 1979; Dimich-Ward et al., 1996; Mocarelli et al., 1996).
Impairments to male-producing fertility are found with both fathers and mothers, leading to a hypothesis that the involved toxic agents impact hormone levels related to sex determination and/or pregnancy outcomes (Toppari et al.,
1996; Toppari and Skakkebaek, 1997).

The Right to Expect Our Children to Have Healthy Bodies and Pregnancies

Given the suspected insidious and sometimes minute but important alterations
induced by EDCs, the identification of changes in health and factual proof of
such change presents a major dilemma. There is a scarcity of baseline data prior
to the introduction of toxic chemicals on which to base the actual occurrence
of possible EDC-induced aberrations. For example, birth defects are the leading
cause of infant death in Florida, although a birth-defect registry, aimed at track-
ing the problem and looking at the causes, was not approved by the state
government until May 1997 (*Gainesville Sun*, 1997). Florida is a state with a
history of heavy agricultural and residential pesticide use. The rate of defects
prior to the introduction of pesticides will never be accurately known.
Although 34 other states have a similar registry, a national registry is still
lacking. Such problems should not be viewed as deterrents for documenting
changes in health status but used to enlarge the scope of recognizable steps that
must be taken to promote better documentation and recognition of the health
changes found in association with EDC exposure. One step that must be
undertaken rapidly is the procurement of broad-based physical and mental

health baseline data on both adults and children living in the few lesser contaminated areas of the world, for EDC exposure will eventually increase in amount and complexity with modernization processes.

The process of growth and development during fetal life and childhood are reflections of health. While the foundations for body growth are laid down during fetal life, the human infant is compositionally immature at birth. Physical growth is a continuous process. Tissue organization and cellular maturation continues until adulthood. It has been demonstrated that infants exposed to high levels of PCBs or herbicides transplacentally are small for gestational age at birth (Munger et al., 1997). An enigma exists in regards to this question: Does *in utero* EDC exposure continue to disrupt postnatal growth? Children exposed transplacentally to PCBs can be used in this debate. Jacobson and Jacobson (1990, 1996) found that children with *in utero* PCB exposure were small for gestational age and remained small at 4 years of age. The studies on a prenatally PCB-exposed group of Yu Cheng children read that they may or may not continue to have continued growth retardation (Gnu et al., 1994; Lai et al., 1994). Cultural and social factors, some of which were considered as variables in the various studies, can account for some of the differences. In addition, one must consider the usual outcome of small infants for gestational age. In a 1972 study, occurring prior to the large-scale recognition of EDCs, babies who were born small were evaluated at 4 years of age. Of these children, 35% remained below the third percentile for both length and head circumference, and only 8% rose above the 50th percentile markers for their age group (Fitzhardinge and Steven, 1972). These data provide hints that other factors besides EDC-induced growth disruption may be involved with the continuation of the exposed fetus's failure grow to a normative level following birth. At the same time, it does not refute correlation between EDCs and limited growth. One must ask if there are any accompanying inborn genetic and/or physiological alterations due to EDCs that accompany below-average growth. This appears to be so. Disorders of ectodermal and neurological tissue are present in children with *in utero* PCB exposure (Rogan et al., 1998).

One of the most important postnatal maturation processes occurs within the central nervous system. Rapid neurological development, particularly learning capabilities, occurs during the first 5 years of life and ends with complete myelination of the peripheral and spinal cord nerve tracts at adulthood. Research has documented that children with high levels of transplacental exposure to PCBs have hypotonia and hyporeflexia at birth, indicating that the central nervous system (CNS) has been affected prior to birth (Rogan and Gladen, 1992). Other signs of defective CNS function that exhibit themselves later in childhood include slowed motor development, with deficits in gross and fine eye-hand coordination (Chen and Hsu, 1994; Cherr and Hsu, 1994; Guillette et al., 1998). The capacity for intellectual abilities also increases during these early years (National Research Council, 1993). Findings suggest that prenatal exposure to PCBs and pesticides tend to affect high cortical function rather than the sensory pathway, resulting in a lower IQ (Chen and Hsu, 1994; Jacobson and Jacobson, 1996). Many of these identified deficits, including behavioral problems and increased activity levels, persist over time (Cherr and Hsu, 1994). Both human and animal research are also providing correlative evidence that prenatal exposure to heavy metals induces varied mental and psychomotor disturbances, including learning, behavioral, and memory disorders (Liu and Elsner, 1995). We do not know if the identified learning/behavioral deficits ever occur with postnatal exposure to an additional compound or if the prenatal deficits are exacerbated by postnatal exposure to similar EDCs. These questions are difficult to answer because of the multiplicity of extraneous factors affecting growth and development in any child, including genetics, diet, ethnic practices, and cultural opportunities for mental stimulation and the overt expression of abilities.

Ethnic and regional differences in thought processes do exist and will continue to exist (Polednak, 1989). Such differences must be taken into account with the mental evaluations of children living in various areas of the world. American children are presented with many opportunities to recall a series of numbers (zip codes, social security numbers, and telephone numbers). In underdeveloped areas, the need to recall a number series is usually absent, making any test item involving this task difficult for the child to comprehend. Revision of the method is often necessary, as done with the children of the Yaqui tribe of Sonora, Mexico, under study for pesticide exposure (Guillette et al., 1998). Only when asked to repeat vowel sounds in abstract order, can the child grasp the task, eventually moving into number repetition. Acceptable childhood play behaviors also vary among cultural groups. Most American preschoolers are encouraged to engage in standing on one foot, which represents a sense of balance. When this same task was asked of Yaqui preschoolers, the children either refused to perform the task or managed to stand on one foot momentarily, usually holding onto an object. Only after questioning the parents did cross-cultural differences regarding the activity emerge. Children had been taught that standing on one foot was dangerous and results in injury (Guillette et al., 1998). Therefore, any claim that low scores on this activity reflected disruption of a sense of balance would have been invalid. Cross-cultural research studies are increasing. Interpretation of findings must always account for social and cultural factors and their implications in regards to neurological and mental performance.

Body functioning also includes the response to disease. The incidence of all cancers in children up to 14 years old rose 7.6% from 1973 through 1989 (Miller et al., 1992). The largest increases were for cases of acute lympocytic leukemia (23.7%), cancers of the brain and nervous system (28.6%), and cancers of the kidney and renal pelvis (25.9%). During the same time interval, other childhood cancers decreased (bone and joints, -15%; Hodgkin's disease, -1.5%; non-Hodgkin's lymphomas, -0.9%). Total cancer incidence for the entire U.S. population increased approximately 16.1% during this period (Miller et al., 1992).

EDCs have also been correlated with a depressed immune response (Colborn et al., 1996; see Chap. 7). The number of T-helper cells is known to be decreased in mice when exposed to DES prior to birth, raising questions with regard to humans (Palmlund et al., 1993). One study on the Yu-Cheng children, with *in utero* exposure to polychlorinated biphenyls and dibensofurans, demonstrates altered T-cell function and increased rates of sinopulmonary infection (Luster, 1996). Immune system depression, believed to be induced by PCB-contaminated food, is at the point where Inuit children have chronic ear infections and fail to produce antibodies in response to the usual childhood vaccinations (Colborn et al., 1996). Pesticides appear to create a similar immune system depression. Over half the families residing in the agricultural regions of Sonora, Mexico, experience seven or more bouts with infectious disease per year, in addition to autoimmune ailments of allergies and asthma, compared to incidence of none to two episodes of infectious disease and no autoimmune symptoms in the reference group. Most common are upper and lower respiratory infections, with adults similar to children in disease incidence (Guillette, 1997). The long-term impact of a compromised immune system gains greater importance when viewed in terms of social and environmental change. Already the more common infectious agents show resistance to new and powerful antibiotics. Looking to the future, will these children be more susceptible to certain diseases of adulthood, including the sexually transmitted diseases and such immune disorders as rheumatoid arthritis, for which there is no known cure?

Evaluating Risk

Risk assessment is usually approached in the context of the probability of a particular compound producing undesirable health outcomes, usually cancer. Risk is generally determined from the extrapolation of data derived from highly exposed subpopulation groups and/or data based on the chemical's effect on rodents, and then applied to adult humans (May 1996). Several problems exist with this approach. First is the assumption that only the heavily exposed subpopulation is at the greatest risk. Little consideration is given to the fact that the majority of all children are exposed to unknown doses of contaminants, including heavy metals, carcinogens, and multiple EDCs. For instance, background levels of TCDD up to 20 ng/kg have been found in the general population, with no identifiable specific exposures (Peterson et al., 1993). Adults and children are also unknowingly exposed through the foods we eat and water that we drink, in addition to the dust of our environment (National Research Council, 1993). Opportunities for children to become contaminated are even greater than parents may suspect. Play leads to direct contact with pesticide residues in yards, schools, and homes (Calabrese, 1997; Stanek, 1995). Other sources of contamination include poorly ventilated classrooms and the arts and crafts supplies at schools (Fields, 1997). In addition to the hidden sources of exposure is the fact that the child can be absorbing more toxic material than an adult in the same area. The child inhales and absorbs lead at a level 2 to 3 times that of an adult due to the child's higher metabolism and higher level of activity (Schell, 1991). One can assume that other airborne EDCs enter at a comparatively similar increased rate.

The universality of contamination places all children at some degree of risk, with the possibility of having cellular disorganization during fetal life and the later development of endocrine-related dysfunctions. The interrelationships between body size, time of exposure, and amount of exposure are not considered. This interrelationship is most important for the developing fetus and the young child (Bern, 1992). "Weak" estrogen, or EDCs that bind to the estrogen receptor, have a far more potent effect on unborn mice than on adult animals (Bern, 1992). There are also critical developmental periods during which exposure can induce modifications in cell function and structure (Bern, 1992; Guillette, 1994). Although these studies involve research on nonprimates, the applicability of findings to human fetal life should not be denied. As described by Bern (1992), the treatment of mice with diethylstilbestrol (DES) during the time period of development of the reproductive tract results in the same vaginal and uterine cell dysplasias as found in women whose mothers received DES during the third month of pregnancy. Such specifics are good to know, but the situation of the world today means that developing embryos are exposed to multiple specifics, many of which remain unknown.

Risk assessment for children needs to be considered both in terms of interurterine exposure and continuing exposure throughout childhood. It is now believed that many EDCs are able to pass the placental barrier and enter the fetus. Fetal blood and breast milk have a high lipophilic content and appear to absorb lipid soluble EDCs. The transfer of the contaminants to the fetus and child is well known (Rogan and Gladen, 1990; Ahlborg et al., 1992). Developing countries, which do not have controls on the use and types of chemicals as strict as those in developed countries, have a fetal blood and breast milk EDC concentration that meets or exceeds levels found in the developed world (Autrup, 1993). In human populations, the average levels of DDT in breast milk range from 70 to 170 mg/l, with highs of 830 (Wolff, 1983). Assorted pesticide residues have been found in such diverse areas as Australia, Uruguay, Spain, Italy, Mexico, and Guatemala (Thomas and Colborn, 1992). Therefore, it seems reasonable to assume that all children born today have experienced *in utero* exposure to some form of EDCs and continued exposure if breast feeding was undertaken. Hopefully, the time will arrive when child risk assessment considers the maternal body load of EDCs prior to pregnancy but not based exclusively on such data. Exclusive use of the toxic equivalency approach may underestimate the risk of deleterious effect, because of the many independent mechanisms involved with these effects and the number of factors involved, including the amount and timing of fetal exposure and possibly the mixture of transferred compounds. For these same reasons, the evaluation of children must extend beyond the typical disease incidence approach to include the endpoints of growth and development, including varied physical maturation process, cognitive abilities, neuromuscular performance, and behaviors.

Both the role of Darwinian evolution and cultural evolution must be incorporated into any evaluative method of growth and development. Genetic differences among children and among racial groups are increasingly recognized as being meaningful in terms of susceptibility to actual disease. Facemire (Chap. 3) discusses racial differences in the adipose tissue composition. The most rapid deposition of total body fat occurs during infancy and reoccurs later during puberesence, especially for the female (National Research Council, 1993). Questions exist if the rapid deposition of fat serves to protect EDC target organs of the neonate when exposed to lipophilic contaminants. The issue becomes paramount with breast feeding, as the intake of varied contaminants via breast milk can be exceedingly high and involve over 250 chemical contaminants (Thomas and Colborn, 1992). The anticipation that rapid fat deposition protects the infant's organs from high concentration of dioxins and feuans in breast milk is included in the 1990 Canadian Environmental Protection Act (Anonymous, 1990). Others claim that the magnitude of the safety margin cannot be determined, and the available information does not rule out the possibility that there is no safety margin for the weight-gaining infant (Ahlborg et al., 1992).

Cultural evolution has created circumstances in which the safety margin is compromised. Social-economic conditions in particular produce outcomes similar to the mental deficits identified with EDC exposure. Undernutrition is known to affect cognitive functioning, including poor scores on IQ tests, decreased intersensory perception, and increased propensity towards illness (Cravioto, 1966; Kamphaus, 1993). Nutrition is not the only social economic variable related to mental and neuromuscular achievement. It has long been known that poor sanitation, inadequate health care, limited and/or low-quality educational and recreational facilities, all interact to play a major role in childhood development (Krogman, 1972). Social-economic inequality is frequently correlated with environmental inequality, with the poor and minorities residing in the more highly contaminated areas (Johnson, 1997). The presence of environmental EDCs may well be the straw that breaks the camel's back, placing the children of these families at extreme risk.

In summary, actual risk assessment should not be based on single factors. Assessment is complicated. The child, from conception onward, is exposed many times to many compounds. The varied mechanisms of action, in conjunction with the varied times of doses and varied time lines of possible adverse effects, add additional confusion to determining risk, as children are not just little adults. They have different exposure, metabolism, and physiological processes. The total problem is compounded by sociocultural factors that create their own risk factors and possibly multiply those of EDC exposure. In addition, childhood risk assessment does not take into account the possibility of delayed effects that may not be expressed until early adulthood or later (Bern, 1992; Guillette et al., 1995).

The Future

The course of the future depends on action taken today. Such action includes two important segments: that of limiting our exposure to EDCs released into the environment and that of integrating EDC research to present a valid and realistic picture of what is actually happening. These two segments apply to all living species, as the physiological and endocrine parameters, although species specific, also share a great degree of similarity.

Scientists involved with the study of EDCs tend to use a categorical approach in their research. Investigation centers on such areas as the impact on biochemistry (i.e., binding properties), cell responses (i.e., mutations), specific organs (i.e., ovarian function), or the general population (i.e., risk, disease incidence). Such research is important in that it provides new knowledge. At the same time, the treatment of these factors as separate entities carries overtones of artificiality, in that research addresses issues that are related causally and conceptually but fails to give a total picture. The building blocks that result from compartmentalized research are seldom erected in total to provide a total view of what may be occurring with all children. The findings give the impression that there are pockets of children with intellectual deficits and other separate pockets with children exhibiting hormonal dysfunction or gross birth deformities or growth retardation. While pockets with extremes do exist, one cannot, and should not, come to the erroneous conclusion that EDCs are not affecting all children to some degree. The unification of specific knowledge from each category is necessary to prevent a heightened state of environment-induced vulnerability for parenting and reproducing, especially with our children and our children's children.

Research involving children must be approached holistically, extending beyond one specific area of interest or expertise. The range of possible outcomes and their endpoints are largely unknown, as are appropriate methods to assess possible probabilities (Weiss, 1998). The broad-based assessment of the normal play behaviors of 4- and 5-year-old pesticide-exposed Yaqui children showed that all areas of play behavior, ranging from ball catching to jumping, from drawing a stick figure to remembering a gift of a balloon, were compromised (Guillette et al., 1998). Such broad-based investigations, delving into unknowns, not only point out the scope and multitude of possible environment-induced deficiencies but also point out a need for more in-depth research in areas not previously recognized as being affected. The holistic approach calls for an interdisciplinary approach involving social, medical, and natural scientists working together without the artificial separation of topical components. Secondly, the need for more international investigation must be recognized, particularly in developing countries. Contamination is not just a problem in industrialized and western nations. Many published reports of birth defects, correlated with maternal and paternal EDC exposure, do not receive recognition because they are usually in lesser-read publications, such as Rupa et al.'s (1991) findings of a 300% increase in congenital defects and a 4-fold increase in neonatal death of children born to pesticide-exposed men in India. Such reports—plus verbal reports by nurses, midwives, agronomists, and others—indicate that children worldwide are exhibiting syndromes consistent with EDC exposure. For instance, a South-African midwife asked me for help in explaining "a strange new disease of newborns" in a particular agricultural area. The symptoms she described fit the syndrome of hermaphroditism. Many countries are those that contain sites with maximum and minimal exposure, providing valid reference groups for research.

Lastly, as research identifies an increasing array of pathological and physiological changes hypothesized to be associated with EDC exposure, consideration must be given to the acceptable and nonacceptable trade-offs that accompany technological advances. Evolution of flora and fauna, including *Homo sapiens*, continues to go onward as life continues. Clean air, water, and sufficient food is needed for all life. Providing these basics involves an integrated plant-based, animal-based, and human-based political economy. Short-term advantages that maintain the political economy must be weighed against long-term disadvantages, as should short-term disadvantages against long-term advantages.

Similar choices must be made for proposing and selecting intervention for protecting children. There are no simple answers. Mothers have been advised to cut away fatty portions of contaminated meat and fish where bioaccumulation is greatest. The removal of fat is a stopgap at best, for where is such tissue discarded? I have observed it being fed to other meat-producing livestock, including goats and hogs. At other times it ends up in a garbage heap, where it reenters the earth. Agricultural workers are advised to wash pesticide-contaminated clothing separate from other articles. Yes, this does decrease skin absorption of these pesticides by others. But where does the contaminated water flow? The possibility of its reentering the water system is present, particularly in rural areas served by shallow wells and drainfields.

Other family-based interventions for decreasing exposure pose similar decision-making problems, balancing economics and health. A mother's decision in regards to breast or bottle feeding frequently reflects the social and economic status of the family in society (Frayser, 1985). Only recently have the possible relationships of decreased maternal breast cancer risk and immunological advantages for the infant play a strong influencing role on the lay person's decision-making process for infant feeding. The maternal cumulative EDC load is a new facet to be considered in the decision-making process. With regard to all EDCs, the estimated intakes for neonates could be exceedingly high, and may exceed the permissible daily intake (Colborn et al., 1996). One point of view is that breast feeding occurs only for a relatively short period of the life span, with exposure reduced below the guidelines during the remainder of the life span (Anonymous, 1990; Ahlborg et al., 1992). Also, the supposition is that with the rapid deposit of fatty tissue during neonatal life, EDC concentration occurs in the adipose tissue rather than the target organs (Anonymous, 1990). The question whether breast feeding should be advocated or not remains a serious matter for scientists to resolve. There should be concern for the transference of EDCs, but considerations must also be given to the positive benefits for the mother and infant.

In all instances, the choice that must be made by the individual involves choosing between short-term and long-term options that will affect their health and their environment. The question all of us must face is: Should EDC production and use be restricted? If so, what will be the outcome in terms of global quality of life and for public health? There are no easy answers to these questions. Advances in knowledge, technology, and policy must provide avenues that will protect both the environment and the people, now and in the future. Until adequate means are found to substitute for present technology, we are left with the question: "Is my child all right?"

References

Ahlborg, U.G., Hanberg, A., and Kenne, K. (1992). Risk assessment of poly-chlorinated biphenyls (PCBs). *Nord* 26: 1–99.

Anonymous (1990). Polychlorinated dibenzodioxins and polychlorinated dibenzofurans. *Canadian Environ. Protection Act* 56.

Autrup, H. (1993). Transplacental transfer of genotoxins and transplacental carcinogenesis. *Environ. Health Perspec.* 101: 33–38.

Bancroft, J. (1993). Impact of environment, stress, occupational and other hazards on sexuality and sexual behavior. *Environ. Health Perspec.* 101 (Suppl. 2): 101–116.

Benedek, T.C., and Kiple, K.F. (1993). Concepts of cancer. In *The Cambridge World History of Human Disease.* (K.F. Kiple, ed.). Cambridge University Press, Cambridge, UK.

Bern, H.A. (1992). The fragile fetus. In *Chemically-Induced Alterations in Sexual and Functional Delopment: The Wildlife-Human Connection.* (T. Colborn and C. Clement, eds.), pp. 9–16. Princeton Scientific Publishing, Princeton.

Calabrese, E., Stanek, E.J., James, R.C., and Roberts, S.M. (1997). Soil ingestion: A concern for acute toxicity in children. *Environ. Health Perspec.* 105: 1354–1358.

Chen, Y.J., and Hsu, C.C. (1994). Effects of prenatal exposure to PCBs on the neurological function of children: A neuropsychological and neurophysiological study. *Develop. Med. Child Neurol.* 36: 312–320.

Cherr, Y.J., and Hsu, C.C. (1994). Effects of prenatal exposure to PCBs on the neurological function of children: A neuropsychological and neurophysiological study. *Develop. Med. Child Neurol.* 36: 312–320.

Colborn, T., Dumanoski, D., and Myers, J.P. (1996). *Our Stolen Future.* Dutton, New York.

Cravioto, J., DeLicardie, E.R., and H.G. Birch. (1966). Nutrition, growth and neurointegrative development: An experimental and ecologic study. *Pediatrics* 38: 319–372.

Dimich-Ward, H., Hertzman, C., Teschkl, K., Hershler, R., Marion, S.A., Ostry, A., and Kelly, S. (1996). Reproductive effects of parental exposure to chlorophenate wood preservatives in the sawmill industry. *Scandinavian J. Work Environ. Health* 22: 267–273.

Epstein, P.R., Dobson, A., and Vandermeer, J. (1997). Biodiversity and emerging infectious disease: Integrating health and ecosystem monitoring. In *Biodiversity and Human Health.* (F. Grifo and J. Rosenthal, eds.), pp. 60–87. Island Press, Washington, D.C.

Fields, S. (1997). Exposing ourselves to art. *Environ. Health Perspec.* 105: 284–289.

Fitzhardinge, P.M., and Steven, E.M. (1972). The small-for-date infant: Later growth patterns. *Pediatrics* 49: 671–681.

Gainesville Sun (1997). New birth-defect registry. May 27, p. 26.

Gnu, Y.L., Lin, C., Yau, W.J., Ryan, J.J., and Hsu, C.C. (1994). Musculoskeletol changes in children prenatally exposed to polychlorinated biphenyls and related compounds (Yu-Chong children). *J. Toxicol. Environ. Health* 4: 83–93.

Gray, L.E. (1991). Delayed effects on reproduction following exposure to toxic chemicals during critical periods of development. In *Aging and Environ. Toxicology: Biological and Behavioral Perspectives.* (R.L. Copper, J. Goldan, and T. Harbin, eds.), John Hopkins University Press, Baltimore.

Guillette, E.A. (1997). Environmental factors and the health of women. In *Second Meeting of National Leaders in Women's Health in Gainesville, FL.* (S.V. Rosser and L.S. Lieberman, eds.), pp. 81–96. Custom Copies, Gainesville.

———, Meza, M.M., Aquilar, M.G., Soto, A.D., and Garcia, I.E. (1998). An anthropological approach to the evaluation of preschool children exposed to pesticides in Mexico. *Environ. Health Perspec.* 106: 347–353.

Guillette, L.J., Jr. (1994). Endocrine-disrupting environmental contaminants and reproduction: Lessons from the study of wildlife. In *Women's Health Today: Perspectives on Current Research and Clinical Practice.* (D.R. Popkin and L.J. Peddle, eds.), pp. 201–207. Parthenon, New York.

————, Crain, D.A., Rooney, A.A., and Pickford, D.B. (1995). Organization versus activation: The role of endocrine-disrupting contaminants (EDCs) during embryonic development in wildlife. *Environ. Health Perspec.* 103 (Suppl. 7): 157–164.

———— and Guillette, E.A. (1996). Environmental contaminants and reproductive abnormalities in wildlife: Implications for public health? *Toxicol. & Indust. Health* 12: 337–350.

Guo, Y.L., Lai, T.J., Ju, S.H., Chen, Y.C., and Hsu, C.C. (1993). Sexual developments and biological findings in Yucheng children. *Chemosphere* 14: 235–238.

Iguchi, T. (1992). Cellular effects of early exposure to sex hormones and antihormones. *Internat. Rev. Cytol.* 139: 1–57.

Jacobson, J.L., and Jacobson, S.W. (1996). Intellectual impairment in children exposed to polychlorinated biphenyls *in utero*. *New England J. Med.* 335: 783–789.

————, ————, and Humphrey, H.E.B. (1990). Effects of exposure to PCBs and related compounds on growth and activity of children. *Neurotox. Terat.* 12: 319–326.

Johnson, B.R. (1997). Life and death matters at the end of the millennium. In *Life and Death Matters: Human Rights and the Environment at the End of the Millennium.* (B.R. Johnson, Ed.), pp. 9–22. AltaMira, Walnut Creek.

Kamphaus, R.W. (1993). *Clinical Assessment of Children's Intelligence.* Allyn and Bacon, Boston.

Karmaus, W., and Wolf, N. (1995). Reduced birthweight and length in the offspring of females exposed to PCDFs, PCP, and Lindane. *Environ. Health Perspec.* 103: 1120–1125.

Knave, B., Gamberale, F., Bergstrom, E.E., Birke, E., Iregen, A., Kolmodin-Hedman, B., and Wennberg, A. (1979). A long-term exposure to electric fields: A cross-sectional epidemiologic investigation of occupationally exposed workers in high-voltage substations. *Scandinavian J. Work & Environ. Health* 5: 115–125.

Krogman, W.M. (1972). *Child Growth.* University of Michigan Press, Ann Arbor.

Lai, T.J., Guo, Y., Yu, M.L., Ko, H.C., and Hsu, C.C. (1994). Cognitive development in Yucheng children. *Chemosphere* 29: 2405–2411.

Leke, F.J., Oduma, A.J., Basson-Mayagoitis, S., and Grigor, K.M. (1993). Regional and geographic variation in infertility: Effects of environmental, cultural and socioeconomic factors. *Environ. Health Perspec.* 101: 64–73.

Liu, G., and Elsner, J. (1995). Review of the multiple chemical exposure factors which may disturb human behavioral development. *Preventive Med.* 40: 209–217.

Luster, M.I. (1996). Immunotoxicology: Clinical consequences. *Toxicol. & Indust. Health* 12: 533–535.

May, M. (1996). Risk assessment: Bridging the gap between prediction and experimentation. *Environ. Health Perspec.* 104: 1150–1151.

Miller, B.A., Reis, L.A.G., Hankey, C.L., Kosary, C.L., and Edwards, B.K. (1992). *Cancer Statistics Review 1973–1989*. NIH Pub. No. 92–2789. National Institutes of Health, Bethesda.

Mocarelli, P., Brambilla, P., Gerthous, P.M., Patterson, D.G., and Needham, L.L. (1996). Change in sex ratio with exposure to dioxin. *Lancet* 14: 348–409.

Munger, R., Isacson, P., Hu, S., Burns, T., Hanson, J., Lynch, C.F., Cherryholmes, K., Van Dorpe, P., and Hausler, W.J., Jr. (1997). Intrauterine growth retardation in Iowa communities with herbicide-contaminated drinking water supplies. *Environ. Health Perspec.* 105: 308–314.

National Research Council (1993). *Pesticides in the Diets of Infants and Children*. National Academy Press, Washington, D.C.

Palmlund, I., Apfel, R., Buitendijk, S., Cabau, A., and Forsberg, J. (1993). Effects of Diethylstibestrol (DES) medication during pregnancy: Report from a symposium at the 10th International Congress of ISPOG. *J. Psychosomatic Obstetric. Gynaecol.* 14: 71–89.

Peterson, R.E., Theobald, H.M., and Kimmel, G.L. (1993). Developmental and reproductive toxicology of dioxins and related compounds: Cross-species comparisons. *Crit. Rev. in Toxicol.* 23: 283–335.

Polednak, A.P. (1989). *Racial and Ethnic Differences in Disease*. Oxford University Press, New York.

Popline (1997). Fertility decline reported. *World Population News Service Popline*. Vol. 19: May–June, p. 33.

Rogan, W.J., Gladden, B.C., Hung, K.L., Shish, S.L., Taylor, J.S., Wu, Y.C., Yand, D., Ragan, N.B., and Hsu, C.C. (1988). Congenital poisoning by polycholorinated biphenyls and their contaminants in Taiwan. *Science* 241: 334–336.

———— and ———— (1990). Perinatal exposure to polychlorinated biphenyls (PCBs) and child development at 18 and 24 months. *Pediatric Resident* 27: 97A.

———— and ———— (1992). Neurotoxiciology of PCBs and related compounds. *Neurotoxicol.* 13: 27–35.

Rupa, D.S., Reddy, P.P., and Reddy, O.S. (1991). Reproductive performance in populations exposed to pesticides in cotton fields in India. *Environ. Res.* 55: 123–128.

Schecter, A., and Ryan, J. (1993). Exposure of female production workers and their children in Ufa, Russia, to PCDDs/PCdFs/planar PCBs. In *13th International Symposium on Chlorinated Dioxins and Related Compounds in State University of New York, Binghamton*, pp. 55–58. State University of New York.

Schell, L.M. (1991). Effects of pollutants on human prenatal and postnatal growth: Noise, lead, polychlorobiphenyl compounds and toxic wastes. *Yearbook Phys. Anthropol.* 34: 157–188.

——— (1997). Culture as a stressor: A revised model of biocultural interaction. *Am. J. Phys. Anthropol.* 102: 67–78.

Sherman, J.D. (1995). Chlorpyrifos (Dursban)-associated birth defects: A proposed syndrome, report of four cases, and discussion of the toxicology. *Internatl. J. Occ. Med. & Toxicol.* 4: 417–431.

Stanek, J.I., and Calabrese E.J. (1995). Daily estimates of soil ingestion in children. *Environ. Health Perspec.* 103: 276–285.

Thomas, K.B., and Colborn, T. (1992). Organochlorine endocrine disruptors in human tissue. In *Chemically-induced Alterations in Sexual and Functional Development: The Wildlife/Human Connection Vol. XXI.* (T. Colborn and C. Clement, eds.), pp. 365–394. Princeton Scientific Publishing, Princeton.

Toppari, J., Larsen, J.C., Christiansen, P., Giwercman, A., Grandjean, P., Guillette, L.J., Jr., Jegou, B., Jensen, T.K., Jouannet, P., Keiding, N., Leffers, H., McLachlan, J.A., Meyer, O., Muller, J., Rajpert-De Meyts, E., Scheike, T., Sharpe, R., Sumpter, J., and Skakkebaek, N.E. (1996). Male reproductive health and environmental xenoestrogens. *Environ. Health Perspec.* 104 (Suppl. 4): 741–803.

——— and Skakkebaek, N.E. (1997). Response to James, W.H. *Environ. Health Perspec.* 105: 162.

Weiss, B. (1998). A risk assessment perspective on the neurobehavioral toxicity of endocrine disruptors. *Toxicol. & Indust. Health* 14: 341–359.

Wolff, M.S. (1983). Occupationally derived chemicals in breast milk. *Am. J. Ind. Med.* 4: 259–282.

World Cancer Research Fund (1997). Food, Nutrition and the Prevention of Cancer: A Global Perspective. American Institute for Cancer Research, Washington, D.C.

INDEX

(Italic letter *f* after a page number means that an illustration is on that page; *t* means table.)